AIRCRAFT ENGINES

DRAKE'S AIRCRAFT MECHANIC SERIES

AIRCRAFT WOODWORK

AIRCRAFT WELDING

AIRCRAFT SHEET METAL

AIRCRAFT ENGINES

AIRCRAFT ELECTRICAL SYSTEMS,
       HYDRAULIC SYSTEMS,
       AND INSTRUMENTS

AIRCRAFT MAINTENANCE AND SERVICE

AIRCRAFT ENGINE MAINTENANCE AND SERVICE

AIRCRAFT ENGINES

by COLONEL ROLLEN H. DRAKE, B. S., M. A.

RESEARCH ENGINEER, PILOT AND GROUND SCHOOL INSTRUCTOR
FORMERLY: INSTRUCTOR, VOCATIONAL AND AVIATION SUBJECTS, LOS ANGELES CITY SCHOOLS; CHIEF, AIR AGENCY UNIT AND CIVILIAN PILOT TRAINING SPECIALIST, CIVIL AERONAUTICS ADMINISTRATION; SUPPLY SPECIALIST, OFFICE OF THE QUARTERMASTER GENERAL

Originally published in 1948

ISBN: 1940001374

The Aviation Collection
by
Sportsman's Vintage Press
2015

This book is dedicated to
MY BELOVED WIFE
*without whose inspiration and help
this book could not have been written*

PREFACE

For many years man was hindered in his conquest of air by the lack of a light, powerful engine to pull his aircraft through the air. It was not until the Wright Brothers built their great four cylinder gasoline engine, which developed approximately 12 horsepower and weighed more than 13 pounds per horsepower, that man's real conquest of the air started.

It is difficult to say whether the airplane has kept pace with the aircraft engine or whether the aircraft engine has kept pace with the developments brought about by the aircraft builders. It has been the aim, however, of engineers, scientists and manufacturers of aircraft engines to continually better their product. Under no circumstances could the airplane exceed the capabilities of the aircraft engine. It is a long step from the original engine of the Wright Brothers to the new 28 cylinder, four row radial engines developing several thousand horsepower and the jet engines which carry man through the air faster than he has ever travelled before.

This book is written for the layman, teacher and the airplane mechanic, to explain the fundamentals of structure and the theory and operation of aircraft engines. While it is impossible to treat in detail every type of aircraft engine in a book of this scope, it is believed that the fundamental principles, modern construction methods and theory of engine operation given in this book can be applied profitably by the layman and the aircraft mechanic.

This book should furnish all material necessary in the way of theory to qualify a person for the Civil Aeronautics Administration's aircraft engine mechanic certificate. It should meet text requirements for formal classes studying Aircraft Engines.

Much information is contained on metals and alloys used in aircraft engines including jet engines. Though this book has been definitely designed for use as a classroom text it can be used to good advantage

PREFACE

for reference as a practical manual and guide to the mechanic in the field. It is designed for use in College Vocational Courses, Trade Schools, Junior Colleges, High Schools, Aviation Ground Schools and Technical Institutes.

Perhaps the most outstanding feature is the non-technical, simple language in which it is written. The author has avoided the use of unnecessary formulas, graphs, confusing tables, obscure footnotes and other material which cannot be clearly understood by everyone. Particular attention has been given in this text to the theory underlying the successful operation of the aircraft engine.

The author wishes to express his grateful appreciation to the following, who have so kindly furnished material which has been of assistance in the preparation of this book: the U. S. Office of Education; Civil Aeronautics Administration; the Departments of Education of the various States, particularly New York, Pennsylvania, Utah and Virginia; AC Spark Plug Division, General Motors Corporation; Allison Division, General Motors Corporation; Bell Aircraft Corporation; Bendix Products Division, Bendix Aviation Corporation; Candler-Hill Corporation; Continental Motors Corporation; Delco-Remy Division, General Motors Corporation; Eaton Manufacturing Company; Fairchild Engine and Airplane Corporation; The Electric Auto-Lite Company; Jack & Heintz, Inc.; Jacobs Aircraft Engine Company; Kinner Motors, Inc.; Kollsman Instrument Division of Square D Company; The Leece-Neville Company; Lycoming Division, the Aviation Corporation; The Glenn L. Martin Company; Norma-Hoffmann Bearings Corporation; Scintilla Magneto Division, Bendix Aviation Corporation; S K F Industries, Inc.; Sperry Gyroscope Company, Inc.; The Timken Roller Bearing Company and Torrington Company.

The author wants particularly to thank the following who not only furnished material for this text but who read and edited a large part of the manuscript: Aeroproducts Division, General Motors Corporation; Curtiss-Wright Corporation; Eclipse — Pioneer Division of Bendix Aviation Corporation; Eisemann Corporation; General Electric Company; Hamilton Standard Propellers United Aircraft Service Corporation; Marvel-Schebler Carburetor Division, Borg-Warner Corporation; Ranger Aircraft Engines, Division of Fairchild Engine and Airplane Corporation; Thomson Industries, Inc.; and Wright Aeronautical Corporation, Division of Curtiss-Wright Corporation.

And lastly the author wishes to thank his many friends who have so

PREFACE

generously contributed their advice and assistance. He also wishes particularly to express his gratitude to Earle R. Hough for his excellent work in the preparation of many of the drawings in this book, and to Mildred Pickrel and Alma Franklin for their untiring patience and cooperation in the preparation of the manuscript. The frontispiece shows the Shooting Star, and is reproduced by courtesy of the Lockheed Aircraft Corporation.

R. H. D.

TABLE OF CONTENTS

I	Introduction	1
II	Fundamentals of Internal Combustion Engines	8
III	Metals and Alloys Used in Aircraft Engines	17
IV	Glossary of Terms Used in Aircraft Engines	25
V	Types of Engines	32
VI	Cylinders	43
VII	Piston Assembly	52
VIII	Crankshafts	64
IX	Crankcases	75
X	Valves and Cams	86
XI	Electrical Fundamentals	99
XII	Ignition Systems	112
XIII	Fuel and Fuel Systems	137
XIV	Carburetors and Superchargers	148
XV	Lubricants and Lubricating Systems	174
XVI	Propeller Fundamentals	193
XVII	Controllable-Pitch Propellers	205
XVIII	Engine Installation and Cooling Systems	255
XIX	Engine Instruments	266
XX	Starting and Starting Systems	300
XXI	Engine Theory and Operation	309
XXII	Jet Propulsion	322
	Index	343

INTRODUCTION TO AIRCRAFT ENGINES

The men who first built life into cold metal so that it would carry future generations not only over the highways and railroads, but under the seas and through the air, must truly have been inspired. They gave to us the aircraft engine.

The earliest experimenters, the alchemists of the middle ages, and the ancients, who first extracted metal from its ore and formed it into tools which would do the will of man, paved the way for the building of all modern machinery. Progress was slow. Thousands of years lie between the crude tools hammered by primitive man from native metals and the

Fig. 1. A bomber powered by two radial engines. (Courtesy The Glenn L. Martin Company)

AIRCRAFT ENGINES

perfection of the complicated aircraft engine, so highly finished and reliable that it seems a living thing, doing the bidding of the pilot at the mere touch of the engine controls.

Perhaps the accidental explosion of some crude experiment first gave man the idea that the motion caused by this explosion might be controlled and made to work for man. As history has it, James Watt invented the

Fig. 2. The original Wright engine. This four-cylinder engine developed approximately 12 horsepower and weighed about 165 lbs. (Courtesy Wright Aeronautical Corporation)

steam engine as the result of watching steam lift the cover of his mother's teakettle. For many years the expanding power of steam was the only motivating force used in engines.

The use of aircraft was delayed nearly half a century because of the lack of a light, powerful engine. The Wright brothers could find no internal-combustion engine weighing less than approximately 33 pounds per horsepower, so they built one which developed approximately 12 hp. and weighed about 13 lb. per hp. This engine whirled two crudely constructed propellers which lifted man from the surface of the earth in his first successful flight in an aircraft that was heavier than air.

Metallurgy, which is the science of metals, has played a most important part in the development of the aircraft engine. Again we think of the alchemist in his smoky laboratory with the crudest kind of equipment, melting and alloying the various metals in his search for the

INTRODUCTION TO AIRCRAFT ENGINES

magic formula which would make gold from base metals. Later, the working of metals and their alloys became a science, and man began to experiment with metals to obtain the type of alloy best suited for his purpose.

Without alloys, the present aircraft engine would be an impossibility. Iron, aluminum, copper, vanadium, tungsten, tin, zinc, lead, and many

Fig. 3. Aircraft engines on the assembly line being prepared for installation. (Courtesy Douglas Aircraft Company, Inc.)

other metals are combined in alloys to produce the modern miracle of the aircraft engine. Even the semiprecious metal, silver, and the precious metal, platinum, have found their places in this wonderful machine.

Hand in hand with the development of the aircraft engine have gone years of research in proper fuels and lubricants. The products from the bodies of prehistoric monsters, as well as the myriads of smaller creatures that roamed the earth millions of years ago, have contributed much to making possible the travel of man through the air. Many scientists believe that petroleum is formed from the products of the bodies of these many animals that lived on the earth in the prehistoric ages. Aircraft fuels and lubricants, refined from the crude petroleums found thousands

AIRCRAFT ENGINES

Fig. 4. A flying boat powered by two radial engines. (Courtesy The Glenn L. Martin Company)

Fig. 5. The first Lawrance radial engine. This was the earliest type of radial engine. (Courtesy Wright Aeronautical Corporation)

of feet below the surface of the earth, make the efficient and dependable performance of the aircraft engine possible. Many skilled scientists have spent their lives in developing and perfecting modern aircraft fuels and lubricants. Many are working day and night to improve these products

INTRODUCTION TO AIRCRAFT ENGINES

in order to keep pace with others who are working on the improvement of alloys and on new developments so necessary to the aircraft engine industry.

The aircraft engine is one of the mechanical and engineering wonders of the present age. It is as delicately built as a watch, yet it is as powerful as a locomotive. It is built of the metals of the earth, but enables man to exceed the birds in flight.

Fig. 6. The R-1 was the first Wright radial engine. This was a nine-cylinder Lawrance radial engine produced in the early 1920's. (Courtesy Wright Aeronautical Corporation)

The first aircraft engines were crude, lacked flexibility, and were subject to failure at critical times. Engine instruments were unknown. Even in some of the engines still in use today, they are simple and few, consisting primarily of a tachometer, oil pressure and oil temperature gauges, and, if liquid cooled, a liquid temperature gauge. The engine controls on some of the aircraft used during World War I consisted only of a switch. The engine ran at full speed or not at all. The present-day engine has a great number of controls and instruments.

AIRCRAFT ENGINES

Engine power has increased from the small 12-horsepower engine of the Wright brothers, whose first flight was shorter than the distance between the wing tips on our largest aircraft of today, to engines devel-

Fig. 7. Three views of a light four-cylinder horizontally opposed aircraft engine. (Courtesy Lycoming Division, The Aviation Corporation)

INTRODUCTION TO AIRCRAFT ENGINES

oping several thousand horsepower. Injection carburetors, superchargers, exhaust gas analyzers, electrical temperature gauges, and many other developments have made this possible.

Early engines had to be almost completely overhauled every 50 to 75 hr., while the modern engine may give excellent performance for 500 hr. or more with only the routine service operations being performed upon it. However, the aircraft-engine mechanic is still largely responsible for the safety of the pilot and his passengers.

The automobile mechanic, if he is not sure of his skill or knowledge, can attempt to make a repair and then take the automobile out and drive it around the block. If it does not work properly, he can drive it back to the garage and try something else. The aircraft-engine mechanic cannot use this fix-and-try method for, if the aircraft engine fails, the plane may be wrecked and the pilot and his passengers may lose their lives.

The aircraft-engine mechanic must know when he turns the aircraft over to the pilot that it is "right" and that it will carry him safely to his destination. This mechanic must have a high degree of skill combined with a complete knowledge of what is right and what is wrong, and the proper method of correcting the wrong.

The small percentage of aircraft accidents due to engine failure testifies to the high skill, intelligence, and knowledge of the licensed aircraft-engine mechanic. The aircraft engine, with its hundreds of parts and microscopic adjustments, seems to the unskilled a complicated piece of machinery. It is like a foreign language to a person without a knowledge of that language. The aircraft engine is actually made up of many simple parts. When one learns these parts and their uses and adjustments, the whole becomes as simple as everyday speech.

The information contained in this text, while not complete, will be of assistance to anyone who wishes to contribute his part to this great field of aviation.

II FUNDAMENTALS OF INTERNAL-COMBUSTION ENGINES

Webster defines engine as "any of numerous machines by which physical power is applied to produce a desired mechanical effect, especially one for converting a physical force such as heat into mechanical power."

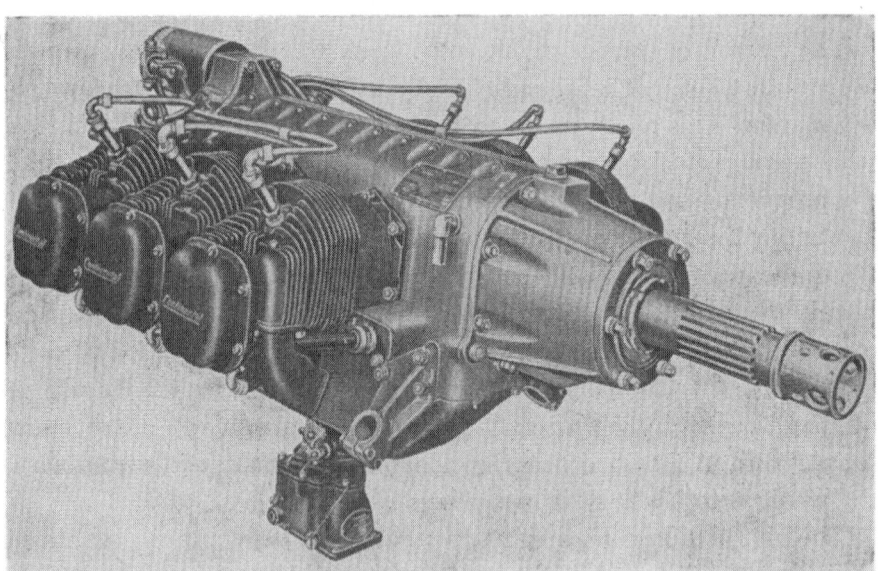

Fig. 8. A six-cylinder opposed light internal-combustion aircraft engine. (Courtesy Continental Motors Corporation)

Engines are commonly grouped into external-combustion and internal-combustion engines. In the external-combustion engine, of which a railroad engine is an example, the heat supply is produced by fuel burned separately from the engine. In the steam engine, fuel is burned under a closed boiler which generates steam under pressure. This steam is led through pipes to the cylinders. The steam is admitted into the cylinder through a valve arrangement and allowed to expand

FUNDAMENTALS OF INTERNAL–COMBUSTION ENGINES

against the piston, pushing it back and forth in the cylinder. The heat generated by the burning fuel is transferred to the cylinder by means of the steam generated in the boiler. The cylinder and its attachments are the steam engine. The combination of the boiler and engine is known as the locomotive.

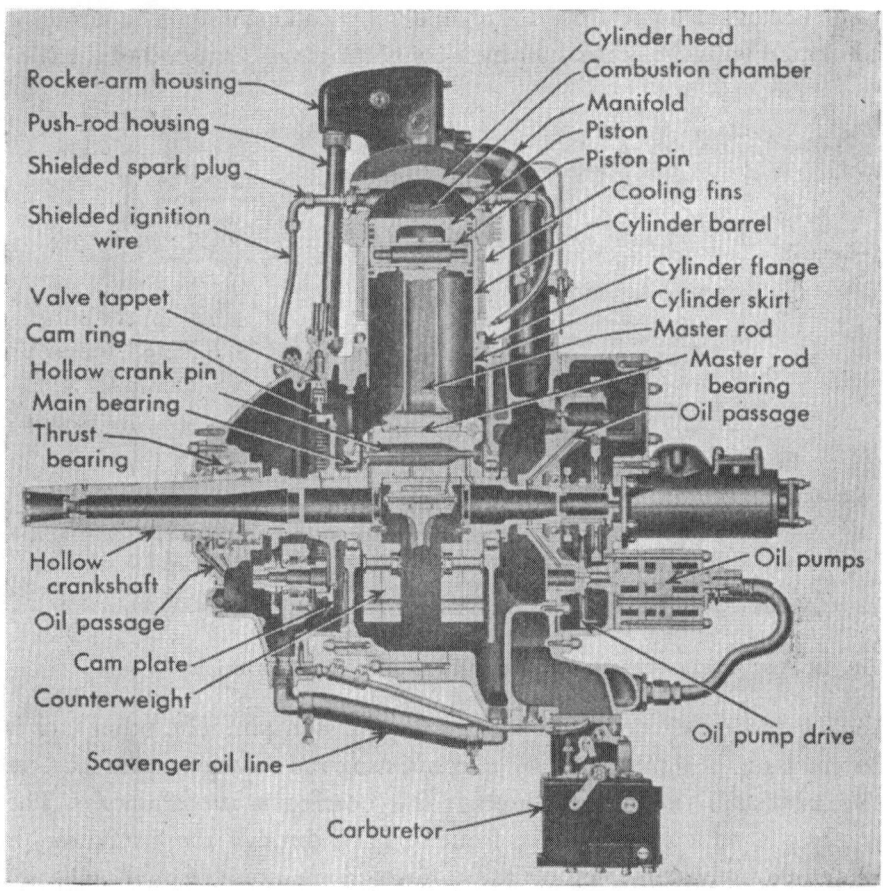

Fig. 9. A cutaway view of a radial engine showing its many parts. (Courtesy Jacobs Aircraft Engine Company)

In the internal-combustion engine, the fuel is burned within the cylinder, and the heat energy in the form of expanding gases is used to push the piston downward in the cylinder. The expression, "explode," is often used to describe the burning of the fuel in the cylinder of an internal-combustion engine. The charge in the cylinder, which is made up of a mixture of air and fuel, does not explode but burns at a high rate of speed. This speed of burning is only a fraction of the speed of burning which would take place if the charge actually exploded.

AIRCRAFT ENGINES

The main parts of an internal-combustion aircraft engine are shown in Fig. 9. The cylinder in which the power is generated is open at one end. The other end is closed by the cylinder head. The piston, which travels up and down in the cylinder, is equipped with piston rings to help make a gastight joint between the piston and the cylinder wall. Heat is carried away from the cylinder by cooling fins. The piston is connected with the crankshaft by a connecting rod. One end of the con-

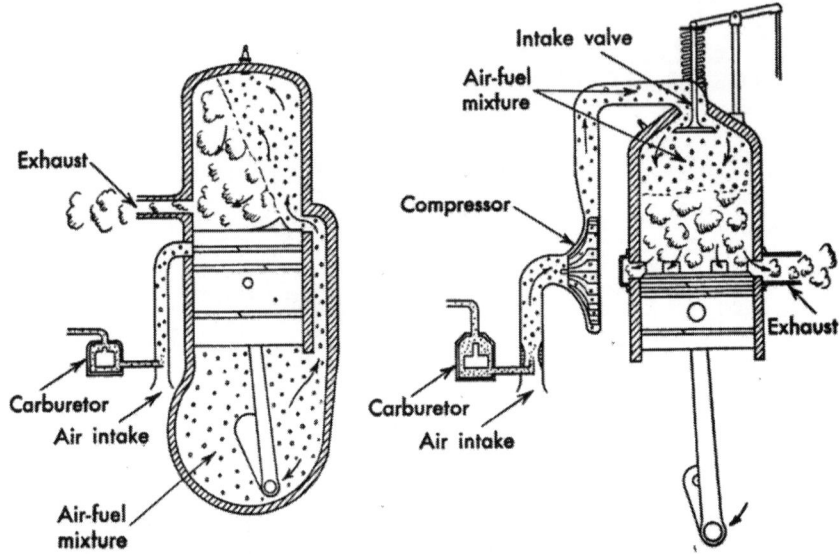

Fig. 10. A schematic drawing showing the two-cycle engine principle.

necting rod is fastened to the piston by a piston pin. The other end is fastened to the crankshaft by a connecting-rod bearing. The part of the crankshaft that passes through this bearing is the crankpin. The crankshaft, which passes from front to back through the crankcase, is supported by bearings resting on webs which are a part of the crankcase. The counterweights help balance the crankshaft. The oil pumps, oil lines, carburetor, manifolds, and ignition system help to make up the complete engine.

Practically all aircraft engines operate on the four-cycle principle. This is also known as the Otto cycle, being named after the man who first worked it out. In the four-cycle engine, a power stroke occurs once in every two revolutions of the crankshaft. The two-cycle aircraft engine has been used to some extent. One advantage of the two-cycle engine is that a power stroke occurs for each revolution of the crankshaft. The older type of two-cycle engine was subject to starting difficulties and

FUNDAMENTALS OF INTERNAL-COMBUSTION ENGINES

was not considered to be as dependable as the four-cycle engine. In this two-cycle engine, the air-fuel mixture was compressed in the crankcase by the downward stroke of the piston, and this compressed mixture was used to force the exhaust gases from the cylinder after the power stroke. It was necessary in this type of engine to use a lubricant which was insoluble in gasoline. This was the reason for the use of castor oil in many of the World War I aircraft engines. In some two-cycle engines, the oil is dissolved in the fuel. Later developments of the two-cycle engine make use of a separate compressor or charger which forces the air-fuel mixture directly into the cylinder (Fig. 10). This does away with the necessity of the fuel's mixing with the lubricant in the crankcase or the fuel tank.

The two-cycle and four-cycle principles are so called from the number of strokes the piston makes in the cylinder for each power stroke. These strokes may be numbered in any order, but the intake stroke is usually No. 1. This stroke starts when the piston is as far in the cylinder as it ever

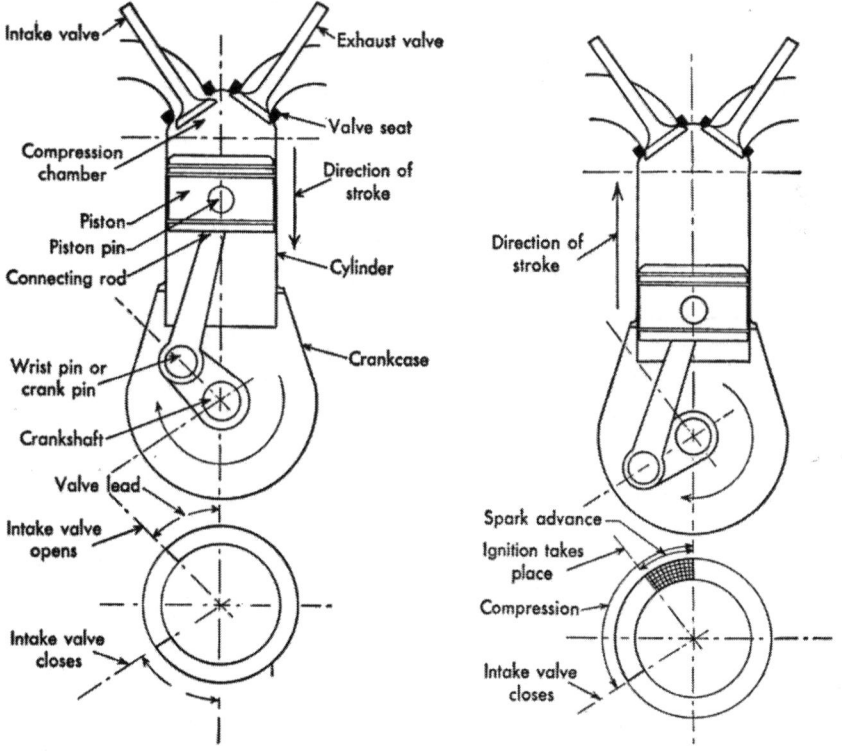

Fig. 11. A schematic drawing showing the intake stroke of an internal-combustion engine.

Fig. 12. A schematic drawing showing the compression stroke of an internal-combustion engine.

11

goes. When the piston is in this position, it is at "top dead center." When the piston is at the bottom of the stroke, it is at "bottom dead center." On the intake stroke, with the piston at top dead center and the intake valve open, the piston starts downward, creating a partial vacuum in the cylinder. The intake valve opens into the intake manifold which is connected with the carburetor. Atmospheric pressure forces the air-fuel mixture from the carburetor and intake manifold into the cylinder to fill the space in the cylinder left by the downward travel of the piston.

Shortly after the piston has reached bottom dead center and starts upward in the cylinder on stroke No. 2, which is the compression stroke, the intake valve closes. The exhaust valve remains closed during this stroke. As the piston starts upward in the cylinder on the compression stroke, the air-fuel mixture is compressed until, by the time the piston reaches top dead center, the entire charge is compressed into the space between the piston and the cylinder head. This space is the combustion or compression chamber. Slightly before the piston reaches top dead center, this mixture is ignited by means of an electric spark which jumps between the points of the spark plugs. The mixture burns rapidly, creating high temperatures which cause the gases to expand and exert high pressures on the piston head. The combustion time is approximately $\frac{1}{300}$ sec.

This pressure forces the piston downward in the cylinder on stroke No. 3, which is the power stroke. Before the piston reaches bottom dead center, the exhaust valve opens.

On stroke No. 4, which is the exhaust stroke, the piston again travels the length of the cylinder from bottom dead center to top dead center, forcing the burned gases out through the exhaust valve opening into the exhaust manifold.

There is only one power stroke for each four strokes of the piston in the cylinder. During the other three strokes — intake, compression, and exhaust — power is used up by the engine. In addition to the usable power developed by the engine, enough momentum must be built up in the moving parts of the engine during the power stroke to carry the engine through the other three strokes. In order to have this momentum available, single-cylinder, internal-combustion engines must be equipped with a heavy flywheel. The momentum of this flywheel, which is obtained from the single power stroke, continues to rotate the engine through the three strokes during which no power is developed. On a two-cylinder,

FUNDAMENTALS OF INTERNAL-COMBUSTION ENGINES

four-cycle engine, there is one power stroke for each revolution of the crankshaft. On a four-cylinder, four-cycle engine, one power stroke occurs every half revolution of the crankshaft. As the number of cylinders is increased, the weight of the flywheel may be decreased, because of

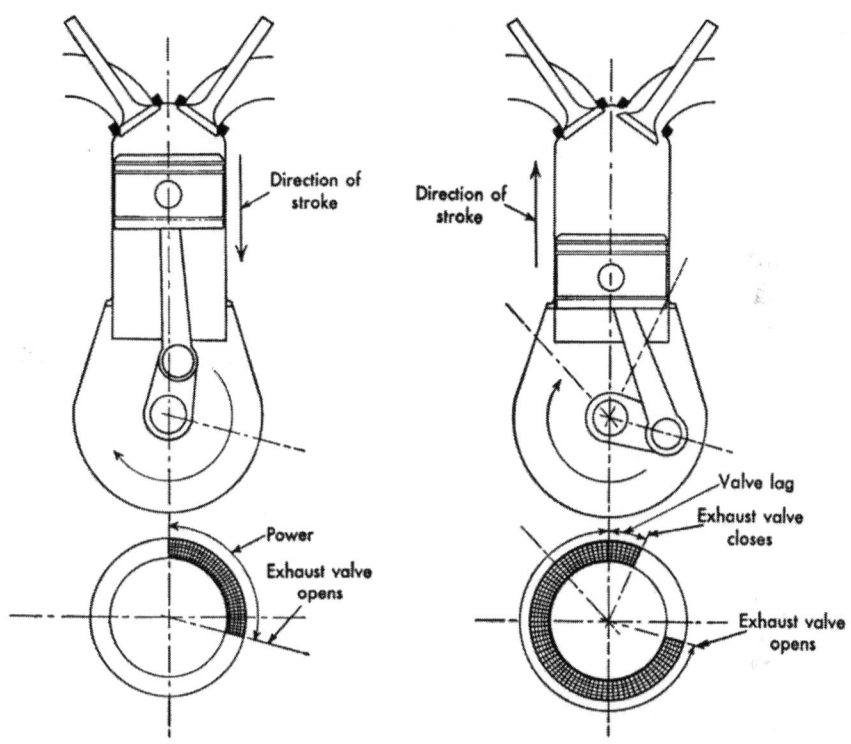

Fig. 13. A schematic drawing showing the power stroke of an internal-combustion engine.

Fig. 14. A schematic drawing showing the exhaust stroke of an internal-combustion engine.

the more frequent power strokes. Since there is not a continuous flow of power, it is necessary to have some kind of flywheel to absorb the regular impulses of the power strokes. On the automobile engine this is accomplished by means of a light flywheel, but on the aircraft engine the propeller acts as the flywheel. Counterweights are placed on the crankshaft in most aircraft engines to assist in balancing the engine to prevent excessive vibration. In order to transmit the power from the piston to the propeller, the piston is connected to the crankshaft by means of a connecting rod. The connecting rod turns a crank, built into the crankshaft, similar to the crank on a grindstone. The propeller is mounted on the end of the crankshaft.

AIRCRAFT ENGINES

The carburetor mixes the air and fuel in the proper proportions, and an ignition system, consisting of a source of high-tension electric current, is necessary to ignite this air-fuel mixture in the cylinder at the proper time.

The crankcase encloses the crankshaft and connecting rods and forms the main framework of the engine. A camshaft or cam plates are used to open and close the valves at the proper time. The valves open only once

Fig. 15. A radial engine crankshaft equipped with counterweights. (Courtesy Wright Aeronautical Corporation)

during each two revolutions of the crankshaft. The camshaft, therefore, rotates at only one half the crankshaft speed. In radial engines the camshaft is replaced by cam plates which rotate at various speeds depending upon the number of cylinders. On some radial engines the cam plate rotates in the same direction as the crankshaft, at a fraction of the crankshaft speed. On others it rotates opposite to the direction of the crankshaft.

Aircraft engines are equipped with a number of accessories, instruments, and controls.

Low weight per horsepower is one of the fundamental requirements for a successful aircraft engine. If the engines of today weighed 13 lb. per hp., as did the original Wright engine, an engine developing several thousand horsepower could hardly lift its own weight.

A railroad engine must have great weight in order to create enough friction on the rails to haul heavy trains, and a few thousand pounds of extra weight only means added tractive force. Each pound added to an airplane engine means that one pound less pay load may be carried. For many years designers tried to build an engine which would develop one horsepower for each pound in weight. Some of the latest engines have even exceeded this figure.

Weight per horsepower is dependent somewhat on the size of the engine, the more powerful engines having less weight per horsepower than the lighter engines. Modern alloys have made it possible to construct high-powered aircraft engines within reasonable weight limits.

Fig. 16. A cutaway view of a crankcase showing the crankshaft, connecting rods, piston, and bottoms of the cylinders of an inverted engine. (Courtesy Ranger Aircraft Engines)

Some of the smaller engines, striving for low prices, have used cheaper materials, which brings about an increase in weight per horsepower. These engines develop one horsepower for each two or three pounds of weight.

Aircraft engines must be economical to operate. Low fuel and oil consumption are desirable. However, the cost of the fuel and oil is not as important as the added weight when large quantities of fuel must be carried on long flights.

A gallon of gasoline weighs 6 lb. An engine using 40 gal. per hr. would require 720 lb. of gasoline for a 3-hour flight. Oil weighs between 7 and

8 lb. per gal., and 10 gal. of oil adds approximately 75 lb. of weight. These combined weights would displace five average-sized passengers. The modern engine uses approximately ½ lb. of gasoline per horsepower per hour under cruising conditions. A properly maintained engine uses only about 0.05 lb. of lubricating oil per horsepower-hour, but must carry a considerable supply to lubricate and cool properly the many parts of the engine. Aircraft taking off on long endurance flights have actually carried fuel in excess of the weight of the entire airplane.

The aircraft engine must operate satisfactorily through a wide range of temperature changes and atmospheric conditions varying from the humid tropical temperatures of over 100° F. to temperatures far below zero — when these changes take place in a matter of minutes.

It is necessary that the aircraft engine be free from excessive vibration through its entire range of speed and power. Severe vibration could destroy not only the engine, but also the entire aircraft structure. Increasing the number of cylinders greatly assists in smoothing out vibration. Balancing the various parts to prevent vibration is an engineering problem with each individual aircraft engine.

III METALS AND ALLOYS USED IN AIRCRAFT ENGINES

While the aircraft-engine mechanic is not expected to have an extensive knowledge of metallurgy, he should have a general understanding of the many metals and alloys that are used in aircraft-engine construction.

Metals, when pure, are classed as elements. There are 92 elements, roughly divided by the chemist into metals and nonmetals. The metals all have metallic luster on freshly polished surfaces. Most of the metals combine readily with other elements, forming chemical compounds. When a metal combines with oxygen, the compound formed is called an "oxide."

Metals vary widely in their physical properties. Physical properties are those properties which may be determined by the senses, such as color, weight, strength, and hardness.

The weight per unit volume of metals, which is density, varies widely. For example, lithium weighs 33.072 lb. per cu. ft., and a cubic foot of iridium weighs 1397.76 lb.

The specific gravity of any substance is the number of times a volume of that substance is as heavy as an equal volume of water. For example, a cubic inch of lead weighs 11.34 times as heavy as a cubic inch of water. Therefore, the specific gravity of lead is 11.34. The specific gravity of aluminum is 2.7; of gold, 19.3.

Metals vary in their hardness. Lead is one of the soft metals. Sodium is lighter than water and is almost as soft as putty. Some of the metals are so hard that they will scratch glass. Some are brittle and break easily when pounded. Others are soft and may be pounded into any shape without breaking. Gold may be hammered so thin that the sheet is almost transparent.

The melting point of metals, which is the temperature at which they change from the solid to the liquid state, covers a wide range. Lithium melts at about 366° F., and lead melts at 621° F., while the melting point of tungsten is 6098° F. Aluminum melts at 1218° F., and steels have melting points from 2480° to 2786° F., depending upon the alloy.

All metals conduct electricity. Silver has the highest conductivity of all the metals, and its conductivity is taken as 100. The conductivity of bismuth is only 1.8 in comparison with silver.

All metals except copper and gold are silvery white in color when freshly polished; copper is reddish and gold yellowish in color.

Some metals corrode readily, while others, such as gold, silver, and platinum, corrode hardly at all under ordinary conditions. Aluminum corrodes rapidly. A layer of aluminum oxide, which corresponds to iron rust (iron oxide), forms on freshly polished aluminum within a few minutes. Corrosion is sometimes called tarnish.

Metals are sensitive to changes in temperature. All metals expand when heated and contract when cooled. Almost all of the metals contract when passing from the liquid to the solid form. Each metal has its own rate of expansion. The amount that a metal changes in length or volume for each degree change in temperature is the coefficient of expansion. When expansion is measured lengthwise, it is known as linear expansion per inch per degree. The amount the metal expands or contracts for each degree change in temperature is small. For example, iron expands 0.00000648 in. for each degree change in temperature. This means that a piece of iron wire 1 in. long expands or contracts 0.00000648 in. for each degree change in temperature. Aluminum, lead, tin, and zinc have a comparatively large coefficient of expansion. The coefficient of expansion of aluminum is 0.00001234. After metal has been caused to expand by heating, it always returns to its original dimensions when cooled to the original temperature.

The electrical conductivity of metals changes with changes of temperature. The conductivity of a metal is lower at high temperatures than at low temperatures.

Most metals are used in the form of alloys. An alloy is made by melting two or more metals together. Metals dissolve in metals in much the same way that sugar dissolves in water. If sugar is dissolved in water and the water is frozen, a sweet ice is formed, which would represent an alloy of sugar and water. Steel is considered to be an alloy of carbon and iron. A true alloy is a solution of metals within metals or a mixture of metals. Carbon is not a metal. There are a number of elements other than metals, such as carbon, phosphorus, and sulphur, which are commonly

METALS AND ALLOYS USED IN AIRCRAFT ENGINES

found in steels and which affect the properties of the steel. These elements form compounds with iron or may be mixed with iron. While these substances do not form true alloys, the properties of the alloys are often affected to a considerable extent by their presence. Even very small quantities of a metal or of a nonmetallic substance may have a great effect on the properties of the alloys formed. For example, a fraction of 1 per cent of sulphur may make steel unfit for use in aircraft-engine construction.

Steel is iron which contains a varying percentage of carbon. When the carbon content is less than $1/10$ of 1 per cent, the product is considered to be iron.

Wrought iron is an alloy of iron and carbon in which the carbon content is less than 0.10 per cent and the manganese content is less than 0.07 per cent. Wrought iron is tough, ductile, and soft, and is easily worked.

When iron has a carbon content of less than approximately 0.06 per cent, it is low-carbon iron. Low-carbon steels, or mild steels, may have a range of carbon content from $1/10$ to $3/10$ of 1 per cent. Low-carbon steels are easily worked.

Metals and alloys are usually furnished to the industry in the form of castings, forgings, and rolled sheet or plate. Castings are formed by pouring the molten metal into a mold and allowing it to cool. Forgings are formed by hammering or working the cast material while it is at high temperatures, but without remelting it. Aluminum forgings may be formed by applying pressure to the molten metal in the mold. Ingots of cast metal are heated and passed between heavy rolls to form sheets and plates. Rolling has much the same effect as forging.

Some of the more common alloys of steel are molybdenum steel, chromium steel, chromium molybdenum steel, chromium vanadium steel, manganese steel, manganese vanadium steel, nickel steel, nickel chromium steel, chromium molybdenum vanadium steel, copper nickel steel, and molybdenum nickel steel. Most of these alloys have trade names and are used widely throughout the aircraft industry.

Stainless steels have come into use in the last few years. Stainless steels are alloys of iron, chromium, carbon, nickel, and other metals. Stainless steels are known to the trade by their number classification, such as 18–8 or 25–20. One of the most common stainless steels is 18–8, which contains approximately 18 per cent chromium, 8 per cent nickel and from 0.08 per cent to 0.20 per cent carbon alloyed with iron. Stainless steel is used for valve seats, exhaust rings, and oil tanks.

The properties of steel are affected by both its physical and chemical composition. The heat treatment which steel receives affects both its physical and chemical properties. The physical constituents of steel may be observed under a microscope. The following are some of the substances present in steel: ferrite, cementite, austenite, pearlite, primary troostite, secondary troostite, martensite, and sorbite. These are all compounds of iron and carbon. The material to be examined must be carefully polished and etched with an acid solution which corrodes the metal and brings out the grain structure. This corrosion or etching indicates the grain or crystal boundaries and assists in their identification. Etching will also show the presence of nonmetallic impurities.

Cast steel is iron containing from 0.05 per cent to 2 per cent carbon and is cast from the molten state. This steel is forged at comparatively high temperatures, depending upon the amount of carbon present. Carbon, manganese, silicon, sulphur, and phosphorus are the elements usually present in cast steel. Nickel, chromium, vanadium, tungsten, and molybdenum are the metallic elements most commonly used in the making of steel alloys.

Copper, manganese, magnesium, zinc, nickel, and silicon are the common elements added to aluminum to make the commercial alloys. Aluminum to which these elements have been added shows a decided increase in both hardness and strength. Commercial metals are usually considered pure when other elements are not present in sufficient amounts to change the properties of the metal.

Strong alloys of aluminum depend mainly for their properties on the amount of the alloying substance which is combined with the aluminum. However, many alloys of aluminum depend for their strength and hardness upon proper heat treatment.

The strong alloys of aluminum are used in aircraft engines in cylinder heads, crankcases, connecting rods, pistons, fuel and oil tanks, oil lines, and propellers.

Gray cast iron is used in the construction of piston rings.

Chromium nickel steel is used where strength, ductility, toughness, and shock-resistant properties are required, such as for the crankshaft, cylinders, connecting rods, wrist pins, and camshafts.

Chromium molybdenum steel is used in aircraft-engine parts where toughness and strength are desired, such as connecting rods, cylinders, and crankshafts.

Chromium steel is especially noted for its stainless and noncorrosive

METALS AND ALLOYS USED IN AIRCRAFT ENGINES

qualities and is used in aircraft engines where these qualities are necessary, such as in valves and piston pins.

Chromium vanadium steel is strong and tough and is used for wrist pins, pinion gears, and bearing races.

Tungsten steel is noted for its hardness and heat-resistant qualities. The low-tungsten grade in the annealed condition is used for permanent magnets. The high-tungsten grade is used principally for exhaust valves. It retains its tensile strength at high temperatures.

Commercially pure copper is used in electrical equipment, wiring, and oil lines. Pure brass is an alloy of copper and zinc. There are a number of different brasses, depending upon the amount of the alloying material added to the copper. Brass is used in bearings, bushings, and fittings where a soft bearing surface is desired.

Pure bronze is an alloy of copper and tin. The various bronzes depend upon the proportion of copper and tin present in the alloy. Bronze may be modified by the addition of phosphorus, forming phosphor bronze, or aluminum, forming aluminum bronze. Bronze is used for bushings, fittings, bearings, valve seats, and valve guides.

The alloys of nickel are used in aircraft fabrication in the form of Monel metal and Inconel. Inconel is an alloy containing about 75 per cent nickel, 12 to 15 per cent chromium, and 9 per cent iron. It also has small amounts of carbon, copper, manganese, and silicon. Monel metal contains approximately 75 per cent nickel and 25 per cent copper, with small amounts of iron, manganese, and silicon. It is used for collector rings and exhaust stacks.

Lead and tin are alloyed with various substances to form different types of bearings. They are also combined with silver to make solders of varying degrees of hardness. Soft solder is an alloy of lead and tin.

Many magnesium alloys, which contain magnesium in percentages varying from 89.9 to 98.5, are furnished to the industry under the Dowmetal products. The magnesium is alloyed with manganese, aluminum, silicon, zinc, and tin. These alloys are coming into use rapidly because of their high strength-to-weight ratio. The magnesium alloys are used for such parts as crankcases and oil sumps.

Alloys for connecting-rod bearings may contain cadmium, silver, zinc, tin, lead, copper, and antimony. Platinum may be used in breaker points.

Heat treatment consists of heating and cooling under various conditions. Heat treatment not only affects the grain structure of the metal, but may affect its chemical composition. As a rule, the longer the cooling time of a metal or alloy which has been heated above its critical temperature, the larger the grain structure. A few of the metals become softer when cooled rapidly, but most become harder. The critical point is the temperature to which a metal must be heated to change the crystalline structure. At ordinary temperatures, changes in crystalline structure may take place, but when it does the process is slow. Most critical temperatures require heating to some degree of redness. At these temperatures, crystalline structures may change rapidly.

With proper heat treatment, such characteristics as hardness, toughness, springiness, ductility, and workability may be controlled. Usually the heating is brought about gradually, and the metal is held at the required temperature for varying lengths of time. Annealing, hardening, tempering, and normalizing are forms of heat treatment.

Annealing is a form of heat treatment which usually consists of heating the metal slightly above the critical temperature for a predetermined length of time and then cooling slowly. Cooling is sometimes carried out in a furnace by a gradual lowering of the temperature. Cooling may be carried out by burying the part in slaked lime, sand, mineral wool, glass wool, shredded asbestos, or ashes. Steel and other alloys of iron usually become softer when annealed. Small iron and steel parts are often annealed by heating to a red temperature with a blowpipe. After the proper temperature has been reached, the part is buried in sand or lime and allowed to cool slowly. This treatment relieves internal strains and softens the metal.

Most steel parts are hardened after forming. This hardening consists of heating the part slowly to a point above the critical temperature. This temperature is maintained until the parts are thoroughly heated and the crystalline structure has had time to stabilize itself. The part is then quenched in brine, water, or oil. Oil is often used because the quenching action is not as severe as that of water or brine. Steel tools, such as cold chisels, when hardened are said to have been tempered. Heavy parts having large cross sections are hardened to a greater extent on the outside surfaces than on the inside, because of the slower cooling beneath the surface. The parts being quenched should be moved around in the quenching solution. This helps relieve stresses and strains on the surface of the part. If a part is allowed to remain quietly in a quenching solution, small cracks or checks may appear on the surface.

Normalizing is similar to annealing. The material to be normalized

METALS AND ALLOYS USED IN AIRCRAFT ENGINES

is heated uniformly throughout to slightly above the critical temperature. It is then allowed to cool in still air. This normalizing treatment will produce a hard, strong metal which is less ductile than an annealed metal. Normalizing relieves internal stresses and produces greater uniformity of structure throughout the piece treated. Aircraft steels are usually normalized. Annealed parts are not commonly used in aircraft structures because of their softness and lack of strength. Normalizing reduces the chances of failure due to fatigue. Chromium alloys are normalized before the regular hardening operations and heat treatments are carried out.

Case hardening is a process by which the outer layer of a steel part is hardened. This is done where the surface will be subjected to severe wear. Case-hardened material has an extremely hard outside which tends to be brittle. This outside is supported by a softer, tougher core which gives strength to the part. The part being case-hardened is submerged in either a liquid or finely divided solid from which extra carbon is absorbed. While in contact with the liquid or finely divided solid, the part is heated to the necessary temperature. The parts may be submerged in a molten bath of cyanide, packed in scraps of leather, or other substances containing carbon. After case hardening, the part may be heat-treated to bring about the proper condition of hardness and temper.

Carburizing is a form of case hardening and consists of packing the part in a steel box surrounded by a mixture of some such substance as ground bone, charred leather, coke, wood charcoal, or even sugar. These substances may be used alone or in combination. Small amounts of barium and/or sodium carbonate are often added to assist in the reaction. A typical carburizing mixture contains approximately 60 per cent charcoal, 25 per cent powdered coke, 14 per cent barium carbonate, and 1 per cent sodium carbonate. The parts are packed in an alloy-steel box made of a corrosion-resistant material. The packing material is arranged so as to be in close contact with the part to be hardened. Copperplating will protect any portion of the part from the hardening process. The box is sealed with moist fire clay. If the fire clay tends to crack, a little salt may be added. The whole is then heated in a furnace to a temperature between 1600° and 1700° F. The temperature is maintained at this level until the desired results are obtained. The longer the parts are held at this temperature, the deeper the carburizing will penetrate. A case-hardened layer $\frac{1}{64}$ in. in depth is usually obtained in 1 hr. of heating at the above temperatures. Another method of carburizing

AIRCRAFT ENGINES

makes use of a melted salt to which amorphous carbon has been added. Carbon must be added at intervals to replace that driven off by the heat. A depth of from $1/10$ to $3/10$ in. of hardening per hour may be obtained by this method. Hardening in a furnace may be carried out by passing methane gas through the furnace while the part is held at the desired temperature. Small areas may be hardened by the application of the oxy-acetylene flame, having an excess of acetylene.

Flame hardening is hardening the outside layer of iron and steel alloys by use of the oxy-acetylene flame. The surface to be hardened is heated rapidly to the desired temperature and then quenched in a bath or stream of cold water. If oil is used for the quenching, it should be cold. The high temperature of the oxy-acetylene flame makes it possible to heat the surface of the metal to a high temperature before the metal at greater depths has been affected.

Cyaniding is a form of case hardening in which the material to be hardened is heated in contact with a cyanide salt. All cyanide salts are extremely poisonous, and great care must be used to avoid coming in contact with the salt or breathing the fumes. The parts to be hardened are heated and dipped into the melted salts.

Nitriding is a form of surface hardening used for special alloys containing steel in combination with chromium or molybdenum. The parts to be hardened are heated to temperatures from 900° to 1000° F. in a special furnace into which ammonia is introduced. At this temperature, the ammonia breaks down and releases an active form of nitrogen which combines with the outside layer of the part being treated. This outside layer is extremely hard and brittle. The nitrided layer is harder than the surface obtained from treating with carbon. The nitriding process does not cause cracking or distortion of the part. No quenching is required, and usually no scaling occurs. The nitrided area is resistant to corrosion. Any part of the surface may be protected from the nitriding operation by coating it with tin or solder.

IV GLOSSARY OF TERMS USED IN AIRCRAFT ENGINES

acceleration. Change in velocity.

aircraft. Any weight-carrying device designed to be supported by the air, either by buoyancy or by dynamic action.

airfoil. Any surface, such as an airplane wing, aileron, or rudder, designed to obtain a reaction from the air through which it moves.

airplane. A mechanically driven fixed-wing aircraft, heavier than air, which is supported by the dynamic reactions of the air against its wings.

alloy. A metal made up of two or more metals usually formed by melting them together.

alternating current (ac). Electric current which flows in one direction, then in the opposite direction, reversing at regular intervals.

alternator. A machine for generating alternating current. An alternating-current generator.

altitude. Height above some given level. Ceiling is given in feet above ground; altimeters are generally set to give height above sea level.

ampere. The unit for measuring the rate of flow (current) of electricity.

aneroid. A metallic cell from which the air has been partially exhausted, arranged to contract or expand with changes in gaseous pressure.

annealing. Heating a metal to a temperature above its critical temperature and then allowing it to cool slowly. This usually softens the metal.

anode. A positive or plus terminal of a battery or other electrical source.

antiknock value (antidetonation). The octane rating of a fuel.

arcing. The jumping of an electric current from one conductor to another through the air.

armature. The revolving core of an electric generator or magneto.

articulated rod (link rod). One of the connecting rods of a radial engine which is attached to the master rod by means of knuckle pins.

atmospheric pressure. The normal pressure of the atmosphere at any given elevation.

autosyn. One of the remote indicating systems.

axial motion. Motion along the axis of a revolving body.

25

AIRCRAFT ENGINES

baffle, cylinder. Metal deflectors used to direct air around cylinders to ensure proper cooling.

battery. A cell or device for supplying an electric current by chemical means.

battery, storage. A cell which is restored by driving an electric current backward through the electrolyte.

British thermal unit (Btu). A unit of heat measure. The amount of heat necessary to raise the temperature of 1 lb. of water 1° F.

cam. A projecting part on a shaft, rim, plate, or wheel to impart a desired movement to a follower in contact with that part of the shaft, rim, plate, or wheel.

camber. The convex or concave curvature of the surface of an airfoil.

cam plate. A flat circular plate having cam lobes around its circumference.

camshaft. A shaft which carries cam lobes.

carburetor. A device for measuring and mixing the proper amount of fuel with air to form the air-fuel mixture.

cathode. The negative electrode of a source of electric current.

cell, dry. An electric cell having all of its parts sealed against the atmosphere. The materials in this cell are not dry.

cell, wet. An electric cell in which the electrolyte is entirely liquid and not sealed from the atmosphere.

centrifugal force. The tendency of a body to move in a straight line when forced to move in a curved path.

centrifugal pump. A pump which makes use of centrifugal force in its operation.

combustion chamber (compression chamber). The total space within the cylinder and cylinder head when the piston is at top dead center.

compression ratio. The ratio between the space within the cylinder and combustion chamber when the piston is at bottom dead center, and the space within the cylinder and combustion chamber when the piston is at top dead center.

congeal. An increase in viscosity of a liquid, usually due to change in temperature.

coulomb. A unit used to measure quantities of electricity.

counterweight. A weight attached to a revolving part such as a crankshaft to bring about conditions of balance.

crankcase. The part of an internal-combustion engine that encloses the crankshaft and connecting rods.

crankpin (wrist pin). The part of the crankshaft about which the connecting-rod bearing fits.

crankshaft. The main rotating shaft of an engine by means of which the power developed in the cylinders is changed to rotary motion.

cycle. One of a complete series of recurring events.
four-cycle principle (Otto cycle). One power stroke is developed for each four strokes of the piston in the cylinder.
two-cycle principle. One power stroke is developed for each two strokes of the piston in the cylinder.

cylinder. A chamber in an engine in which a piston slides back and forth.

density. Actual weight per unit volume of any substance.

GLOSSARY OF TERMS USED IN AIRCRAFT ENGINES

detonation. In an engine, a rapid combustion replacing normal combustion. When detonation occurs, there may be loss of power, engine overheating, and noise.

diesel engine. An internal-combustion engine in which the air-fuel mixture is ignited by heat of compression within the cylinder.

diffuser. A device for reducing the speed of the air-fuel mixture leaving the supercharger impeller and thereby permitting a charge of uniform density to enter each cylinder.

direct current (dc). Current that flows in one direction.

distill. A process by which a liquid is changed to vapor and then condensed back to a liquid.

dynamic damper. A short pendulum in the form of a counterweight, designed to damp out engine vibrations set up by the power strokes.

dynamotor. Combination of generator and motor.

electrode. Either terminal of an electric source. An electrode may be a wire, plate, or other conducting object.

electron. A negative particle of electricity.

element. A simple substance made up of only one kind of atom, such as iron, aluminum, or oxygen.

engine, internal-combustion. Any engine in which the energy is produced in the engine cylinder.

engine, external-combustion. A heat engine which derives its heat from fuel consumed outside the engine cylinder.

exhaust port. The opening from which gases escape from the cylinder after combustion.

exhaust valve. A valve for opening and closing the port through which the exhaust gases escape from the cylinder.

feathering. Rotating a propeller blade so that the leading edge meets the air in such a manner that the air stream is approximately parallel to the chord of the blade.

flywheel. A wheel on an engine crankshaft to absorb the variable force of the power stroke of the piston and to carry the engine over the dead centers.

foot-pound (ft.-lb.). A unit of energy, or work, being equal to the work done in raising 1 lb. avoirdupois against the force of gravity a distance of 1 ft.

forging. Metal fabricated by the blows of a hammer or a die forcing the material to conform to the contours of the forming die.

fuel. Any substance used to produce heat by burning.

fuselage. The main structure of an airplane, approximately streamlined in form, to which are attached the wings and tail units.

generators. A general term applied to machines which are used for the transformation of mechanical energy into electrical energy.

glide. The normal flight of an airplane without power.

governor. A regulating device.

graduations. Marks setting off intervals such as inches or degrees on a measuring instrument.

gyroscope. A rapidly revolving, heavy wheel.

gyroscopic. Pertaining to an action brought about by the use of a gyroscope.

AIRCRAFT ENGINES

horsepower. That power necessary to raise 1 lb. 33,000 ft. in 1 min., or that power which will raise 33,000 lb. 1 ft. in 1 min. against the pull of gravity.

hydraulic. Pertaining to liquids in motion.

hydrometer. An instrument for measuring the density of liquids.

idling. The speed of rotation of an engine with the throttle closed.

ignition. The act of igniting; subjection to the action of fire or intense heat.

impeller. The vaned rotating part of a supercharger.

induced current. Current set up in a conductor due to the effect of a magnetic field.

induction coil. An apparatus for transforming an ordinary battery current by induction into an alternating current of high potential.

inertia. The tendency of a body to continue in its state of rest or motion.

inflammable. Burns readily.

intake valve. A valve for opening and closing the port through which the charge of air-fuel mixture enters the cylinder.

jet propulsion. Propulsion by means of the reaction of gases under high pressure escaping through an opening.

jig. A pattern, form, or framework, accurately dimensioned and aligned, in which identical structures or parts can be produced to meet a standard.

lapping. Finishing a metal surface to a high polish by the use of a fine abrasive.

linear. Lengthwise.

link rod. (*See* articulated rod.)

loading, power. The gross weight of an airplane divided by the rated horsepower of the engine computed for air of standard density, unless otherwise stated.

lodestone. A naturally occurring iron ore which attracts magnetic substances.

lobe. The rounded projection of a cam.

lubricant. A substance used to reduce friction.

magnetic field. Any space through which magnetic influence is exerted.

magneto. A device for producing electricity dependent on permanent magnets for its magnetic field, commonly used for ignition in internal-combustion engines.

main bearing. A bearing supporting the crankshaft of an internal-combustion engine.

manifold. A pipe or casting with several outlets, used between the carburetor and cylinders of an internal-combustion engine or to carry off the exhaust heat and gases.

meshes. Fits together; for example, two gears.

metallurgy. The science of metals.

metering jet. An opening of predetermined size which regulates the flow of fuel through the carburetor.

molten. In liquid form.

momentum. The energy contained in a body due to its motion.

nacelle. An enclosed shelter for personnel or for a power plant. A nacelle is usually shorter than a fuselage and does not carry the tail unit.

GLOSSARY OF TERMS USED IN AIRCRAFT ENGINES

neutron. An uncharged particle in the nuclei of atoms.

octane. The antiknock rating given in numbers.

octane rating. The percentage of iso-octane in a mixture of iso-octane and normal heptane required to match the performance of the fuel being tested in a special test engine under controlled conditions.

ohm. The practical unit of electrical resistance, being the resistance of a circuit in which a potential difference of 1 volt produces a current of 1 ampere.

piston. The plunger which moves within the cylinder of an engine or pump. The efficiency of compression depends largely on the proper fitting of the piston.

precision instrument. An instrument for accurately measuring quantities.

preignition. The ignition of the compressed gas in the combustion chamber before it is desired.

propeller. Any device for propelling a craft through a fluid such as water or air; especially a device having blades which, when mounted on a power-driven shaft, produce a thrust by their action on the fluid.

adjustable propeller. A propeller whose blades are so attached to the hub that the pitch may be changed while the propeller is at rest.

automatic propeller. A propeller whose blades are attached to a mechanism that automatically sets them at their optimum pitch for various flight conditions.

controllable propeller. A propeller whose blades are so mounted that the pitch may be changed while the propeller is rotating.

geared propeller. A propeller which rotates at a speed different from that of the engine crankshaft due to a train of gears between the crankshaft and the propeller shaft.

pusher propeller. A propeller mounted on the rear end of the engine or propeller shaft.

tractor propeller. A propeller mounted on the forward end of the engine or propeller shaft.

propeller area. The blade area times the number of blades.

propeller efficiency. The ratio of the thrust power to the input power of a propeller.

propeller pitch. The distance the propeller would screw forward in a semi-solid due to the angle of blade setting.

effective pitch. The distance an aircraft advances along its flight path for one revolution of the propeller.

theoretical pitch. The distance a propeller would move through the air if no slip occurred.

static pitch. A condition where the propeller rotates in the same path without forward motion, as when warming up the engine.

propeller radius. The distance of the outermost point of a propeller blade from the axis of rotation.

propeller slip. The difference between theoretical pitch and effective pitch.

propeller thrust. The component of the total air force on the propeller which is parallel to the direction of advance.

propeller tipping. A protective covering of the blade of a propeller near the tip.

AIRCRAFT ENGINES

proton. A positive charge of electricity.

p.s.i. Pounds per square inch.

pulsating direct current. Current which flows in one direction all the time, but is periodically interrupted or changes its intensity at more or less regular intervals.

ram effect. The pressure built up by the velocity of a fluid such as air.

rarefied atmosphere. Atmosphere of less density than standard.

resistor. An electrical resistance such as a resistance coil.

rheostat. A variable resistance. Usually, a radial arm touches a circular resistance so that when it is rotated, more or less of the resistance is included in the circuit.

rocker arm. An oscillating arm borne by a shaft which is usually used to operate valves.

rocket. A device propelled by the burning of fuel contained within the device itself, for example, a sky rocket.

S.A.E. (Society of Automotive Engineers), **or Saybolt, numbers.** These numbers are used to indicate the viscosity of a liquid such as oil.

safetyed. Fastened securely in place.

scavenger oil (or **scavenged oil**). Used oil which is removed from an engine after it has passed through the bearings, etc. Oil removed from a dry sump engine by the scavenger pump.

scavenger pump. A pump for removing used oil from the engine and returning it to the oil tank.

servo unit. A cylinder containing a piston which moves back and forth bringing about a desired movement such as the moving of an aileron by the action of the automatic pilot.

shielding (radio). A method by which radio interference due to the aircraft electrical system is eliminated.

sodium. A metallic element.

solenoid. Operated by means of a magnetic coil which is energized by means of an electric current causing a soft iron core to move due to magnetic attraction.

spark plug. A plug which is used to ignite the air-fuel mixture.

specific gravity. The number of times the weight of a unit volume of a substance is of the weight of an equal volume of water.

spinner. A fairing of approximately conical or paraboloidal shape, which is fitted coaxially with the propeller hub and revolves with the propeller.

splines. A series of lands and grooves cut in a shaft to prevent its turning when inserted into matching lands and grooves.

stellite. A nonferrous, cobalt-chromium-tungsten alloy used for hard-surfacing other metals or alloys.

sump, dry. An engine in which the oil supply is carried other than in the crankcase of the engine.

sump, wet. An engine in which the oil supply is carried in the crankcase.

supercharger. A device, usually a rotating fan, for supplying the engine with a greater weight of air-fuel mixture than would normally be inducted at the prevailing atmospheric pressure.

surging. Irregular flow of a liquid.

synchronize. To cause to agree in time, as to time a magneto armature

GLOSSARY OF TERMS USED IN AIRCRAFT ENGINES

with the distributor so they agree in their timing; to cause an agreement in the timing of two separate ignition systems attached to one engine.

tachometer. An instrument that indicates in revolutions per minute the rate at which the crankshaft of an engine turns.

tappet. A mechanism between the cam and valve or push rod.

tension. A pulling force.

thermal circuit breaker. A circuit breaker operated by means of changes in temperature.

torque. That force which produces or tends to produce rotation.

torsional force. A twisting force.

transformer. Two coils of wire wound on a common iron core used to change the voltage of an electric current by induction.

vacuum. A true vacuum is a space entirely empty of all matter, even traces of gas.

valve. A device for opening or closing a passageway or a port.
　by-pass relief valve. A valve designed to regulate the oil or fuel pressure within the engine by allowing excess oil or fuel to bypass the pump.
　valve guide. A bearing through which the valve stem passes.
　valve tappet. A tappet between a cam and a valve.

vapor pressure. The pressure exerted by a vapor within a confined space.

venturi tube. A short tube of varying cross section. The flow through the Venturi causes a pressure drop in the smallest section, the amount of the drop being a function of the velocity of flow.

volatility. A property of a liquid which determines the temperature at which it passes into the vapor state. Too high volatility in a fuel may produce vapor lock. Too low volatility produces difficulty in engine starting at low temperatures.

volt. That electromotive force which if steadily applied to a circuit whose resistance is 1 ohm will cause a current of 1 ampere to flow.

voltage booster. A device for increasing the voltage of an electric current.

volumetric efficiency. The highest possible filling of a combustion chamber and cylinder with air-fuel mixture when the piston is at bottom dead center.

watt. A unit of electric power found by multiplying amperes by volts.

Wheatstone bridge. A device for the measurement of resistances.

windmilling. The turning of a propeller by the force of the air stream.

wobble pump. A hand fuel pump used for starting or for emergency purposes. The handle may be provided with holes by means of which an extension rod may be attached for operating the pump at some distance from the pilot's seat.

wrist pin. (*See* crankpin.)

V TYPES OF ENGINES

While the fundamental principles of all internal-combustion engines are much the same, many different types of engines have been built. The first internal-combustion engine had a single cylinder and was equipped with a heavy flywheel. As multiple-cylinder engines were developed, most of them were of the in-line type. In this engine the

Fig. 17. A light trainer-type aircraft equipped with a four-cylinder horizontally opposed engine. (Courtesy Piper Aircraft Corporation)

cylinders are arranged in a row along the crankshaft. Quite early in the history of the aircraft engine, attempts were made to arrange the cylinders around the crankshaft, and a number of rotary engines were developed. In this engine, which worked on the two-cycle principle, the cylinders rotated about the crankshaft. The propeller was attached directly to the crankcase, rotating with the engine.

The nonrotating radial engines first built had two cylinders. These cylinders were opposed, one on each side of the crankshaft. The two-cylinder radial engines were soon followed by the three-cylinder, five-

TYPES OF ENGINES

cylinder, seven-cylinder and nine-cylinder radials. The nine-cylinder radial engine was followed by the two-row radials, having seven or nine cylinders in each row. All radial engines have an odd number of cylinders in each row, which simplifies their firing order.

Fig. 18. A seven-cylinder, radial, air-cooled aircraft engine. (Courtesy Jacobs Aircraft Engine Company)

One classification for aircraft engines is based on cylinder arrangement. In the vertical, in-line engine the cylinders are arranged above the crankshaft with the heads upward. Four-, six- and eight-cylinder, vertical, in-line engines have been used successfully. These were followed by the V-type engine which has two banks of four or six cylinders each.

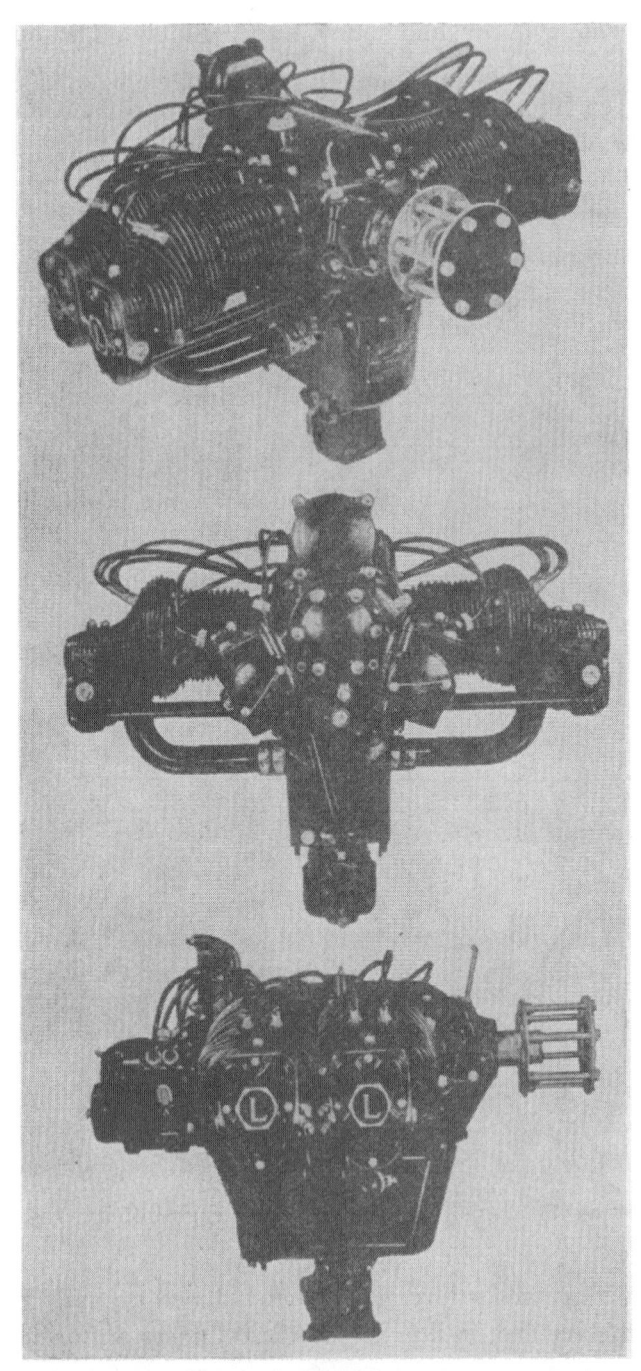

Fig. 19. Three views of a light four-cylinder opposed wet-sump aircraft engine. (Courtesy Lycoming Division, The Aviation Corporation)

Fig. 20. A four-cylinder opposed aircraft engine with the oil tank attached to the engine. This figure shows the two magnetos and ignition wiring. Note the shielded spark plugs. (Courtesy Continental Motors Corporation)

Fig. 21. A six-cylinder horizontally opposed engine showing the generator, magnetos, and carburetor. (Courtesy Continental Motors Corporation)

35

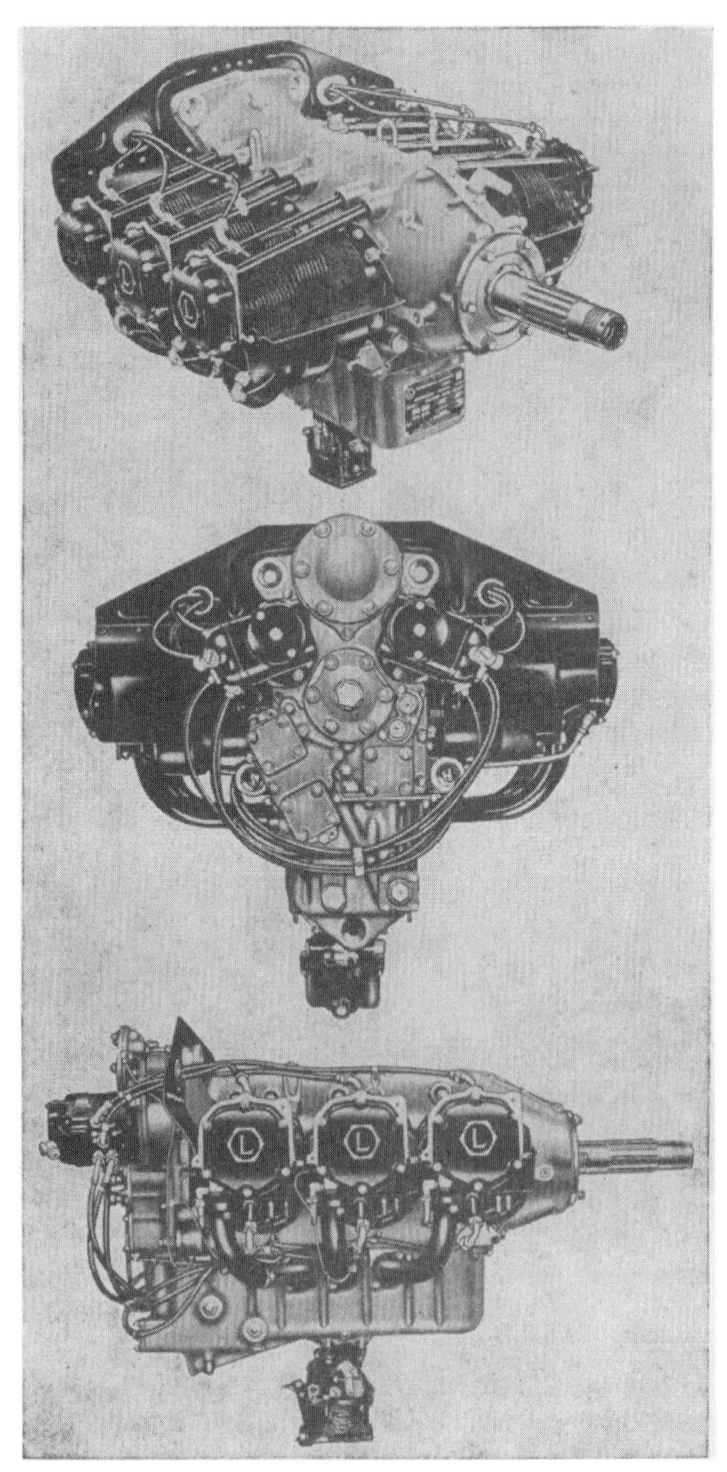

Fig. 22. Three views of a six-cylinder horizontally opposed wet-sump engine. (Courtesy Lycoming Division, The Aviation Corporation)

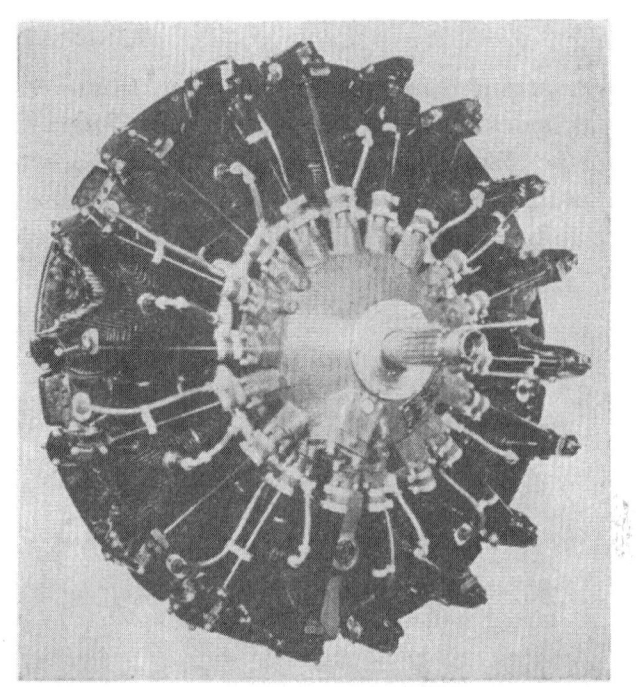

Fig. 23. A nine-cylinder radial aircraft engine developing approximately 300 horsepower. (Courtesy Lycoming Division, The Aviation Corporation)

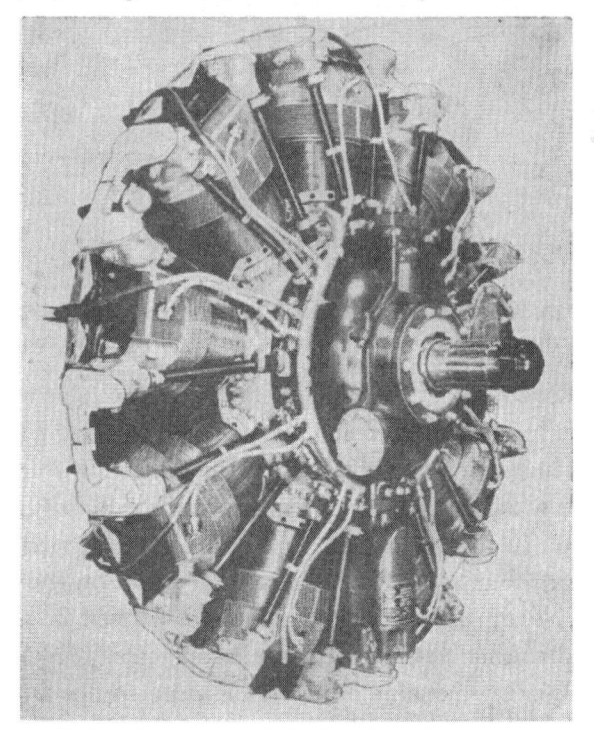

Fig. 24. A nine-cylinder single-row radial aircraft engine developing approximately 1350 horsepower. (Courtesy Wright Aeronautical Corporation)

AIRCRAFT ENGINES

The banks are arranged from 45° to 60° apart. In the vertical, in-line engines the cylinders being above the crankshaft tended to obstruct the view of the pilot. Largely because of the desire for visibility, the inverted in-line engines were developed. In this engine the cylinders are arranged in line below the crankshaft with the head of the cylinder downward.

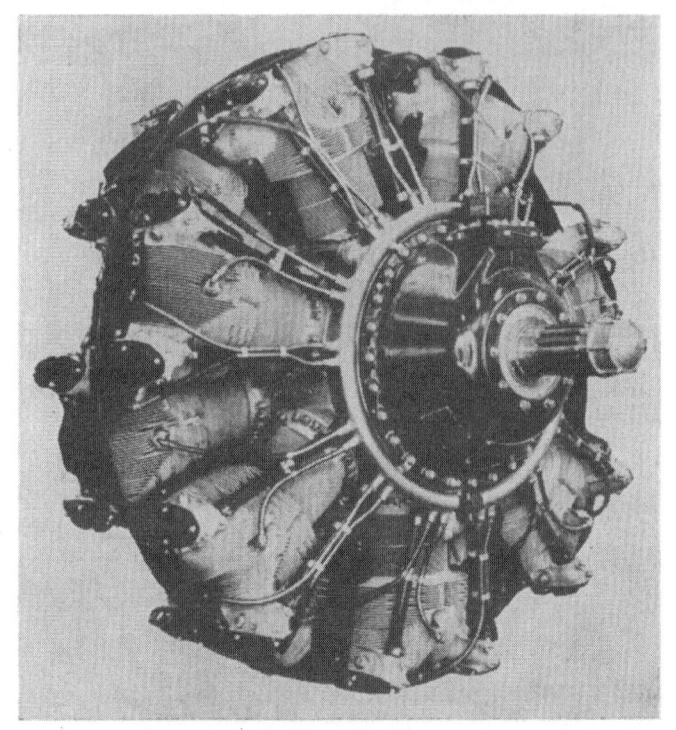

Fig. 25. A 14-cylinder twin-row radial aircraft engine developing approximately 1700 horsepower. (Courtesy Wright Aeronautical Corporation)

This engine led to the need for a type of lubricating system in which the oil would be carried separately from the engine. Two common lubricating systems are used in aircraft engines and are classified as wet sump and dry sump. In the wet-sump engine, the oil is carried in the engine itself, either in the crankcase or in a special sump on the bottom of the engine. In dry-sump engines, the oil supply is carried in a separate tank and the oil is furnished to the engine under pressure. The oil, after passing through the bearings, drains into one or more sumps from which it is returned to the oil tank by means of scavenger pumps. The rapid movement of the pistons prevents the oil from flowing into the cylinders

TYPES OF ENGINES

and gathering in amounts sufficient to interfere with engine operation.

The inverted V-type engine is similar to the vertical V-type, except that the cylinders are below the crankshaft level, sloping downward and outward from the crankshaft at an angle of from 45° to 60°.

Fig. 26. An 18-cylinder twin-row radial aircraft engine developing approximately 2200 horsepower. (Courtesy Wright Aeronautical Corporation)

X-type engines have been built which are a combination of the vertical and inverted V-type. These engines have two rows of cylinders in the form of a V above the crankshaft and two rows of cylinders in the form of an inverted V below the crankshaft.

A W-type engine was built having three rows of cylinders above the crankshaft level.

The opposed type of aircraft engine has come into popular use. In this engine, the cylinders are arranged horizontally in two rows 180° apart on opposite sides of the crankshaft. Opposed-type engines have two or more cylinders on each side of the crankshaft and develop from 40 to well over 1000 hp. The advantage of the opposed-type engine is that it occupies very little space vertically and allows good visibility. Some of the newer engines have been designed to be placed in the wing of the aircraft, leading to much better streamlining.

In radial engines, the cylinders are arranged about the crankshaft in the form of a circle. The cylinders radiate outward from the crankcase

AIRCRAFT ENGINES

in the same manner as do the spokes of a wheel. Because some of the cylinders in the radial engines point downward, these engines use a dry sump. Single-row, radial engines with nine cylinders develop from 175 to 1350 hp. and multiple-row, radial engines having as many as 28 cylinders develop over 4000 hp.

Two types of cooling systems are used in aircraft engines: air and liquid. With the exception of the rotary type of engine, most of the early

Fig. 27. A six-cylinder inverted in-line air-cooled aircraft engine. Three-quarter right rear view. (Courtesy Ranger Aircraft Engines)

engines were water cooled. The water was carried in a radiator and circulated through jackets surrounding the cylinders. Recent liquid-cooled engines use ethyl glycol, a liquid similar to Prestone, which has a higher boiling point and lower freezing point than water.

Engines may also be classified as supercharged and nonsupercharged. Supercharging is usually associated with engines of high horsepower, although engines of comparatively low horsepower are now being supercharged, and this will probably be standard equipment on all except the lightest of aircraft in the near future.

Fuel systems are another method by which engines may be classified.

TYPES OF ENGINES

Fig. 28. A 12-cylinder V-type aircraft engine with a single stage supercharger, developing approximately 1300 horsepower. (Courtesy Allison Division, General Motors Corporation)

The common, float-type-carburetor, fuel system mixes the fuel with the air in the carburetor before it passes into the intake manifold. In this system the fuel is usually fed to the carburetor by gravity. The pressure fuel system is similar to the float-carburetor system except that the fuel is fed to the carburetor under pressure by a pump. The injector-type-carburetor fuel system differs in that the fuel is sprayed into the air stream after the air passes through the Venturi tubes in the carburetor. The injection fuel system does not make use of a carburetor. In this system the fuel is injected by means of pumps into the air stream, passing through the intake manifold near the intake-valve opening.

Considerable experimenting has been done with the Diesel aircraft engine. At present, the Diesel aircraft engine is not used to any great extent in this country. It is expected, however, that this engine will be

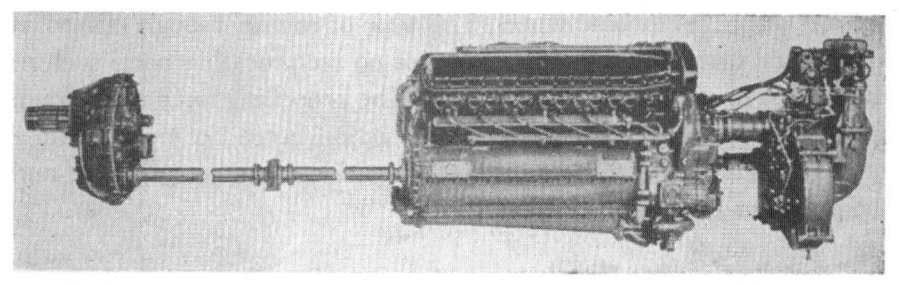

Fig. 29. A 12-cylinder V-type aircraft engine with an auxiliary stage supercharger and a propeller reduction gear box with an extension shaft. (Courtesy Allison Division, General Motors Corporation)

AIRCRAFT ENGINES

developed and used in the aircraft industry in the future, as it is now being used to a considerable extent in other countries.

A propeller may be driven directly by the crankshaft or may be driven by means of a reduction gear. Reduction gears are installed to allow the propeller to rotate at a lower rate of speed than the crankshaft.

Fig. 30. A 12-cylinder V-type engine with an auxiliary stage supercharger. Note the propeller extension shaft connection at left of illustration. (Courtesy Allison Division, General Motors Corporation)

Engines having the propeller attached directly to the crankshaft are direct-drive engines. When the reduction-gear arrangement is used, it is known as a reduction-gear-drive engine.

The jet engine is a radical departure from the conventional reciprocating engine. It lacks the cylinders, ignition system, and superchargers which have been standard parts of high-powered aircraft engines. This engine, which is discussed in greater detail in Chapter XXII, has but one moving part consisting of a rotor shaft to which is attached a compressor and a gas turbine wheel. This type of engine, though classed as an internal-combustion engine, contains no reciprocating parts such as the pistons in the conventional engine. The propelling force in the pure jet type of engine is developed by jet reaction instead of by a rotating propeller. The Propjet engine is a combination of the jet principle and the propeller type of propulsion.

VI CYLINDERS

The cylinders of an internal-combustion engine are the heart of the engine. It is within the cylinder itself that the power is developed. The cylinder must be strong enough to hold the tremendous pressures developed by the burning of the air-fuel mixture. It must also be able to withstand the high temperatures developed.

The cylinder is simply a tube in which the piston slides back and forth. The main part of the cylinder, or cylinder barrel, must be made of an alloy which will resist the wear caused by the continual sliding back and forth of the piston. The cylinder barrel is made of a strong steel alloy. It is usually forged and machined to exact size. Chromium nickel steel, chromium molybdenum steel, and carbon steel are alloys used in cylinder-barrel construction. On engines designed for small airplanes, the cylinders may be of cast nickel iron, while for large airplane engines they are usually forged nickel steel which has been heat-treated. On the higher-priced engines, both the barrel and head are forged, which means that the rough casting is hammered or pressed into shape while in a semiplastic condition at high temperatures.

When an engine is revolving at a rate of 2000 r.p.m. there are 1000 power impulses of each cylinder per minute. Aluminum has approximately three times the heat conductivity of steel, which is a decided advantage in using the aluminum alloys in the construction of the cylinder head. The greatest heat in the cylinder is developed at the head end of the cylinder. Overheating is one of the most frequent causes of engine cylinder failure. All parts of the barrel and head do not expand equally when heated. Therefore, allowance must be made so that the cylinder will be as near the proper shape as possible when at operating temperatures. The temperatures recommended by the manufacturer should be maintained at all times while operating the engine.

Some provision must be made to keep the barrel cool. In some engines the cylinder is surrounded by a jacket through which liquid is circulated

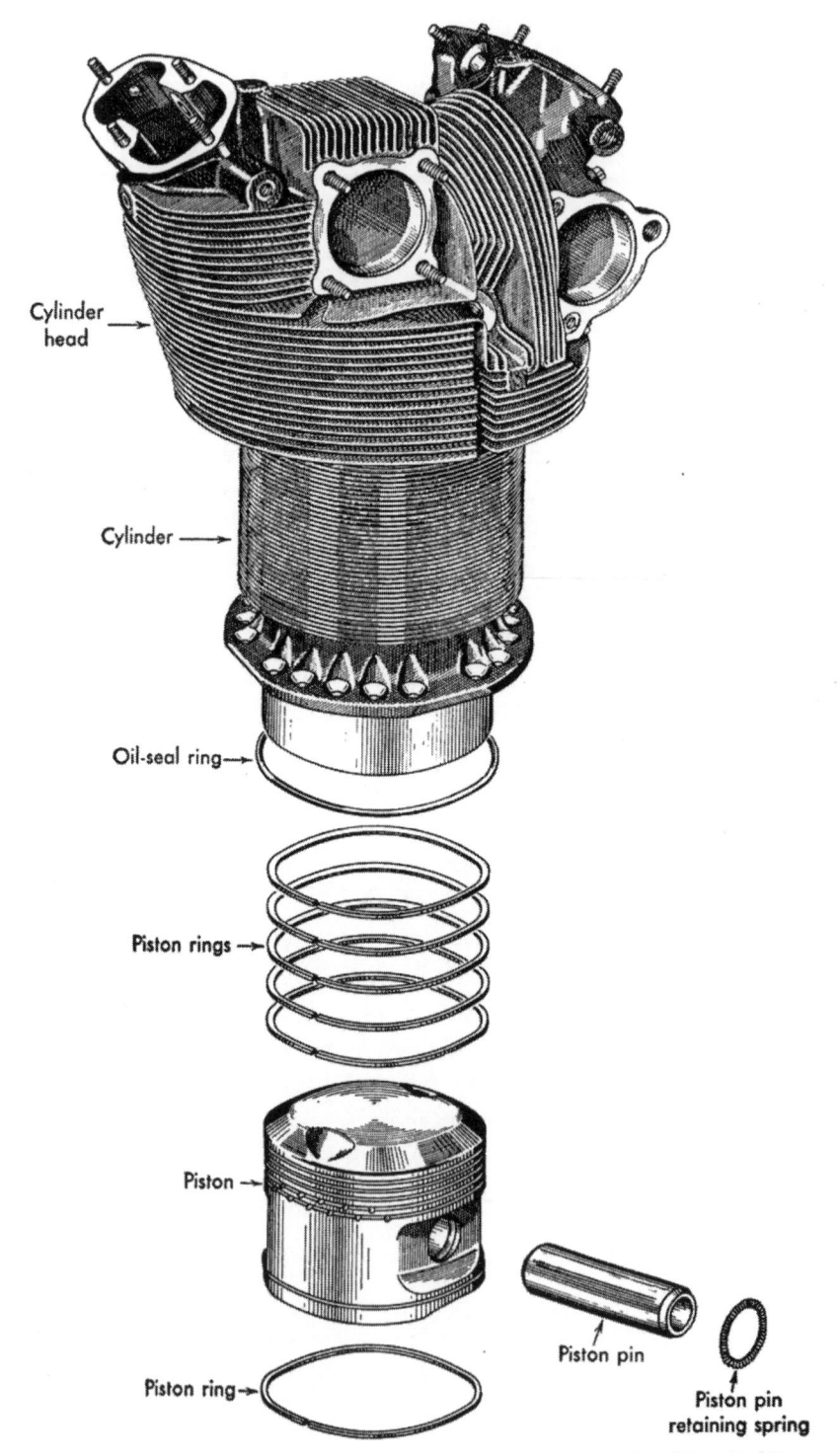

Fig. 31. An exploded view of a cylinder showing piston assembly. (Front.) (Courtesy Wright Aeronautical Corporation)

CYLINDERS

to carry away the excess heat. In the older types of engine this liquid was water. In the more modern engines, a solution having a higher boiling point than water is carried in a sealed radiator and circulated through the cooling jacket. On air-cooled engines, deep narrow fins

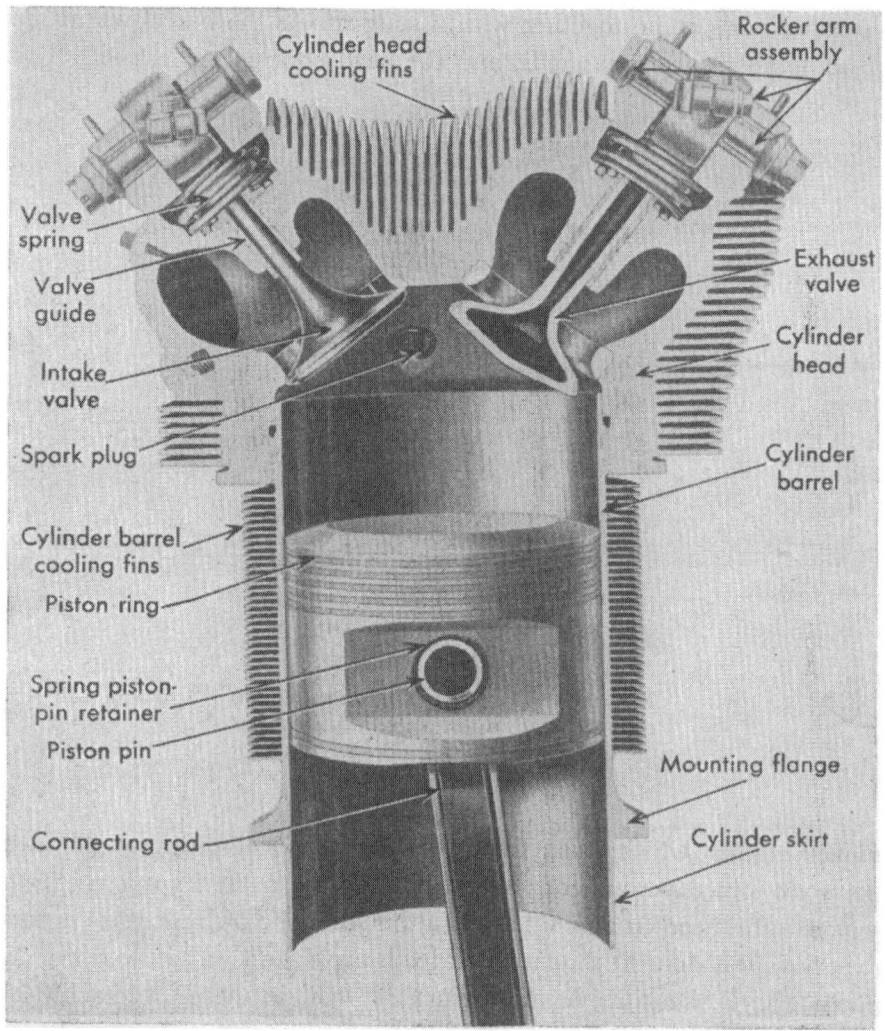

Fig. 32. A cutaway view showing piston and valve assembly. Note cooling fins on cylinder barrel and cylinder head. (Courtesy Wright Aeronautical Corporation)

are machined on the outside of the cylinder barrel to carry away the excess heat. The top end of the cylinder is closed by a cylinder head. This head is usually made of an aluminum alloy which may be either cast or forged and is equipped with fins to assist in cooling. The cylinder head is usually screwed on to the top of the cylinder barrel and shrunk into place. This shrinking-on is done by cooling the barrel to a low temperature and heating the head to a comparatively high temperature. While in this condition, the parts are screwed together. These parts have been so built that when the head is hot and the barrel is cool, the threaded parts come together with a snug fit. As the barrel warms up to air temperature it expands, and the head contracts as it cools. This

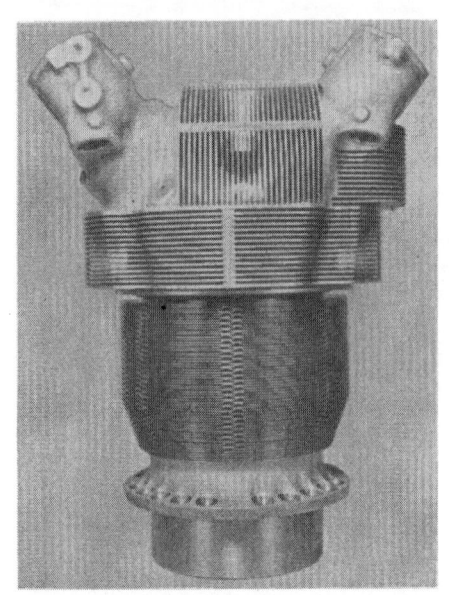

Fig. 33. A late-type forged aluminum cylinder head with inserted cooling fins on the cylinder barrel. (Courtesy Wright Aeronautical Corporation)

makes a joint which is so tight that for practical purposes the two parts are joined permanently together. Usually, no attempt is made to separate the head from the barrel, although some of the large engines may have this done at the factory. Heads are shrunk on to the barrel at about 450° F. On some small engines the cylinder head is detachable.

The bottom or skirt of the cylinder usually projects into the crankcase a considerable distance. The skirt does not need to be as strong as the upper part of the cylinder. The primary purpose of the skirt is to act as a piston guide. The lower part of the piston is in the skirt at the end of the downward stroke. The skirt is cooled by the lubricating oil in the crankcase.

The cylinder head contains the valve openings into which the valve seat is usually shrunk. Valve seats are generally made of aluminum bronze. On large engines the exhaust-valve seat may be formed of heat-

CYLINDERS

Fig. 34. A cutaway view of one cylinder of an inverted in-line engine showing the valve seat, valve arrangement, valve guide, valve spring, and intake and exhaust manifold openings. Note arrangement of cooling fins. (Courtesy Ranger Aircraft Engines)

resistant steel such as stainless steel. The valve guides are shrunk or pressed into the cylinder head and are usually aluminum bronze or steel. It is important that the coefficient of expansion of the valve seat and valve guide be the same as that of the cylinder head.

The spark-plug bushings are usually shrunk into the cylinder head. These are usually formed of aluminum bronze. The bushing must be

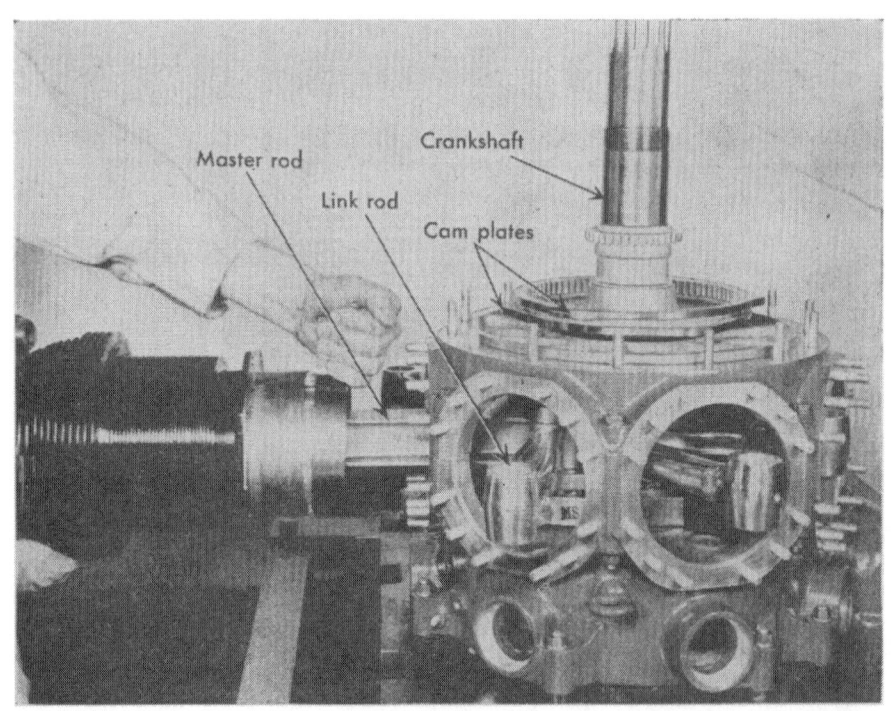

Fig. 35. Installing a cylinder of a radial engine. Note the studs on the crankcase for fastening the cylinder in place. This figure shows the piston rod assembly and cam plates. (Courtesy Jacobs Aircraft Engine Company)

Fig. 36. A six-cylinder inverted in-line aircraft engine. This engine has the camshaft and rocker-arm assembly below the cylinder heads. The camshaft is driven through the vertical drive at the front of the engine. (Courtesy Ranger Aircraft Engines)

CYLINDERS

fastened securely into place as the removal of spark plugs, for service, tends to loosen them. In some light engines, the cylinder head is bolted to the barrel by long bolts which extend past the barrel and fasten to the crankcase. Some cylinder heads are fastened both by shrinking and by means of bolts, or they may be fastened by a combination of screwing the head on the barrel, shrinking, and long bolts.

Fig. 37. I-, or straight-type, cylinder head.

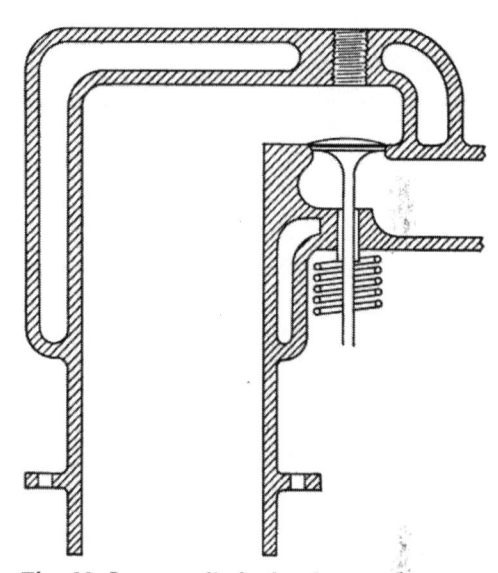

Fig. 38. L-type cylinder head.

The flanges to which the intake and exhaust manifolds are fastened are usually an integral part of the cylinder head. The flange to which the exhaust manifold is fastened generally has more and deeper cooling fins than the intake-manifold flange. This is because of the excessive heat from the exhaust gases.

Cast into the head are the attachments necessary for the rocker arm and rocker-arm housing. The rocker arm operates the valves. Attachments for other external accessories may also be cast into the head.

The pressure within the cylinder is enormous. As much as 15 tons pressure may be developed at each power impulse of the cylinder. This pressure tends to cause expansion and distortion of the cylinder and cylinder head. The pressure within the cylinder tends to force the cylinder away from the crankcase which is its main fastening to the engine. A flange is cast around the base of the cylinder barrel and is fastened to the crankcase by means of bolts or studs. In fastening the

AIRCRAFT ENGINES

barrel to the crankcase, these bolts or studs should be tightened evenly to avoid unequal strains on the crankcase. There is usually a gasket between the flange and the crankcase to prevent oil leakage.

Stresses caused by the recurring power impulses tend to cause metal fatigue and cylinder failure. Insofar as possible, sharp corners or bends

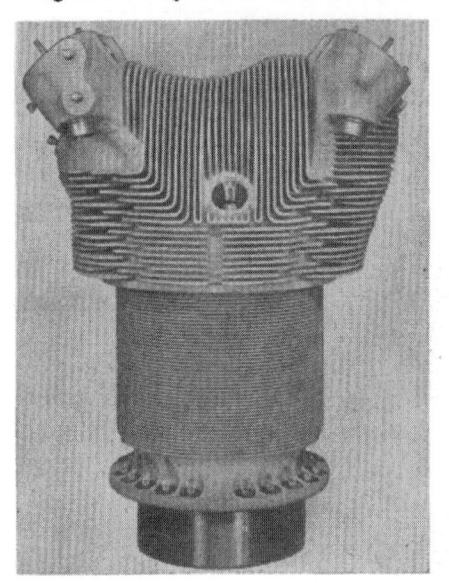

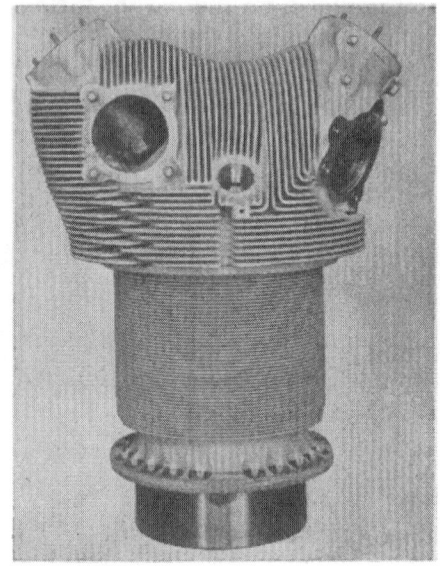

Fig. 39(a). Fig. 39(b).

Figs. 39(a and b). The front and rear view of a forged steel cylinder equipped with a forged aluminum head. Note arrangement of cooling fins. (Courtesy Wright Aeronautical Corporation)

should be avoided in cylinder and cylinder-head construction. Overcoming failure in this part of the engine has been a problem for the engineer as well as for the foundryman who needed to develop a metal of uniform strength throughout.

Most cylinders and heads are now the straight type, having both valves in the top of the head. Some of the older engines had the so-called L-type of head with both valves mounted at one side of the head.

The inside of the cylinder wall must be mirror smooth. After polishing, the inside of the cylinder is toughened or hardened, usually by nitriding. The barrels of forged cylinders are quite thin, and great care should be exercised in honing during overhaul, as there is danger not only of weakening the barrel, but also of removing the hard inside layer. Cylinders may be tested by hydraulic pressure during overhaul to determine indications of failure, but care should be taken not to cause distortion of any part of the cylinder.

CYLINDERS

Fig. 40. Honing a cylinder barrel. (Courtesy Wright Aeronautical Corporation)

The design of the inside of the cylinder head has a considerable effect upon the efficiency of the cylinder. The space in the cylinder between the top of the piston and the cylinder head, when the piston is at top dead center, is the combustion or compression chamber. This chamber is usually dome-shaped and the piston top is usually flat. This design is necessary to obtain proper compression.

VII PISTON ASSEMBLY

The piston assembly consists of a piston, piston rings, piston pin or wrist pin, and connecting rod. The piston moves up and down in the cylinder barrel drawing in the fresh charge of air and fuel, compressing it, and receiving the full force of the burning charge. Pistons transmit the force of the expanding gases to the crankshaft through the connecting rod. The piston cannot be made tight enough to seal the cylinder because the heat causes it to expand. The piston must be strong enough to withstand the high pressures developed in the cylinder and be of such material that it will assist in carrying away the heat of the burning fuel.

Many automobile pistons are made of cast iron. This material is not suitable for most aircraft engines because of its weight and comparatively low heat conductivity. With each revolution of the crankshaft, the piston attains high velocity twice, and twice comes to a complete standstill. This means that when the engine is revolving at the rate of 2000 r.p.m., the piston starts and stops 4000 times per minute. Between starts and stops the piston builds up a velocity of hundreds of feet per minute. If the stroke is $5\frac{1}{2}$ in., the piston travels 11 in. during each revolution of the crankshaft. If the crankshaft is revolving 2000 times per minute, the piston must travel 22,000 in. per min. When one realizes that the piston starts and stops 4000 times per minute while traveling at this high average rate of speed, it is easily seen that the strain on the piston is great. The heavier the piston, the greater the strain.

A cast-iron piston weighs approximately three times as much as an aluminum piston: therefore, most pistons in aircraft engines are made of aluminum alloys. This material may be cast or forged. The pistons are cast in a sand mold or formed in a permanent mold. When formed in a permanent mold, pressure is applied to the molten metal, having the effect of forging.

Pistons are machined to exact size, but are usually left slightly overweight for balancing purposes. The manufacturers usually control the

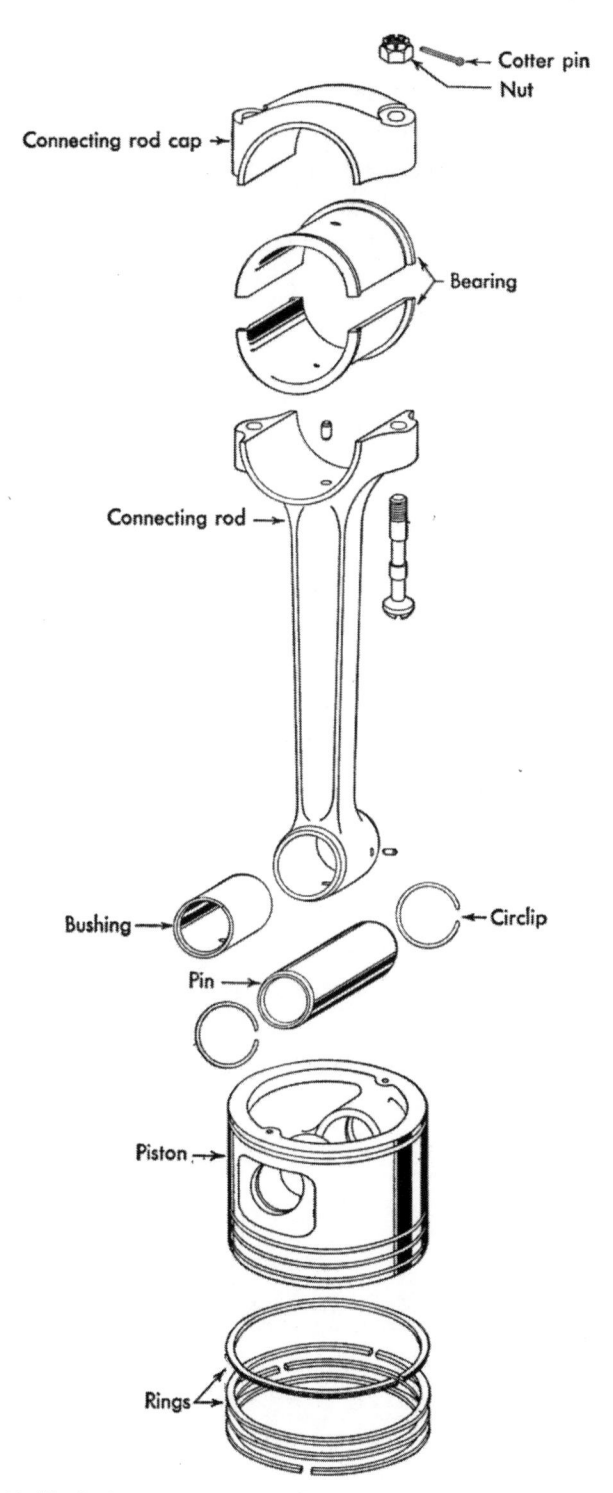

Fig. 41. Typical piston assembly. (Courtesy Ranger Aircraft Engines)

AIRCRAFT ENGINES

weight of pistons to within a few tenths of an ounce for balance in the engine. To balance the piston perfectly, some of the material is removed from the inside of the skirt. Pistons are cup shaped with the top end closed. This top end is called the "head." Some pistons are cast around a steel strut which assists in controlling expansion. This type of piston

Fig. 42. A six-cylinder inverted in-line engine having cylinder No. 1 cut away to show position of the piston in the cylinder. (Courtesy Ranger Aircraft Engines)

allows small clearances within the cylinder of approximately 0.003 in., as against about 0.009 to 0.045 in. for other types. To provide for expansion greater clearances must be allowed between the piston and cylinder wall when large pistons are used.

Most aluminum alloy pistons are heat-treated to increase their strength. There are a number of different types of pistons, classified according to their head shape. In most aircraft engines a flat-head piston is used. This type of piston may have two recesses or hollows cut into the top of the head to allow for valve clearance. On the exhaust stroke, when the piston is at top dead center, both the intake and exhaust valves are open and project into the cylinder.

Some pistons have a dome-shaped head to increase the compression ratio. Other pistons have a concave head, although this type is not used in most of the late engines. The trunk type or straight piston is straight

PISTON ASSEMBLY

up and down on the sides. The slipper type has part of each side cut away to reduce the amount of piston area which may come in contact with the cylinder. The closed end of the piston is the head and the part forming the open end is the skirt. Many pistons are slightly tapered from the bottom of the skirt to the head, the head being smaller than the skirt. The skirt is reinforced on each side by cast-in bosses through which holes are drilled to receive the piston pin.

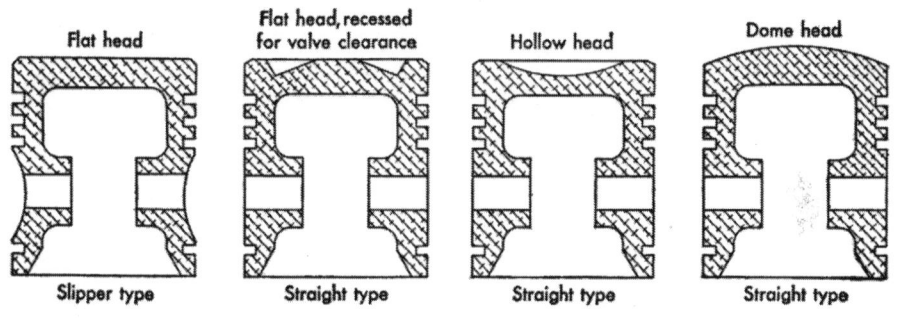

Fig. 43. Cross sections of various types of pistons showing types of piston heads.

Pistons must be so constructed that they will carry away as much heat as possible. Some pistons have cooling fins cast on the inside. These fins transfer heat to the oil which is splashed against the bottom part of the piston.

Cooling of the pistons is one of the limiting factors in engine performance. When the engine is overheated, the piston is one of the first parts to suffer. Aluminum is not only strong, but also has the ability to resist continual vibration. Any failure of a pistonhead is usually indicated by a foglike smoke which is blown out of the breather pipe of the engine. This is caused by the gases in the cylinder blowing through the pistonhead into the crankcase.

Since it is not possible to form a tight seal in the cylinder by the piston itself, piston rings are necessary. These rings keep the gases from blowing past the piston into the crankcase. The rings are inserted in grooves that are machined into the piston parallel to the edge of the head. The rings may vary in number depending upon the horsepower of the engine, and are rectangular or tapered in cross section. The number and location of the grooves vary in different engines. For example, the pistons in one of the light aircraft engines have two grooves above the pin and one below. The pistons in one of the high-powered engines have five grooves above the pin and one below. The metal between the grooves is known as a "land." That is, there are alternately a groove and a land.

Piston rings are usually made of gray cast iron. Gray cast iron has a fine texture and will take a high polish. It is springy enough to exert pressure on the cylinder wall, forming, with the assistance of the oil film, a gastight seal. Alloys of steel, some of which have been coated or plated with other metals, have been tried with varying degrees of success for the manufacture of piston rings. Most rings are heat-treated. The surface of the ring in contact with the cylinder must be highly polished. The sides of the ring which are in contact with the lands should also have a

Fig. 44. A forged aluminum piston of the slipper type. Note the piston pin, coil spring piston pin retainer, and cooling fins on the inside of the piston. (Courtesy Wright Aeronautical Corporation)

high polish to form a gastight joint. There is a trend toward plating the top compression rings with chromium to give longer wear and more effective compression.

Rings may be classified according to their use, such as compression rings, oil control rings, or oil wiper rings. The main function of the rings is to form a gastight joint between the piston and the cylinder wall. The top rings that receive the full force of the expanding gases must be strong. The top rings are compression rings and may be tapered slightly on the side (approximately 15°), which allows a wedging action, increasing the tightness of the fit between the ring, cylinder wall, and land. Tapered rings are usually marked "top" on the narrow side. The portion of the ring in contact with the cylinder wall and the land may have a smooth machined finish or may be lapped to a high polish. The oil control rings may have slightly chamfered edges. These rings carry oil over the sur-

PISTON ASSEMBLY

face of the cylinder. Wiper rings usually have a slightly beveled face and are used to remove excess oil from the cylinder walls. The bottom of the groove back of the wiper rings is usually drilled with oil relief holes which allow the oil scraped from the cylinder wall to flow through the piston wall into the crankcase. Escaping gases may burn the rings, causing rough edges to form. Small particles may break off the rough edges, scoring the cylinder wall and the other rings.

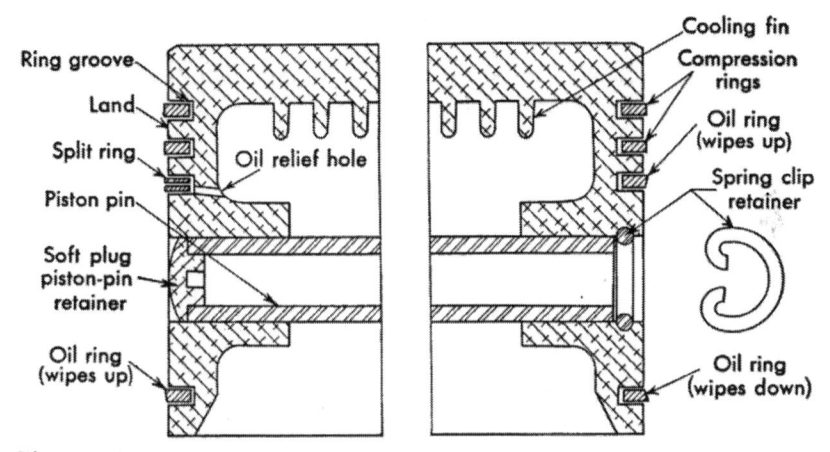

Fig. 45. Cross sections of pistons showing ring grooves, lands, piston pin and piston retainers, compression rings, split rings, and oil rings.

Oil control rings and wiper rings may have their ordinary position reversed when inserted in the bottom cylinders of a radial engine or in the cylinders of an inverted engine. The position of the rings determines whether the amount of oil on the cylinder wall is to be increased or decreased. In the top cylinders of a radial engine, the rings are usually arranged to carry oil upward over the cylinder walls, and in the bottom cylinders they are usually arranged to remove excess oil from the cylinders. All piston rings are split, that is, the ring is not continuous. This must be done to allow for expansion of the ring. The clearance between the two ends of the ring varies with different engines and must be in accordance with the manufacturer's specifications.

Rings are sometimes made thicker on the side opposite the split to equalize the expansion. The clearance between the ends should be measured accurately with a feeler gauge while the ring is mounted in a ring gauge. If a ring gauge is not available, a piston without rings may be inserted in a cylinder and a ring placed in position within the lower part of the cylinder skirt, which is the smallest part of the cylinder. The piston is then drawn back against the ring to align it in the cylinder.

AIRCRAFT ENGINES

While the ring is in this position, the end clearance may be readily measured. It is important that the side clearance of the ring, that is the space between the side of the ring and the land, be accurately determined. If the gap between the ends of the ring is too small it may be enlarged by clamping a small file in a vise and carefully removing some of the material from the end of the ring by sliding the gap over the file. Material from the side of the ring is best removed by the use of a lapping plate and lapping compound. A suitable jig should

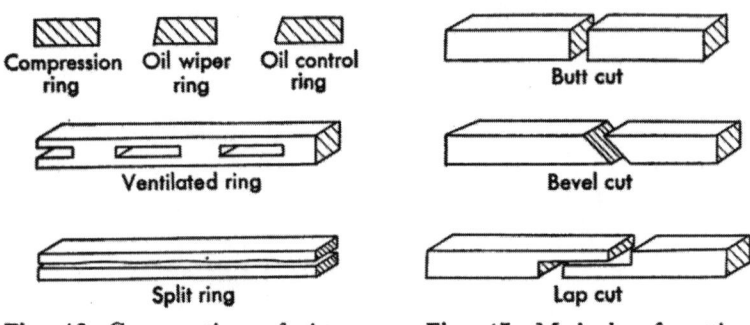

Fig. 46. Cross sections of piston rings, ventilated ring, and split ring.

Fig. 47. Methods of cutting piston ring gaps.

be used to equalize the pressure on the ring when lapping. Light rings are more desirable than heavy rings which tend to pound the sides of the grooves. Two light rings are often more desirable than one heavy ring. Light rings must be carefully adjusted as they have a tendency to flutter. The gap in the ring may be cut in several different ways such as a step cut, bevel cut, or straight butt cut.

The thin part of a step-cut ring is easily broken off, and the small pieces may score the cylinder walls. The thin points of bevel-cut rings have a tendency to break off under certain conditions, also causing damage to the cylinder wall. The plain butt or square-cut ring is the most commonly used as it gives better service. The split ring is made up of two parts that do not fit closely together, allowing the oil to flow between the two parts. Another type, the ventilated ring, has rectangular holes which assist in cooling.

Piston rings may be made by casting a tube of the desired size. The individual rings are cut from the end of the tube, machined to size, and the gap properly cut. Each ring may be cast individually and machined to the correct size. On this type of ring the scale is sometimes left on the inside. The manufacturer states that this increases its springiness and durability. The inside of some rings is hammered or peened to increase

PISTON ASSEMBLY

their springiness and cause them to exert a more nearly even pressure on the cylinder wall.

The ring is always made somewhat larger in diameter than the cylinder into which it is to fit. Rings must always be compressed when they are inserted into the cylinder and should exert a continuous pressure against the cylinder wall.

Fig. 48. A cutaway view of a cylinder and piston to show the full floating piston pin through the connecting rod. Note the spring clip piston pin retainer. (Courtesy Ranger Aircraft Engines)

Pistons have a tendency to rock back and forth around the piston pin. When this rocking motion is pronounced, a decided piston slap may be heard. Due to the noise of the propeller and exhaust this slap is not as annoying as in an automobile engine. Various types of spring expanders have been tried behind the rings to do away with piston slap, but without much success. The high temperatures destroy the temper of the spring expanders. Piston slap is developed as a result of piston or cylinder-barrel wear and, when it becomes excessive, the entire assembly should be overhauled. Pistons come in various oversizes. A worn piston may be replaced by an oversized piston, providing the cylinder barrel has been properly sized.

Piston pins or wrist pins are inserted through the holes drilled in the

AIRCRAFT ENGINES

piston bosses, passing through the opening in the upper end of the connecting rod. These pins form the connection between the piston and the connecting rod, which connects the piston with the crankshaft.

Piston pins are usually made of chromium steel or nitralloy, and are hard, strong, and highly polished. Piston pins are usually hollow.

Piston pins may be clamped to the end of the connecting rod, but more common practice is the full-floating installation. In the full-floating installation, the pin is free to revolve both in the connecting-rod bearing and the opening through which it passes in the piston. To prevent the hard pin from working out and scoring the cylinder walls, two general methods are used. In one method the pin is held in place by a circular coil spring or a spring steel clip which slips into a slot in the piston at the end of the wrist pin. In the second method, a soft alloy plug is inserted into the hollow end of the piston pin which prevents the pin from coming in contact with the cylinder wall.

Since aluminum is a good bearing material, the pin bears directly on the piston bosses. A bronze bushing is usually inserted in the end of the connecting rod to act as a bearing for the pin. Because of the difference in expansion between the aluminum alloy in the piston and the steel in the piston pin, the pin is inserted in the piston with a tight fit. When heat expands the piston, sufficient clearance is obtained to allow the oil to circulate through the bearing. The fit is not so tight in most pistons but that a well-oiled pin may be forced into place with the hands.

Connecting rods form the connection between the pistons and the crankshaft. The power developed by the combustion in the cylinder is transferred to the crankshaft through the connecting rods. The high stresses to which the connecting rods are subjected are not steady, but are applied at intervals by the action of the burning of the air-fuel mixture in the cylinder. The connecting rod is not only subjected to compression stresses developed by having the piston pushed downward in the cylinder, but also to tension stresses which are developed by the connecting rod's stopping the piston at the upper end of each stroke, particularly at the end of the exhaust stroke.

Connecting rods should be as light as possible to reduce the forces on the crankshaft. The connecting rod starts and stops in the same manner as the piston, twice each revolution. The part of the connecting rod connected with the piston moves up and down in the cylinder. The end fastened to the crankshaft travels with a circular motion.

To prevent failure in service, all cracks, checks, tool marks, or other

PISTON ASSEMBLY

blemishes should be carefully removed from the connecting rod. When a connecting rod fails, serious damage to the engine always results. The small end of the connecting rod, which is fastened to the piston by means of the piston pin, is usually in the form of a continuous loop of metal. This loop, which is a part of the connecting rod, is fitted with a bushing through which the piston pin passes. This is usually a solid, press-fit,

Fig. 49. A crankshaft and connecting rod assembly for a 12-cylinder V-type engine. Note the split-type connecting rods, also the counterweight and dynamic dampers on the end of the crankshaft. (Courtesy Allison Division, General Motors Corporation)

bronze bushing. The large end of the connecting rod fastens to the crankshaft. The opening in this end is usually split and made up of two parts in order that it may be fitted around the crankshaft. The loop in the large end of the connecting rod is fitted with a bearing which rests against the crankshaft. These bearings are of several types, usually made of steel or some other backing material and lined with a soft bearing material to prevent wear on the crankshaft. Most engines use a thin pre-fit shell bearing which is lined with the bearing material. This bearing material in the older engines was babbitt, but the more modern engines use cadmium silver, silver cadmium zinc alloys, lead bronze alloys, or alloys which may be lined with a thin layer of lead or other soft material. Copper lead-bearing alloys are sometimes used.

Most connecting rods are machined from heat-treated forgings of alloys, such as aluminum, chromium molybdenum steel, and chromium

AIRCRAFT ENGINES

nickel steel. Connecting rods are often polished or shot-blasted to remove surface blemishes, thereby reducing chances of future failure.

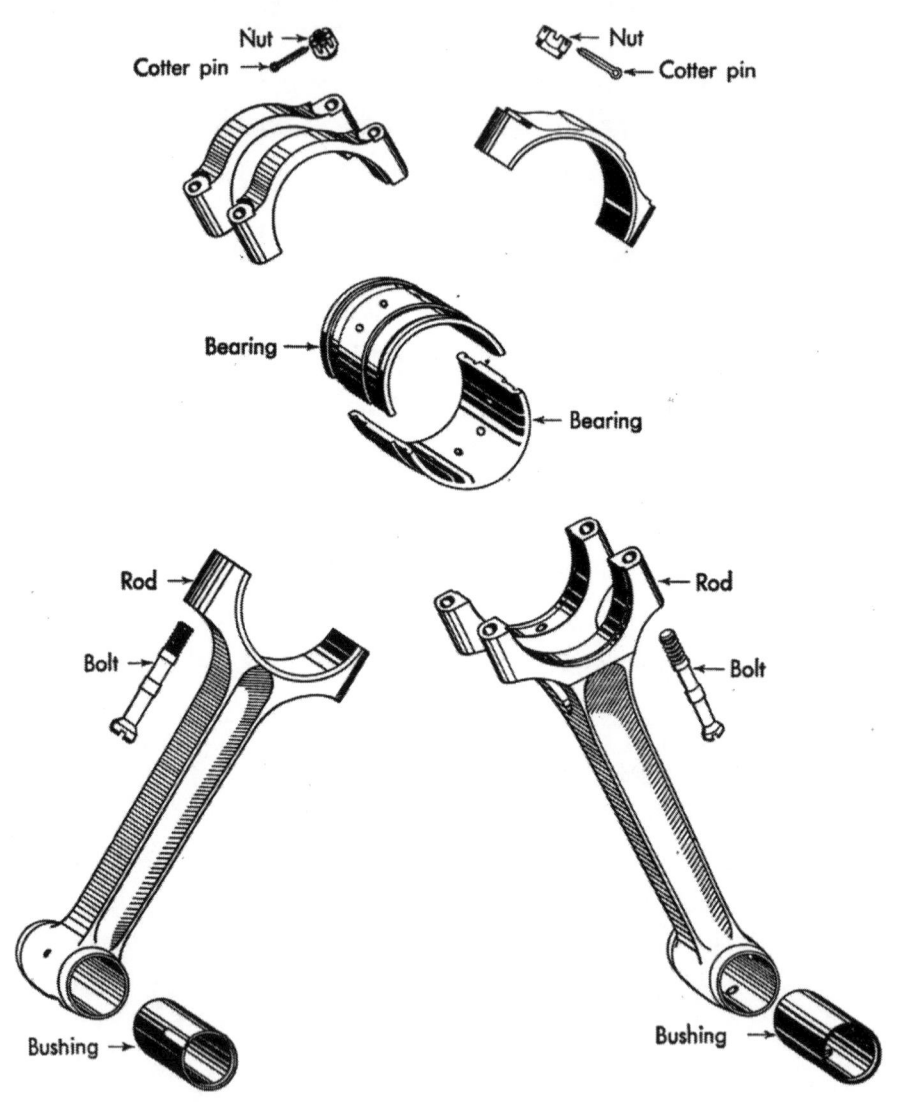

Fig. 50. A split-type connecting rod with bushings and bearings. (Courtesy Ranger Aircraft Engines)

Connecting rods vary from the simple or conventional rod type to the complicated arrangement used in high-powered radial engines.

Connecting rods usually have an I- or H-section. Forked and blade types of rods are used where two connecting rods make use of a single bearing on the crankshaft. One rod has a divided end similar to a wish-

PISTON ASSEMBLY

bone, each part having a separate bearing, as shown in Fig. 50. Its partner rod is of the blade type which has a narrow end containing a bearing which fits between the two bearings of the forked rod when fastened to the crankshaft.

Since all of the pistons in a radial engine in each row are connected with a single bearing on the crankshaft, it is necessary to make use of a master rod and link rods. Link rods are also known as articulated rods.

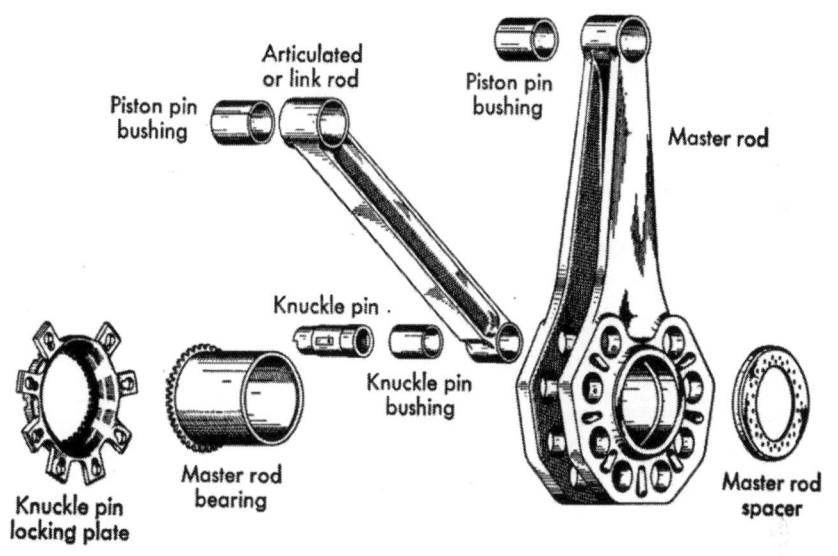

Fig. 51. Master rod and link rods. (Courtesy Wright Aeronautical Corporation)

The large end of the master rod carries the bearing which fits about the crankpin. The other rods are fastened to a flange on the master rod by means of knuckle pins. Some of the early link rods were tubular, and each end contained a pressed bronze bushing. The older type of link rod required replacing at frequent intervals, but modern rods have a much longer life. The large end of the master rod moves in a circular path while the ends of the link rods, which are fastened to it, follow an elliptical path. The knuckle pins must be securely locked. The large end of the master rod may have a solid bearing if the split type of crankshaft is used. When the crankshaft is in one piece, the master-rod bearing must be split and the two halves fastened together by means of bolts which must be securely safetyed.

VIII CRANKSHAFTS

All of the power of the engine is concentrated on the crankshaft. The crankshaft is the means by which the power developed in the cylinders is converted to usable motion. The piston travels back and forth, and the power developed in the cylinder is transferred to the crankshaft through the connecting rods. The crankshaft is the means by which this back-and-forth motion is changed to a circular motion. The crankshaft not only turns the propeller, but also all movable parts of the engine, such as magnetos, oil pumps, cam arrangements, and superchargers.

Fig. 52. A crankshaft for a twin-row radial engine. This crankshaft is in three parts and is equipped with dynamic dampers. (Courtesy Wright Aeronautical Corporation)

The crankshaft is an application of the simple crank such as is used to turn a grindstone or to crank an automobile engine. On a simple crank, the part which is held in the hand corresponds to the crankpin of the crankshaft. This is the part around which the large end of the connecting rod is fastened. The bearings on the crankshaft are also known as journals.

The part of the crankshaft which connects the crankpin to the main part of the shaft is the cheek. A flange to which the propeller may be fastened is sometimes machined onto the crankshaft.

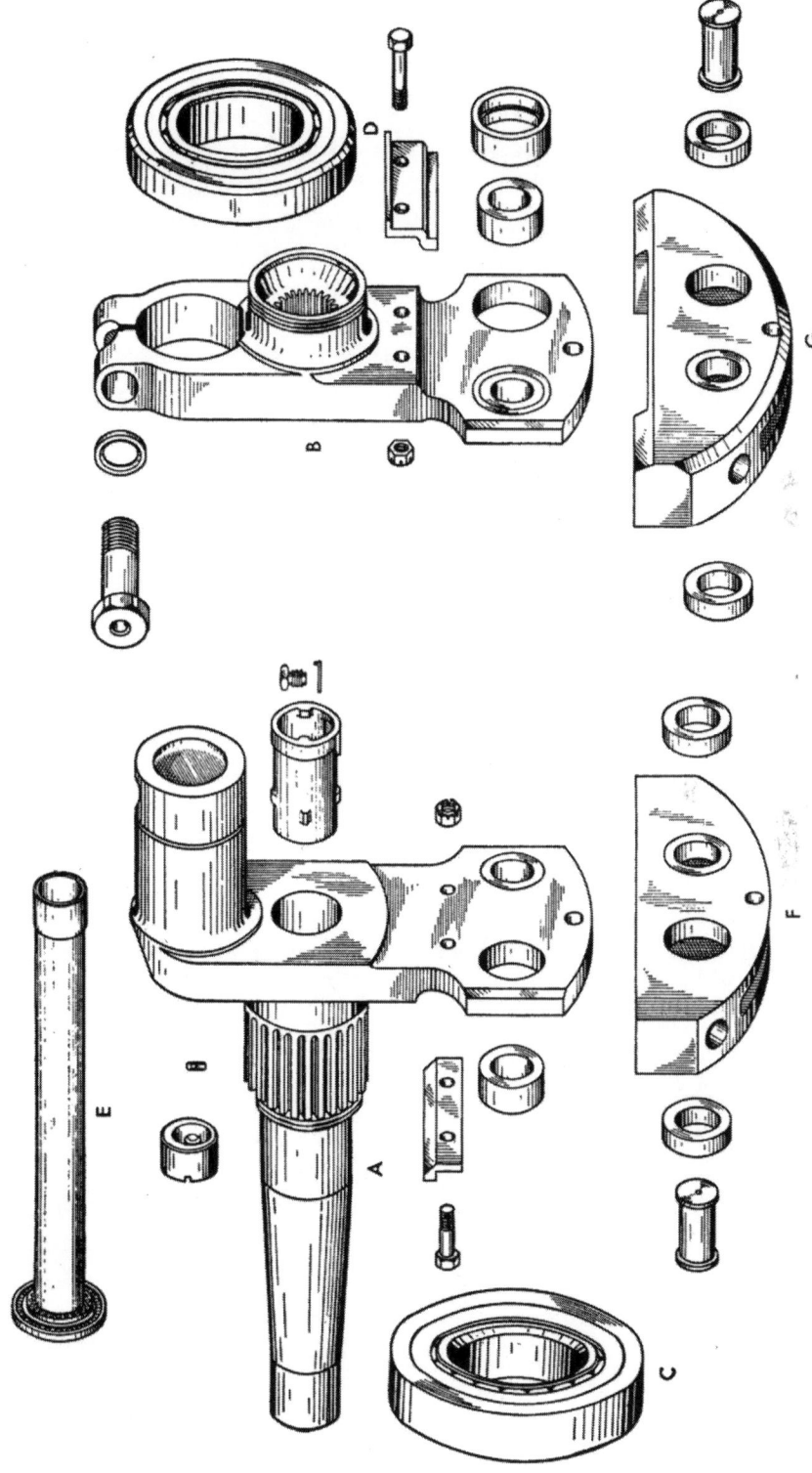

Fig. 53. Exploded view of a split crankshaft for a single-row, in-line, nine-cylinder aircraft engine. A, front half; B, rear half; C, D, main bearings; E, breather tube; F, G, dynamic dampers. (Courtesy Wright Aeronautical Corporation)

65

AIRCRAFT ENGINES

The distance from the center line of the crankshaft to the center line of the crankpin is the crank throw. This distance is the radius of the circle around which the center of the crankpin travels.

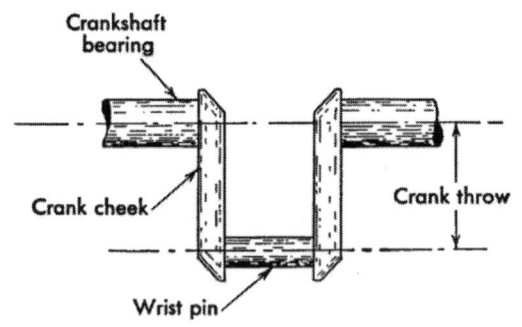

Fig. 54. A schematic drawing to show "crank throw."

Radial engines with one row of cylinders have a single-throw crankshaft. A simple, two-cylinder, in-line engine must have two cranks or two throws, one for each cylinder. All in-line engines have a crank or throw for each cylinder. Twin-row, radial engines have a two-throw crankshaft.

Long crankshafts are subject to excessive vibration. In aircraft-engine construction, the engineers have tried to make the crankshaft as short as possible. The crankshaft is subjected to vibrations caused by the successive power impulses in the cylinders. The power is not applied to the crankshaft evenly by the cylinders as a whole, or by any individual cylinder. The power stroke of each cylinder lasts through about 120° of crankshaft rotation. Since there is one power stroke in each cylinder to every two revolutions of the crankshaft, power is applied only about one sixth of the time. Out of 720° rotation of the crankshaft, the engine uses up power through approximately 600° for each cylinder. The original engine built by the Wright brothers developed only about 3 hp. per cylinder. The OX5 World War I engine developed about 15 hp. per cylinder, the Liberty motor about 35 hp. per cylinder, and one of the light three-cylinder radial engines of the late 1920's developed about 12 hp. per cylinder. Some of the modern engines develop as much as 150 hp. per cylinder, and this tremendous force is applied to the crankshaft suddenly, just after top dead center at the beginning of the power stroke. In low speed engines, the vibration is less, due to the longer interval between each power stroke.

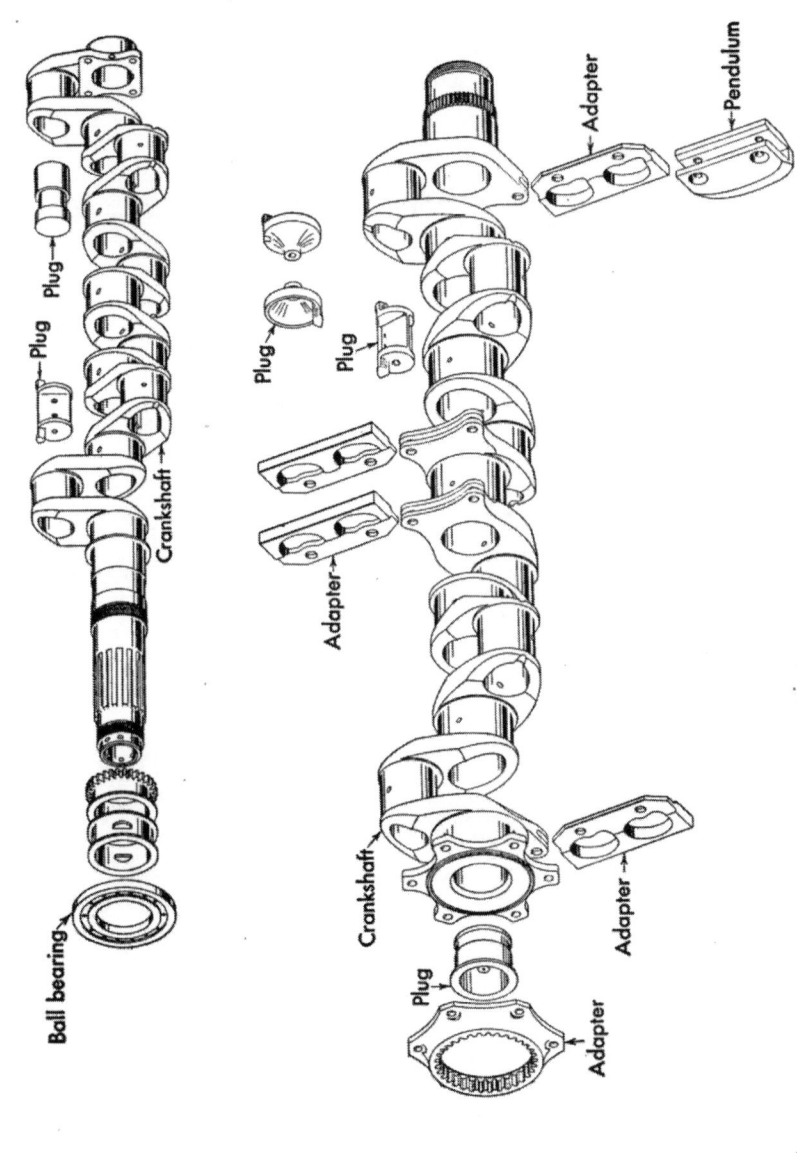

Fig. 55. Top, a crankshaft and its various parts for a six-cylinder, single-row, inverted, in-line aircraft engine. Bottom, a crankshaft and its various parts for a twelve-cylinder, inverted, V-type aircraft engine. (Courtesy Ranger Aircraft Engines)

67

AIRCRAFT ENGINES

In the earlier engines, crankshaft revolutions were not much more than 1000 per min. This speed has more than doubled in the modern engine, greatly increasing the vibration problem.

In order to develop higher horsepower, the propellers have been slowed down by means of reduction gears, allowing increased propeller efficiency.

Fig. 56. Rough steel forgings for a single-throw crankshaft. (Courtesy Wright Aeronautical Corporation)

Crankshafts are machined from high-strength, steel-alloy forgings. Such alloys as chromium nickel molybdenum steel and chromium nickel steel are used. Crankshafts are machined to a high finish, and all tool marks and scratches must be removed. Fillets are also employed to ensure smooth contours. Most crankshafts are heat-treated and surface-hardened by nitriding. No sharp corners are left, since these would lead to failure in the part when vibration forces are concentrated at these points.

Crankshafts are drilled out both to lighten the part and to allow for the flow of oil through the inside of the shaft. When great bearing forces are applied, a large bearing will support much more weight than a small bearing. A tube is stronger than a solid rod which has the same cross section of material. Bearing surfaces are highly polished and finished with extremely small tolerances. These surfaces may be nitrided to increase their wearing qualities.

The crankpin is usually hollowed out, forming an oil reservoir which

CRANKSHAFTS

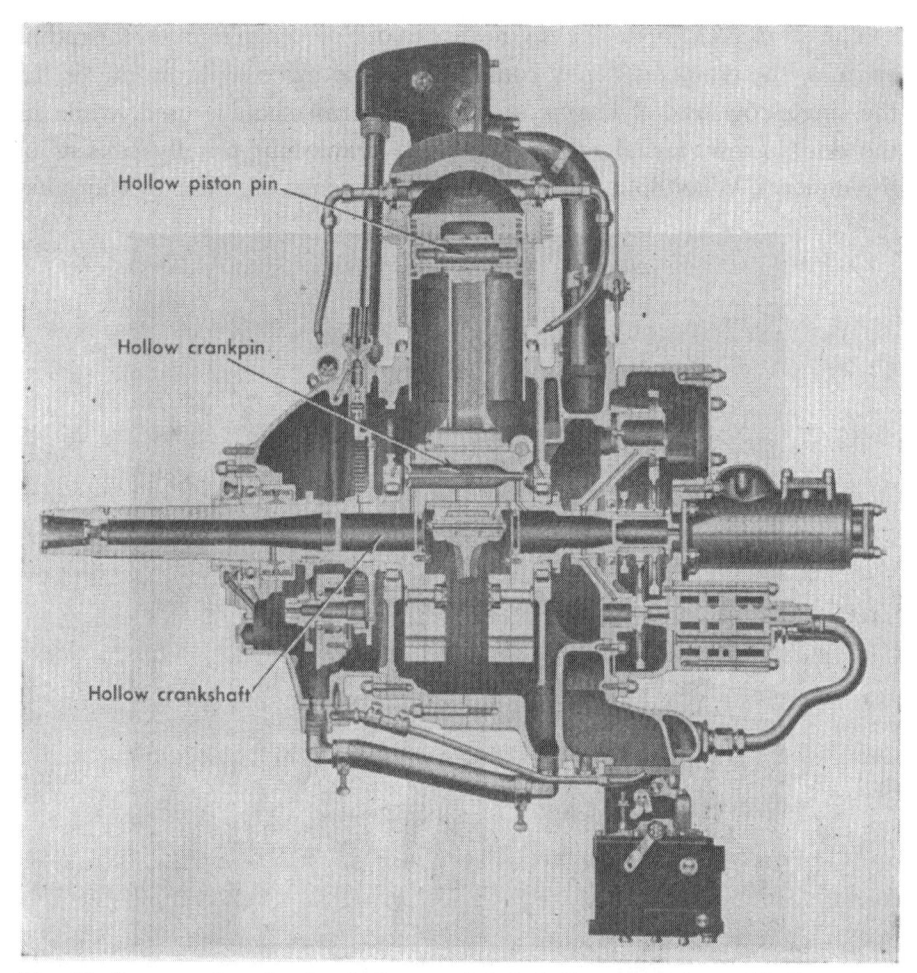

Fig. 57. Cutaway view showing hollow crankshaft, hollow crankpin, and hollow piston pin. (Courtesy Wright Aeronautical Corporation)

also serves as a collection point for sludge or solid particles in the oil. An opening or jet is generally provided from this reservoir through which oil is sprayed onto the underside of the piston and cylinder walls. This oil serves to lubricate the cylinder and cool the piston.

Crankshafts differ somewhat in arrangement. A two-throw crankshaft for a four-cylinder, horizontally opposed or a two-cylinder, in-line engine has the cranks arranged 180° apart. A four-throw, eight-cylinder, V-type engine also has the cranks on each side of the center line. A crankshaft for a six-cylinder, in-line or twelve-cylinder, V-type engine has the cranks arranged in pairs 120° apart. The pairs are arranged 1–2, 3–4, 5–6, as shown in Fig. 60.

69

AIRCRAFT ENGINES

The crankshafts of in-line engines are in one piece. In many of the radial engines, the crankshaft may consist of two or three separate pieces. In the single-row, radial engine, a two-piece crankshaft is used, while in the double-row, radial engine the main crankshaft usually consists of three pieces. When split, the two parts are fastened together by clamping

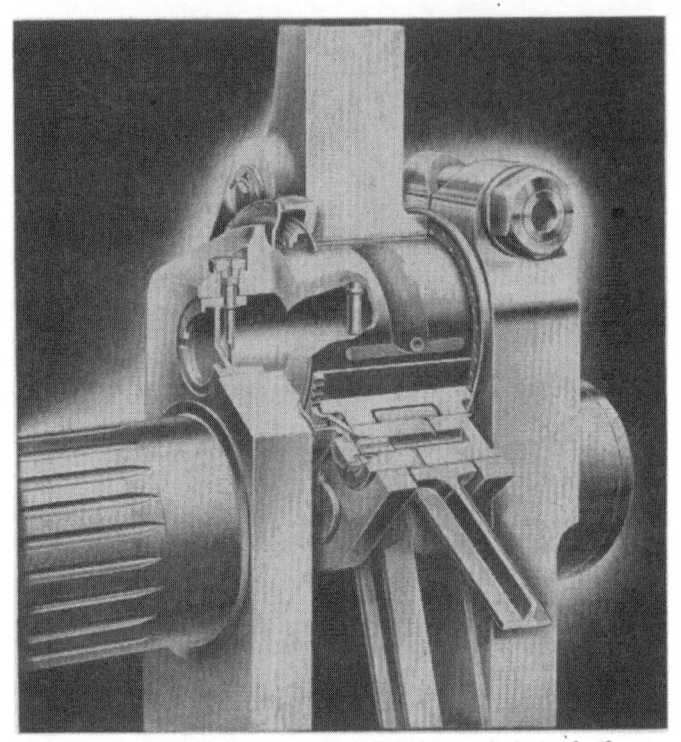

Fig. 58. A cutaway view showing a hollow crankpin and oil passages. (Courtesy Wright Aeronautical Corporation)

one end of the crankpin in a clamp which forms one cheek. The crankpin may be divided, one half carrying a male splined part which fits into a corresponding female part on the other half of the crankpin. The parts are fastened together by means of bolts, the tension on which is measured by the elongation of the bolt itself when the nut is tightened.

One of the large engines has the three parts of the split crankshaft fastened together by the clamp method as shown in Fig. 52. This type of crankshaft has a main bearing on the middle part and a bearing on each side. These bearings rest against bearings in the crankcase. In most of the large radial engines, the crankshaft has extensions or shanks extending to the front and rear. The tail shaft, or rear shank, is usually fastened to the main crankshaft by means of a splined joint. The front shank, or

CRANKSHAFTS

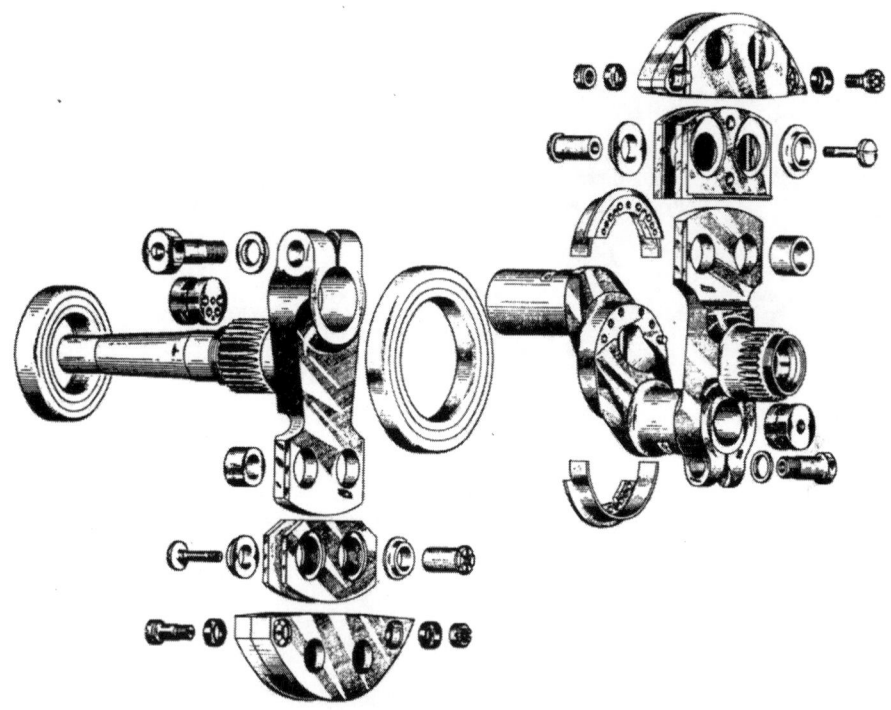

Fig. 59. Exploded view of a split crankshaft for a twin-row, radial engine, showing counterweights and accessories. (Courtesy Wright Aeronautical Corporation)

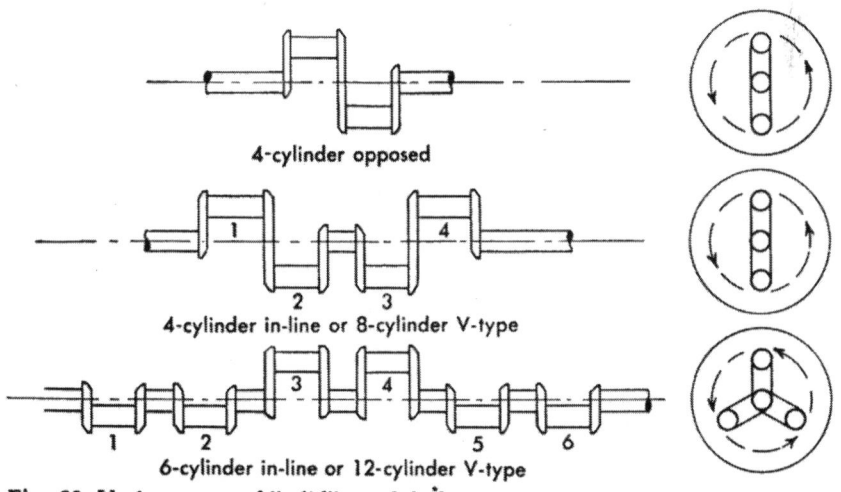

4-cylinder opposed

4-cylinder in-line or 8-cylinder V-type

6-cylinder in-line or 12-cylinder V-type

Fig. 60. Various types of "solid" crankshafts.

propeller shaft, is usually fastened to the main crankshaft in a similar manner. Propeller shafts may be tapered or splined at the point where the propeller is fastened to the shaft. Each end of the crankshaft may have gear teeth machined on it for turning various engine accessories.

71

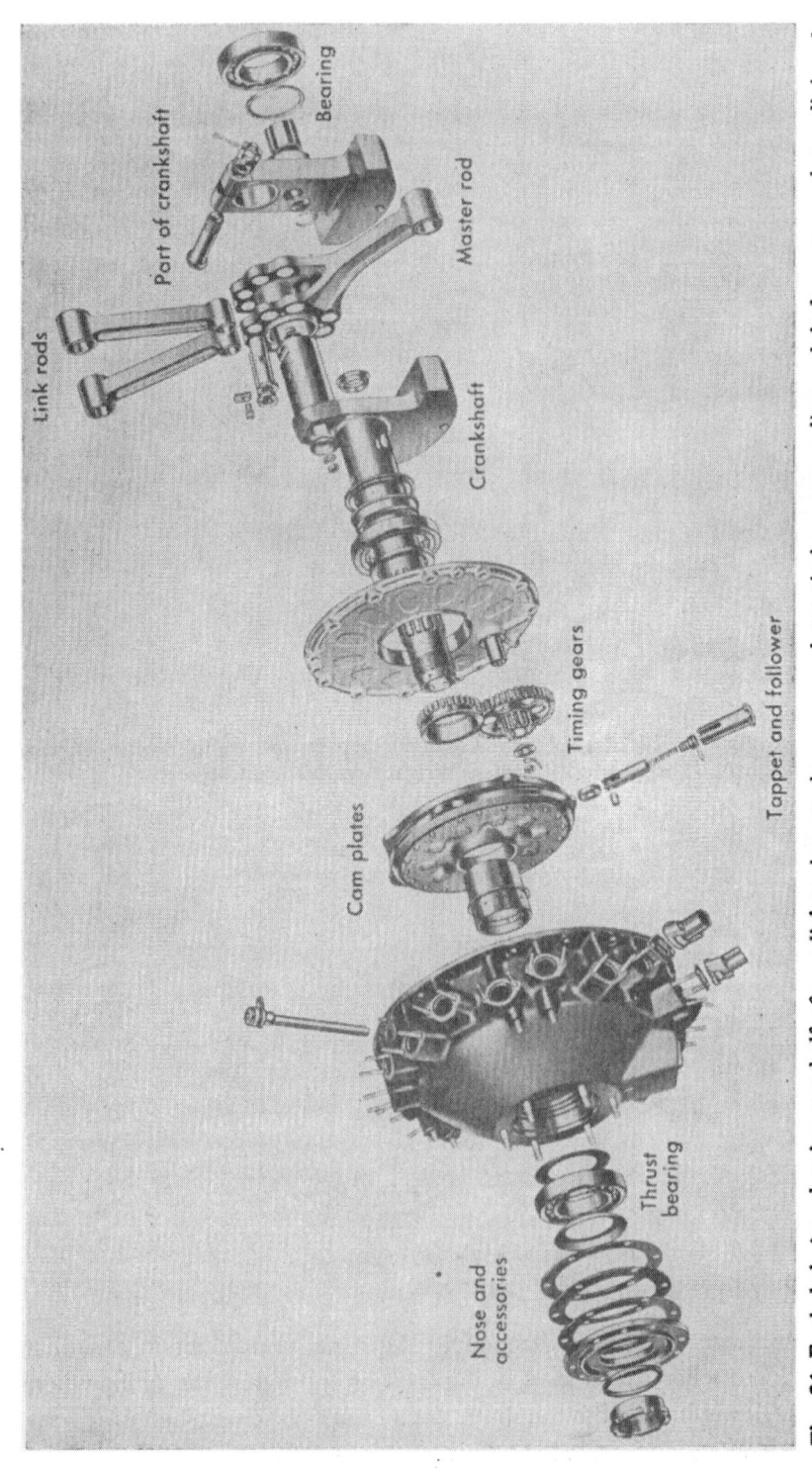

Fig. 61. Exploded view showing one half of a radial engine crankcase: cam plates, timing gears, split crankshaft, master rod, two link rods, and accessories. (Courtesy Lycoming Division, The Aviation Corporation)

CRANKSHAFTS

Fig. 62. A crankcase for an in-line aircraft engine, showing the main bearings. (Courtesy Allison Division, General Motors Corporation)

Due to the forces set up by the weight of the connecting rod and piston, crankshafts are usually balanced by means of counterweights which may be permanently fastened to the end of the cheek opposite the crankpin. Opposed engines do not usually have counterweights, as the pistons in the opposed cylinders are 180° apart and counteract the action of each other. To reduce vibration, dynamic dampers are often

Fig. 63. Accessory drive gears mounted on a crankshaft. (Courtesy Jacobs Aircraft Engine Company)

AIRCRAFT ENGINES

used. These dynamic dampers are heavy weights, in the same location as the counterweights, attached by means of a loose joint. This joint is formed by pins running through holes in the weight which are larger than the pin itself. These weights act as a short pendulum, having the

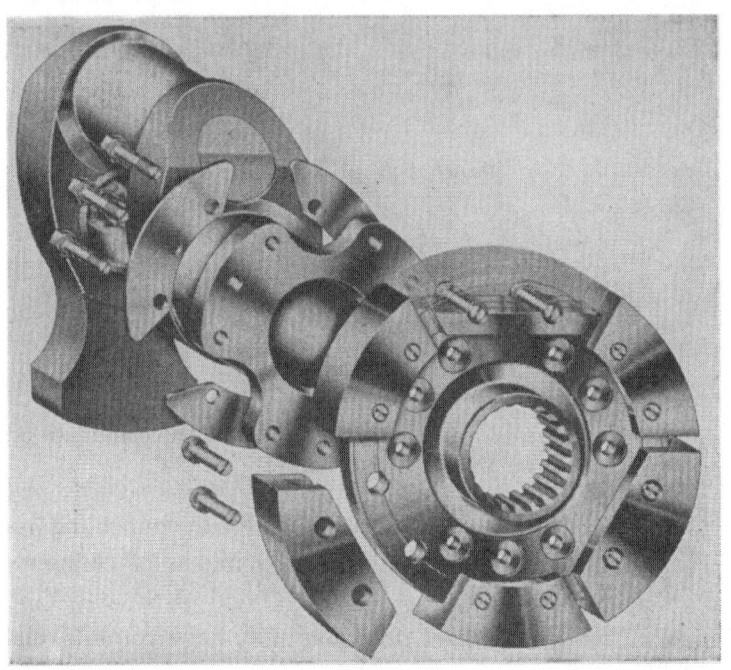

Fig. 64. Dynamic dampers mounted on the end of a crankshaft for a 12-cylinder, V-type aircraft engine. (Courtesy Allison Division, General Motors Corporation)

same period of vibration as that set up by the power impulses in the cylinder. In some V-type, in-line engines, these dynamic dampers are arranged about a flange or flywheel attached to the end of the crankshaft.

Most crankshafts are marked to show the proper location of the propeller blades in relation to the counterweights. The location of the propeller blade may increase or decrease the vibration of the crankshaft.

IX CRANKCASES

The crankcase is the main framework of the engine. It must be rigid and strong. The crankcase not only holds the various parts of the engine together, but in many installations, transmits all of the propeller thrust to the aircraft. It must be constructed to resist the torsional forces between the propeller and the airplane, and also all of the vibrations set up in the engine itself.

The crankcase supports the main bearings which carry the crankshaft. While in operation, most crankcases are filled with a spray of oil and oil vapors. Gases that escape from the cylinders past the pistons enter the crankcase.

A breather arrangement must be made to prevent the building up of pressure in this part of the engine. On small engines, a breather pipe leads from the crankcase to the top part of the engine. On larger engines, this breather pipe may be fitted with baffles to cause the oil vapors to condense and the oil to flow back into the crankcase. The breather pipe may be carried to the oil tank when the tank is separate from the engine. In some engines, breathing is accomplished through the propeller shaft.

In relation to the oil supply of the engine, crankcases are divided into wet- and dry-sump types. When the oil supply for the engine is carried in the crankcase, it is classified as a wet sump. In a dry-sump engine the oil is carried in a tank entirely separated from the engine. In most of the older engines of the in-line type, the oil was carried in the crankcase. With the development of the inverted engines and radial engines, a separate oil tank became necessary. Some of the opposed types of engines still make use of the wet sump.

Cast iron, which is used in many internal-combustion engines, is not suitable for crankcase construction in high-powered aircraft engines. Cast iron is brittle and would soon fail under the stresses and strains set up by the vibration of the aircraft engine. If it were made strong enough to resist these forces, the cast-iron crankcase would be too heavy. The

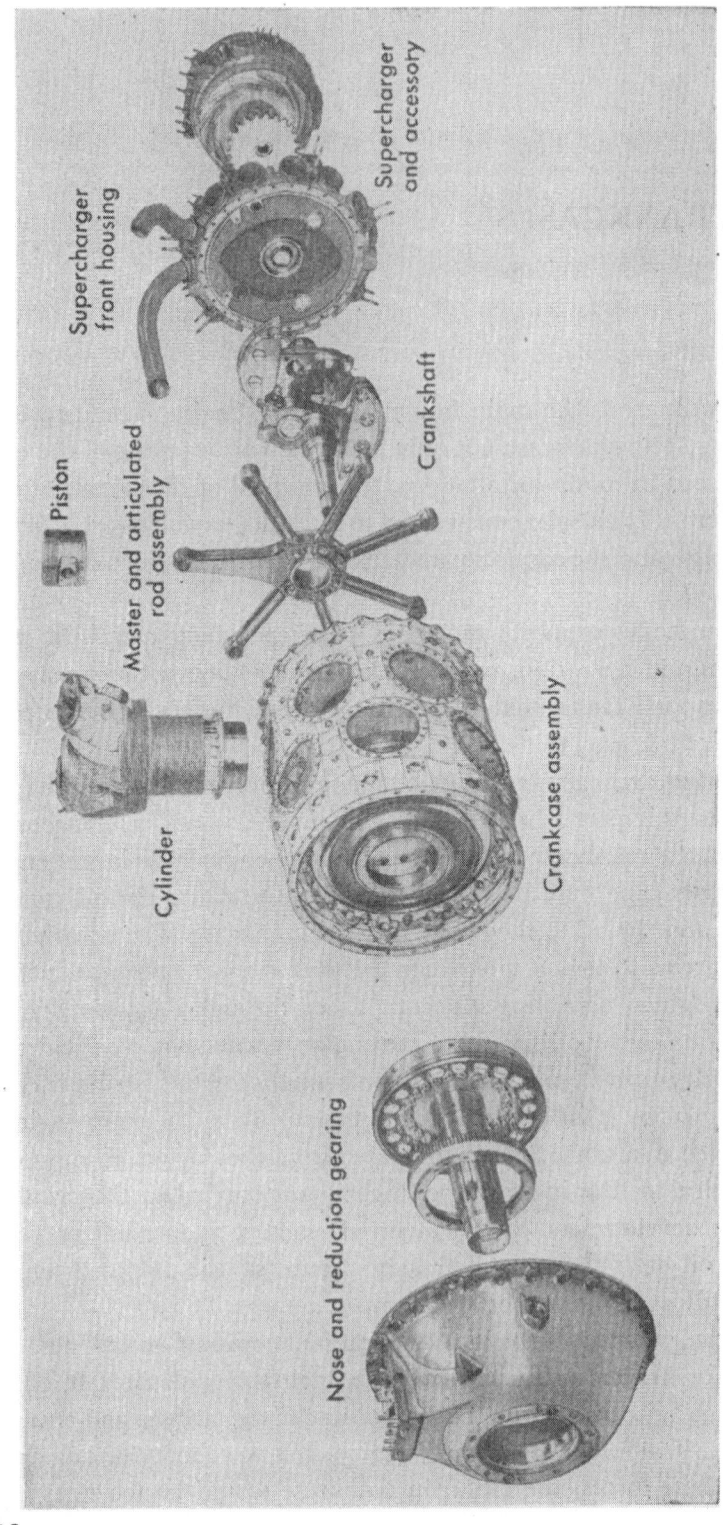

Fig. 65. Exploded view showing some of the parts of a twin-row radial engine. (Courtesy Wright Aeronautical Corporation)

Fig. 66. Crankcase for a 12-cylinder, V-type aircraft engine showing main bearings and supporting webs. (Courtesy Allison Division, General Motors Corporation)

Fig. 67. Casting a magnesium crankcase. (Courtesy Wright Aeronautical Corporation)

most common materials used in the construction of crankcases for aircraft engines are cast or forged steel, aluminum, and magnesium. All of these materials are alloys of the metals and are in the heat-treated condition. Aluminum and magnesium alloys are used because of their lightness. Where extreme strength is required, alloys of steel are used.

Many crankcases are cylindrical or spherical in their general shape. In the vertical, in-line engines, the crankcase usually consists of a top

Fig. 68. The part of a crankcase which carries the oil in a wet-sump aircraft engine. Note cooling fins on bottom of sump. (Courtesy Lycoming Division, The Aviation Corporation)

and a bottom half. One half carries the mounting pads to which the cylinders are fastened, and usually each half carries one part of the main bearings through which passes the crankshaft. In some engines, one half of the crankcase carries the entire main bearing, making it possible to remove the other half without disturbing the bearing adjustment.

The parts of a crankcase may be made oil tight by gaskets or ground joints. The main bearings are usually supported on webs which are a part of the crankcase. These webs serve not only as supports for the main bearings, but add strength to the crankcase as well.

On some of the small, four-cylinder, opposed aircraft engines, the two cylinders on each side are cast as an integral part of one half of the crankcase, which is divided vertically along its center line. Both the top and bottom of the crankcase have openings which are closed on the

CRANKCASES

top by a plate and on the bottom by the oil sump. Bearing adjustment may be made by removing the top plate. The cylinder head is usually removable.

On other opposed-type engines, the crankcase is divided vertically along its center line to form two parts, having the cylinders mounted on pads in a manner similar to the in-line engines.

Most of the engine accessories are attached to the crankcase. Care

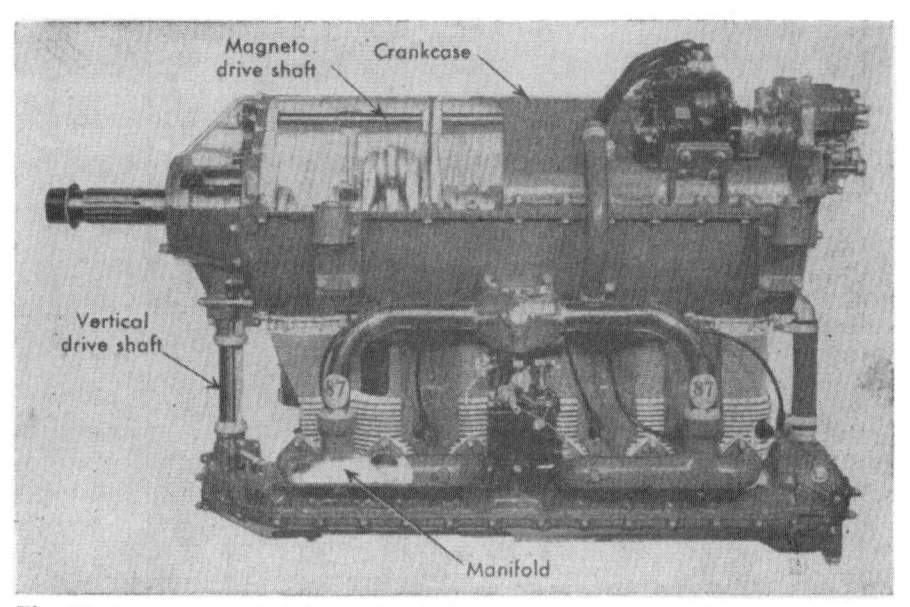

Fig. 69. A cutaway view of the crankcase of a dry-sump, inverted, in-line engine showing the hollow magneto drive shaft which carries oil to the main bearings. (Courtesy Ranger Aircraft Engines)

must be taken in locating these attachments to avoid excessive vibration and strain on critical parts of the crankcase.

On radial engines, the crankcase is made up of from three to seven parts. The main section to which the cylinders are attached may or may not be divided along the center line. The front section closes the opening in the main section, houses reduction gears, and supports the propeller mounting. One 75 hp., four-cylinder, opposed engine uses a semisteel crankcase and cylinder block cast in two units. The crankcase is split vertically and has a ground joint requiring no gasket. The parts are secured with through bolts, studs, and nuts. This crankcase carries three steel-back, copper-lead-lined, main bearings.

One popular, six-cylinder, opposed engine developing from 150 to 175 hp. uses an aluminum crankcase that is split vertically on the crank-

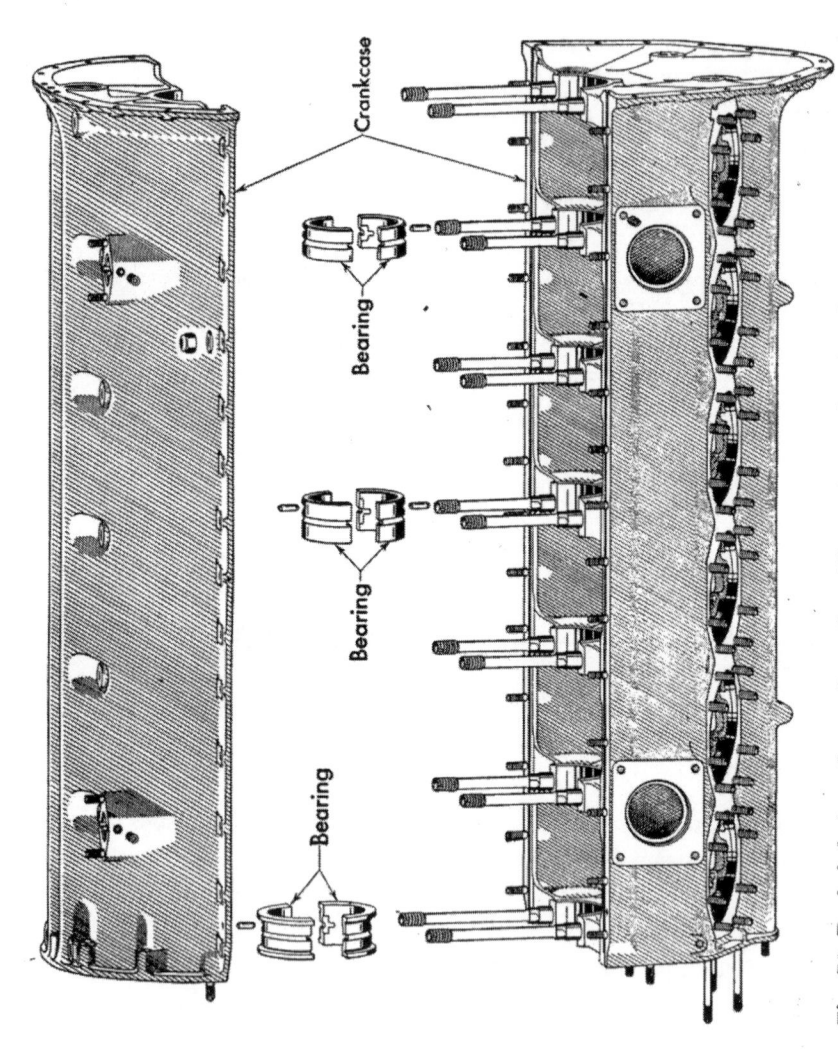

Fig. 70. Exploded view of a crankcase for a V-type, inverted engine showing main bearings and cylinder mounting pads. (Courtesy Ranger Aircraft Engines)

CRANKCASES

shaft center line. The two parts of the crankcase are bolted together with long studs that pass through the cylinder barrel flanges. All oil passages are cast into the crankcase itself.

In some six-cylinder, in-line, inverted engines, the main crankcase is made of aluminum alloy. This case is split into an upper and lower half at the crankshaft center line. Each half has seven webs which support

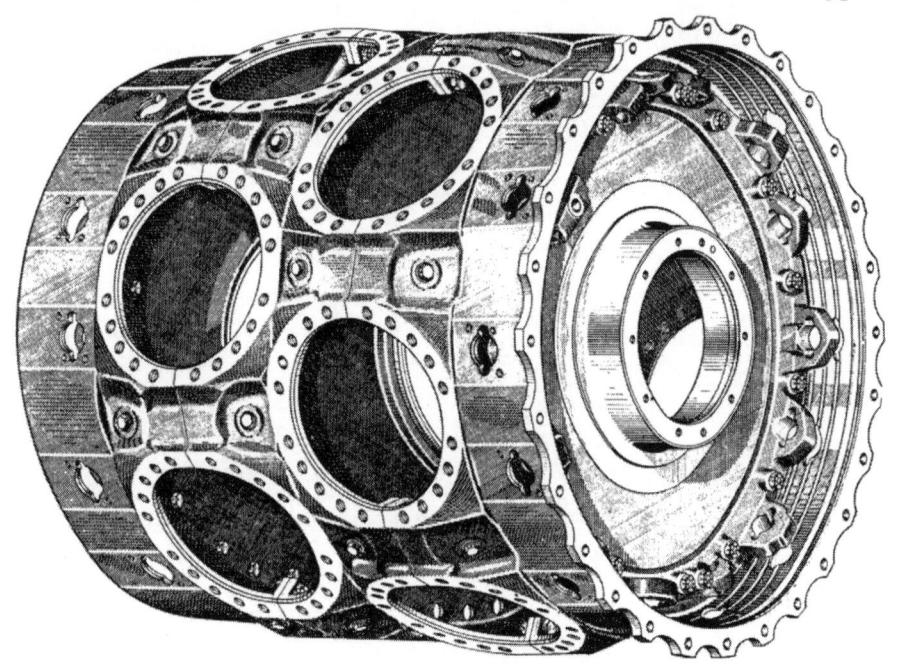

Fig. 71. The main section of a crankcase for a twin-row, 14-cylinder, radial aircraft engine. (Courtesy Wright Aeronautical Corporation)

the main bearings. The halves are held together by 14 main bearing studs that are anchored into the webs of the upper crankcase and extend through the lower half. The upper crankcase in this engine corresponds to the lower crankcase in the vertical in-line engine. The crankcase flanges are sealed with rubber strips and held together by closely spaced bolts. The lower half of the main crankcase has four mounting pads for fastening the engine to the engine mount. In this engine the halves of the crankcase are machined together and cannot be replaced separately. The cylinders are mounted on cylinder pads on the bottom of the lower half of the crankcase.

Some of the nine-cylinder, single-row, radial engines developing from 265 to 300 hp. have a crankcase formed of four sections. These sections are fastened together at flanged surfaces. The main crankcase is an

AIRCRAFT ENGINES

aluminum casting that provides mounting pads for the cylinders and carries the rear main bearing. This part of the crankcase also carries the lugs for attaching the engine to its mount. The thrust-bearing housing is of cast magnesium alloy. This part supports the propeller thrust bearing and contains the cam followers and guides. The front, main-bearing plate

Fig. 72. A six-cylinder, opposed-type aircraft engine with the crankcase split vertically. (Courtesy Continental Motors Corporation)

is formed of an aluminum alloy forging and is attached to the front end of the main crankcase. This part of the crankcase carries the main bearing and the cam, idler-gear assembly. The rear section which forms the accessory drive housing is attached to the rear of the main crankcase. This section supports the accessory equipment, starter drive housing, and accessory drives.

In some of the nine-cylinder radial engines developing approximately 1200 hp., the crankcase is built up of six principal sections. These sections are all of heat-treated alloys. The front section is of cast magnesium alloy. This section contains the main thrust ball bearing and supports the cam followers. It also contains the stationary reduction gear, reduction-gear pinion, and the governor drive assembly. The forward end of the oil sump is attached to the lower part of this front section. The front section contains all oil passages for the oil used in connection with the hydraulically operated propellers. The front and rear

CRANKCASES

Fig. 73. A nine-cylinder radial aircraft engine developing 1200 horsepower. (Courtesy Wright Aeronautical Corporation)

main sections of this crankcase are machined from steel forgings. The main section is divided vertically along the cylinder center line. The supercharger front housing is machined from a magnesium alloy casting. The web or diaphragm of this part forms the forward wall of the diffuser section of the supercharger. The rear supercharger housing is made from magnesium alloys and contains the induction passage to the im-

AIRCRAFT ENGINES

Fig. 74. Placing the crankshaft and connecting-rod assembly into one half of a radial engine crankcase. Note the heavy construction of the crankcase. (Courtesy Jacobs Aircraft Engine Company)

peller entrance and houses all accessory drive gears. This section forms the rear wall of the diffuser chamber. The rear section of this crankcase is the supercharger rear housing cover which is machined from a magnesium alloy casting. This section provides mounting pads for two magnetos, a generator, starter, oil pump, and dual accessory drive housing.

The crankcase of some 14-cylinder, twin-row, radial engines is made up of 7 principal sections: front section, front main section, center main section, rear main section, supercharger front housing, supercharger rear housing, and supercharger rear housing cover. The oil sump on this engine is attached to the outside of the crankcase at its lowest point. The three main sections are of chromium alloy steel forgings. All of the other

CRANKCASES

sections are of cast magnesium alloy. The main section of the crankcase is divided into three parts at the center line of the front and rear cylinder rows. The front section houses the entire propeller reduction-gear assembly. The main crankcase section houses the crankshaft, cam mechanism, valve tappet guides, and mounting pads for the 14 cylinders. The three main roller bearings on the crankshaft are supported by steel retainers shrunk into the center line of the diaphragm of the web of each main section. The supercharger rear housing forms the rear wall of the diffuser section and supports the propeller-shaft and rear oil-seal assembly. The remaining section, the supercharger rear housing cover, provides mounting for several of the engine accessories.

The construction of a suitable crankcase is one of the large problems of the aircraft-engine designer.

X VALVES AND CAMS

The valves are the means by which the ports leading into the cylinder are opened and closed to allow the intake of the air-fuel mixture, commonly known as the charge, and the escape of the burned mixture in the form of hot gases. Most American aircraft engines have two valves for each cylinder: the intake valve and the exhaust valve. Some engines

Fig. 75. Cam and rocker arm arrangement for a 12-cylinder, V-type engine having two exhaust valves and two intake valves for each cylinder. (Courtesy Allison Division, General Motors Corporation)

VALVES AND CAMS

may have two intake valves and two exhaust valves for each cylinder. American designers, by streamlining the passages through which the air-fuel mixture and exhaust gases pass, have made it possible to obtain efficient engine performance by the use of two comparatively large valves in each cylinder.

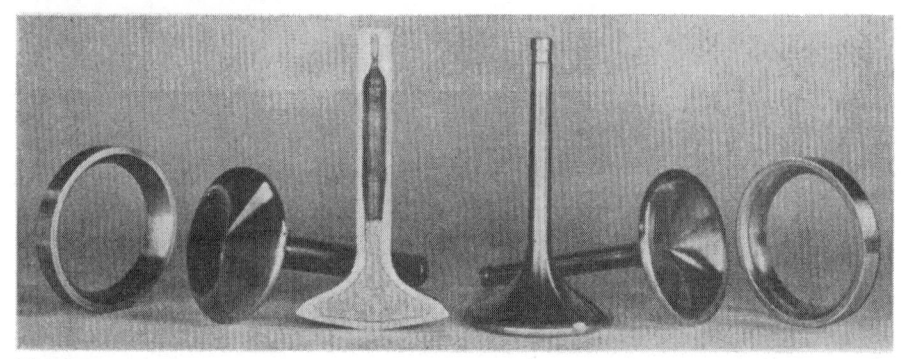

Fig. 76. A mushroom-type, hollow, sodium-cooled exhaust valve, a tulip-type intake valve, and valve seats. (Courtesy Wright Aeronautical Corporation)

A valve has a number of distinct parts. The long, cylindrical part of the valve is the stem. On one end of the stem is a disc-shaped part, which is the head; the opposite end of the stem is the tip. The face of the valve is the beveled outer edge of the head that comes in contact with the valve seat when the valve is closed. The point where the stem joins the head is the neck.

Some of the older engines were equipped with a sliding valve in the form of a sleeve which slides within the cylinder barrel itself. The up-and-down rotary motion of the sleeve opened the valve ports at the proper time. All modern American engines are equipped with poppet valves. There are three general types of poppet valves: the mushroom, tulip, and semitulip. The first poppet valves were of the solid mushroom type. These valves had a tendency to break at the neck because of the weight concentrated in the head of the valve. The semitulip type is used in most modern engines as an intake valve. In the lighter engines, both the intake and exhaust valves are of the full tulip type.

Valve heads are subjected to the full heat of the fuel burning in the cylinder. Cooling the valves is a major problem in an aircraft engine. This is true particularly of the exhaust valve. In large engines the exhaust valve has a considerably larger stem than the intake valve. The head is usually the full mushroom type. The stem and head are of hollow construction and partially filled with metallic sodium. The exhaust

AIRCRAFT ENGINES

valve is subjected to the full heat of the gases as they pass out of the cylinder into the exhaust manifold. At valve-operating temperature, the sodium becomes liquid and splashes back and forth in the valve. This action assists in carrying the heat away from the head of the valve through the stem. The stem loses heat through its close contact with the valve guide. Some heat from the valve is carried away through the valve seat.

Fig. 77. Cam and rocker arm arrangement of an in-line engine having one intake and one exhaust valve in each cylinder. (Courtesy Ranger Aircraft Engines)

The face of the valve and of the corresponding valve seat is usually beveled at either 30° or 45°. When the face is beveled at 30°, the angle between the face and the space through which the gases travel is more nearly streamlined. This shallow bevel, however, tends to trap particles of carbon between the face and the seat which allows gases to escape, causing burning and warping of the valve. There seems to be somewhat more of a tendency for the valve with the shallow bevel to bounce. The manufacturers who use the 45° angle claim that the steeper taper allows a tighter, wedged fit between the face of the valve and the seat. It is also claimed that this fit has more of a tendency to crush solid particles and

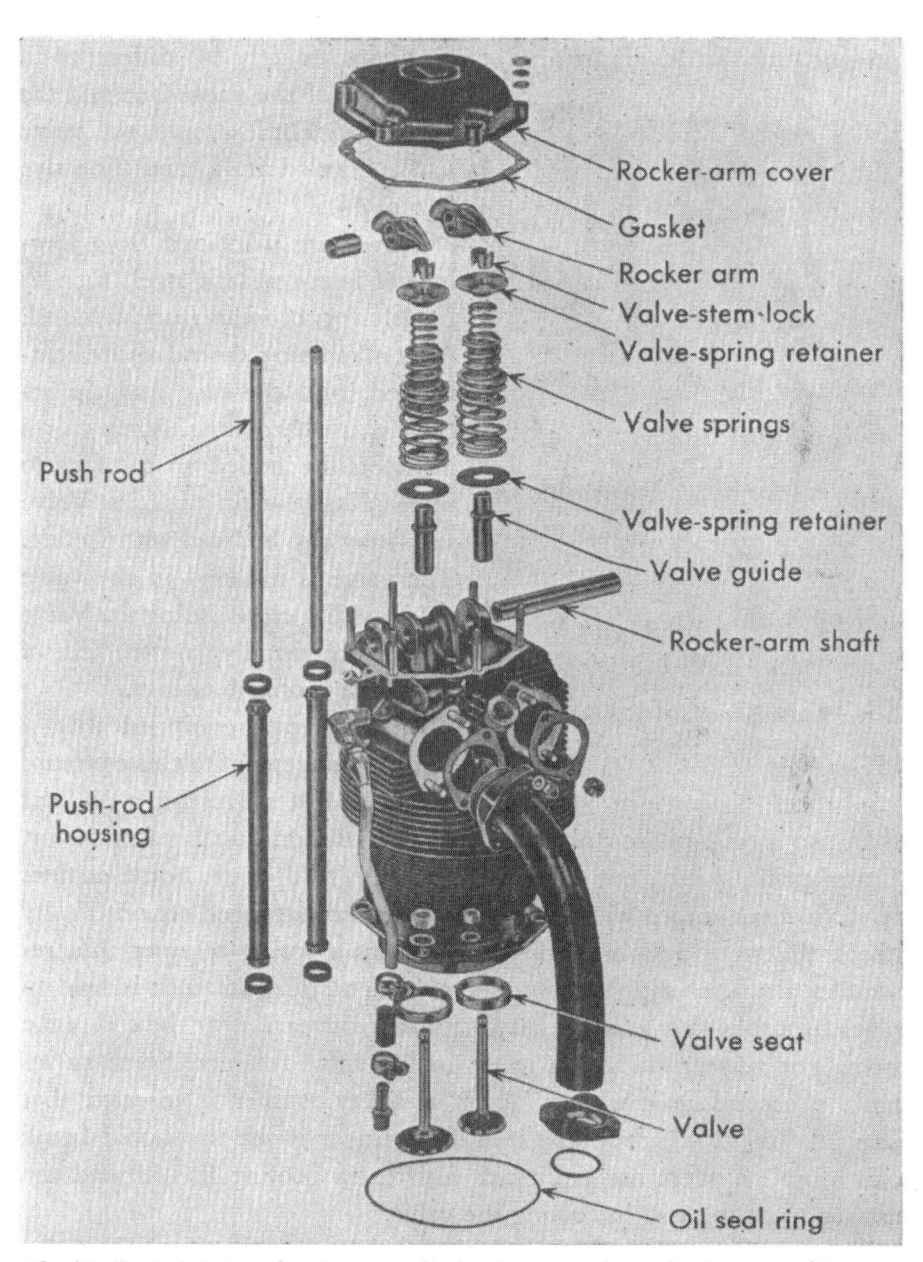

Fig. 78. Exploded view showing one cylinder, its accessories and valve train. (Courtesy Lycoming Division, The Aviation Corporation)

89

close the opening more tightly. The steeper fit reduces the tendency to bounce. Both the 30° and 45° faces are used in aircraft engines at the present time. In some engines there is approximately ½° difference in the bevel of the valve face and the valve seat. For example, the valve seat is beveled at 45° and the valve face at 44.5°.

Valves are machined from forgings of heat-resistant steel, such as tungsten or chromium tungsten steel. After machining, the valves are heat-treated, and the face and tip receive extra hardening. Some valves have Stellite added to the face to increase its wearing qualities. Valve seats may also be faced with Stellite. The valve is installed in the valve guide with a close-sliding fit. Valve guides are usually constructed of shrunk-in bronze bushings.

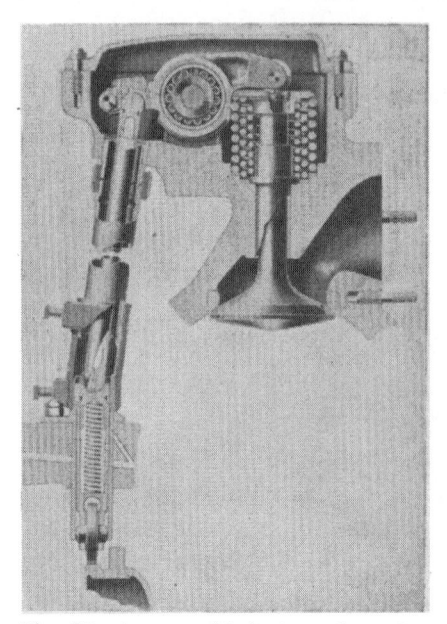

Fig. 79. An assembled view of a valve train. (Courtesy Wright Aeronautical Corporation)

Each valve is equipped with a spring arrangement to cause prompt closing. Most valves have not less than two springs of the coil type which give an additional safety factor. If one spring breaks, the valve will continue to operate. Some engines have valves equipped with three springs that are arranged concentrically about the stem and are held in place by a spring retainer that resembles a washer slipped over the valve stem. This retainer is held in place by a split steel washer that fits into a groove cut near the end of the stem. The washer fits into a recess in the valve retainer. Some valves have a second groove into which a safety washer is inserted that prevents the valve's dropping into the cylinder if the tip should break off. Most cylinders are equipped with extra cooling fins around the exhaust port to assist in cooling the valve.

The valve train is the valve and its operating mechanism. The valve train consists of the valve and its equipment, rocker arm, push rod, tappet, and cam follower. The rocker arm rocks back and forth around the rocker-arm shaft, one end resting on the valve stem and the other end against a push rod or a cam. The rocker arm may be inserted in

VALVES AND CAMS

bronze bushings and have a plain type of bearing, or be equipped with roller bearings. In a few engines the camshaft is located across the cylinder heads, and one end of the rocker arm follows the cam on this

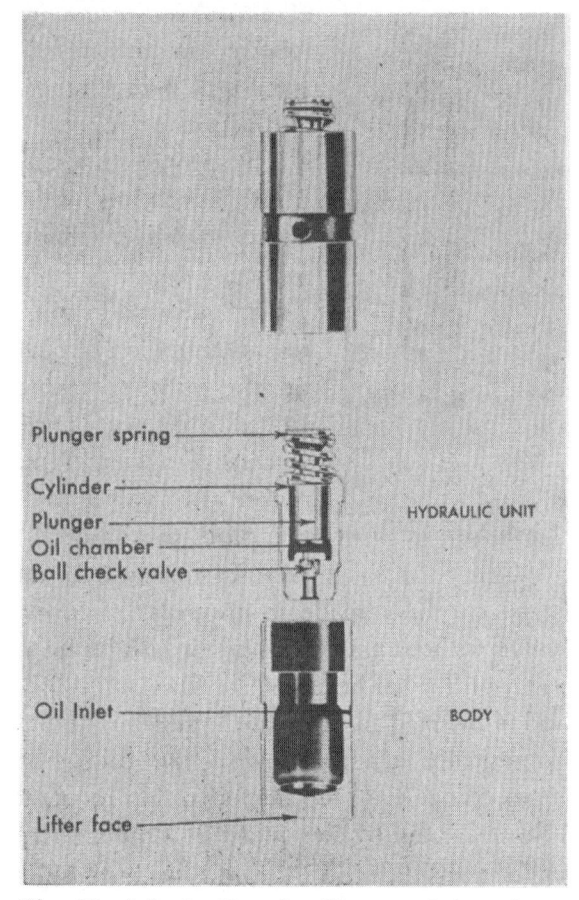

Fig. 80. A hydraulic valve lifter, consisting of two basic parts. (Courtesy Eaton Manufacturing Company)

shaft. In many engines, particularly the radial type, a push rod is necessary to operate the rocker arm.

Push rods are usually of hollow, tubular, alloy construction and have a hardened steel ball or a rounded steel plug pressed firmly into each end. The ends may be drilled with a small hole which allows oil to flow through the push rod, thereby lubricating parts of the valve train. The push rod is usually enclosed in a tube known as the push-rod housing.

In the opposed type of engine and in-line engine, the cam shaft is usually contained in the crankcase, and a push rod extends from the valve

AIRCRAFT ENGINES

tappet to the rocker arm. The valve tappet is the part of the valve train that receives the impulse from the cam and transmits it to the push rod. In some engines the bottom of the tappet is in direct contact with the cam. In most radial engines a cam follower in the form of a roller of some type is in contact with the cam and acts as a tappet.

The proper function of the valve train plays a vital part in the engine's ability to deliver efficient trouble-free service.

The valve train consists of the entire combination of parts which make up the valve-operating mechanism. In the ordinary valve train, which consists of a cam, tappet, push rod, rocker arm, valve, and valve spring, the parts make metal-to-metal contact. In this type of valve train, adjustments must be carefully made to ensure the proper opening and closing of the valve without excessive play or lash in the train.

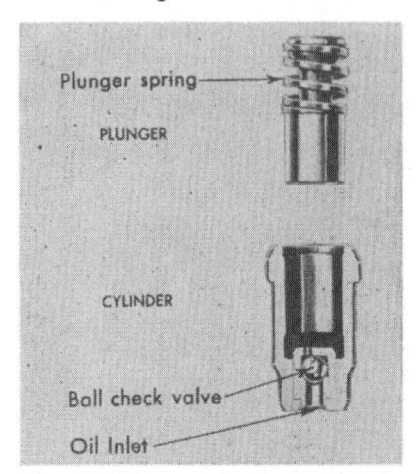

Fig. 81. The two basic parts of the hydraulic unit of the hydraulic valve lifter. (Courtesy Eaton Manufacturing Company)

A hydraulic valve lifter has been developed which is used on some aircraft engines and is inserted in place of the regular style of tappet. This valve lifter automatically adjusts its own length during each revolution of the camshaft to compensate for the expansion or contraction of the valve train. The adjustment takes place in a column of oil which is incorporated in the valve lifter.

The hydraulic valve lifter consists of two basic parts: the lifter body and the hydraulic unit which is inserted in the body. The hydraulic unit consists of a plunger which operates in a cylinder. A light spring holds the plunger in its outermost position against the push rod. This leaves a small chamber at the bottom of the cylinder below the plunger. This chamber is kept filled with oil by pressure from the engine lubricating system. The oil enters the chamber through inlets in the body and in the base of the cylinder. A ball-check valve located at the cylinder outlet controls the flow of oil. When the ball-check valve is closed, the oil in the chamber, being noncompressible, keeps a lifting mechanism against the push rod as positively as though the whole lifter were a single piece of metal.

VALVES AND CAMS

As the valve train expands or contracts with changes in engine temperature, the lifter adjusts its own length to compensate for the change. Accurate clearance is provided between the plunger and the cylinder

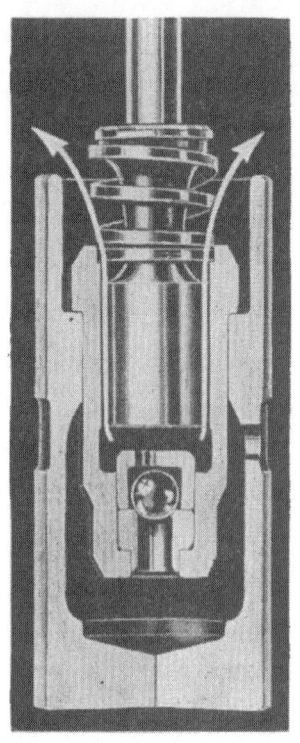

Fig. 82. Hydraulic valve lifter showing the path of oil to take care of "leak down." Note ball valve is closed. (Courtesy Eaton Manufacturing Company)

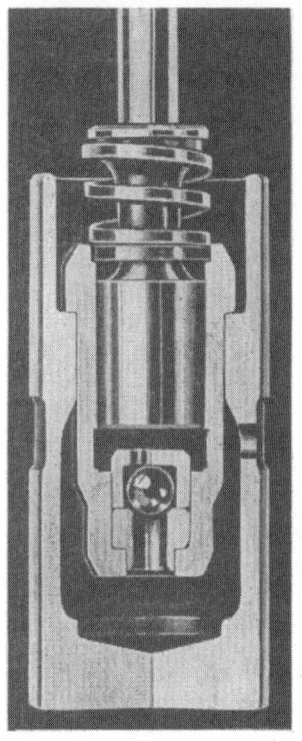

Fig. 83. Hydraulic valve lifter showing the ball valve open. (Courtesy Eaton Manufacturing Company)

wall, which permits the escape or leak down, as it is commonly called, of a small amount of oil from the chamber. This leakage automatically compensates for any expansion in the valve train, allowing positive valve seating. When the valve train contracts, the plunger spring holds the plunger outward. This relieves the pressure on the oil in the chamber and on the ball-check valve. The ball moves from its seat and permits the intake of oil from the engine lubricating system. This hydraulic lifter corrects its length each time the valve closes and thus maintains a zero clearance.

Cams are eccentrics on either a shaft or plate. A camshaft is equipped with cams. On an in-line engine there must be one cam for each valve.

93

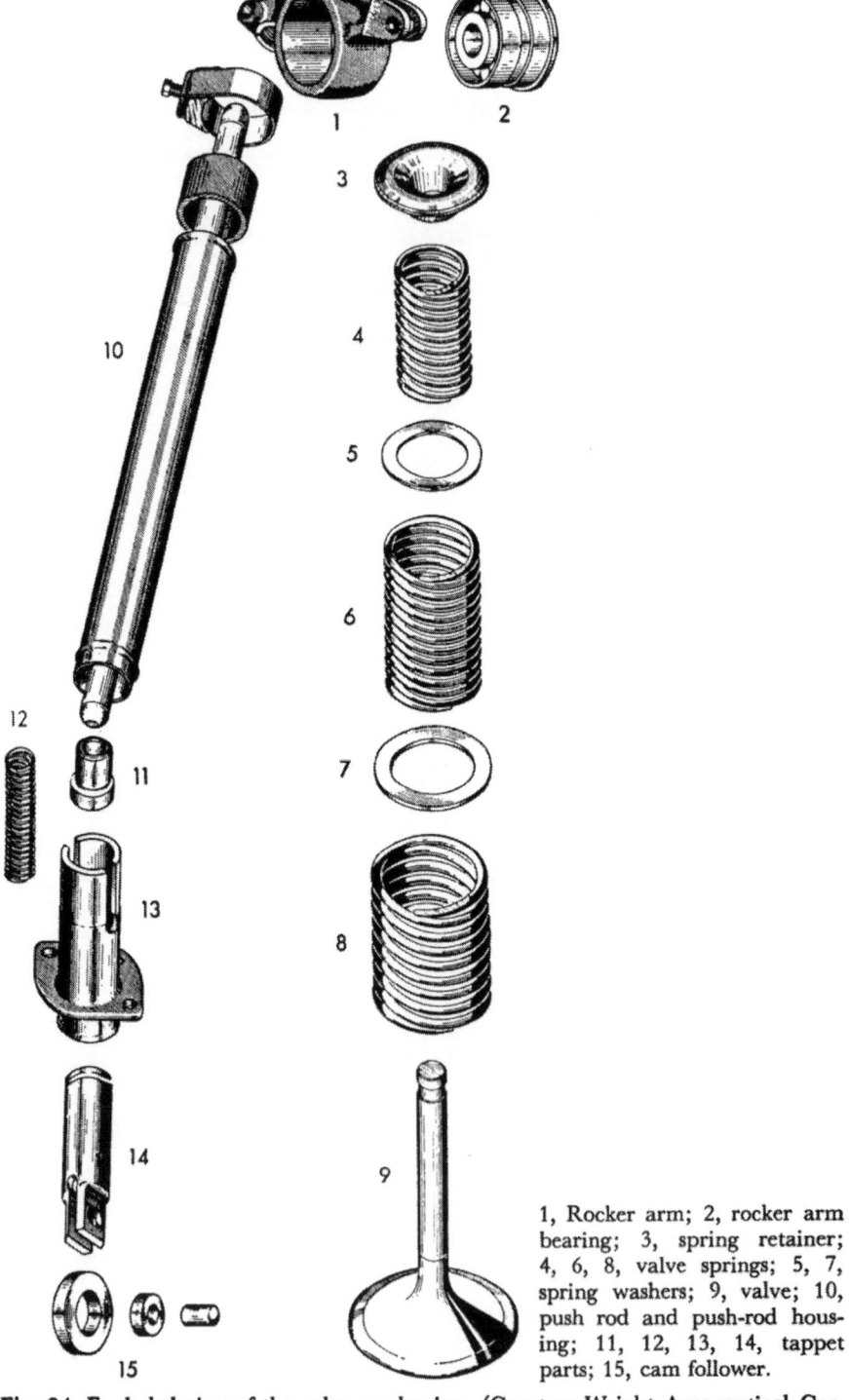

Fig. 84. Exploded view of the valve mechanism. (Courtesy Wright Aeronautical Corporation)

1, Rocker arm; 2, rocker arm bearing; 3, spring retainer; 4, 6, 8, valve springs; 5, 7, spring washers; 9, valve; 10, push rod and push-rod housing; 11, 12, 13, 14, tappet parts; 15, cam follower.

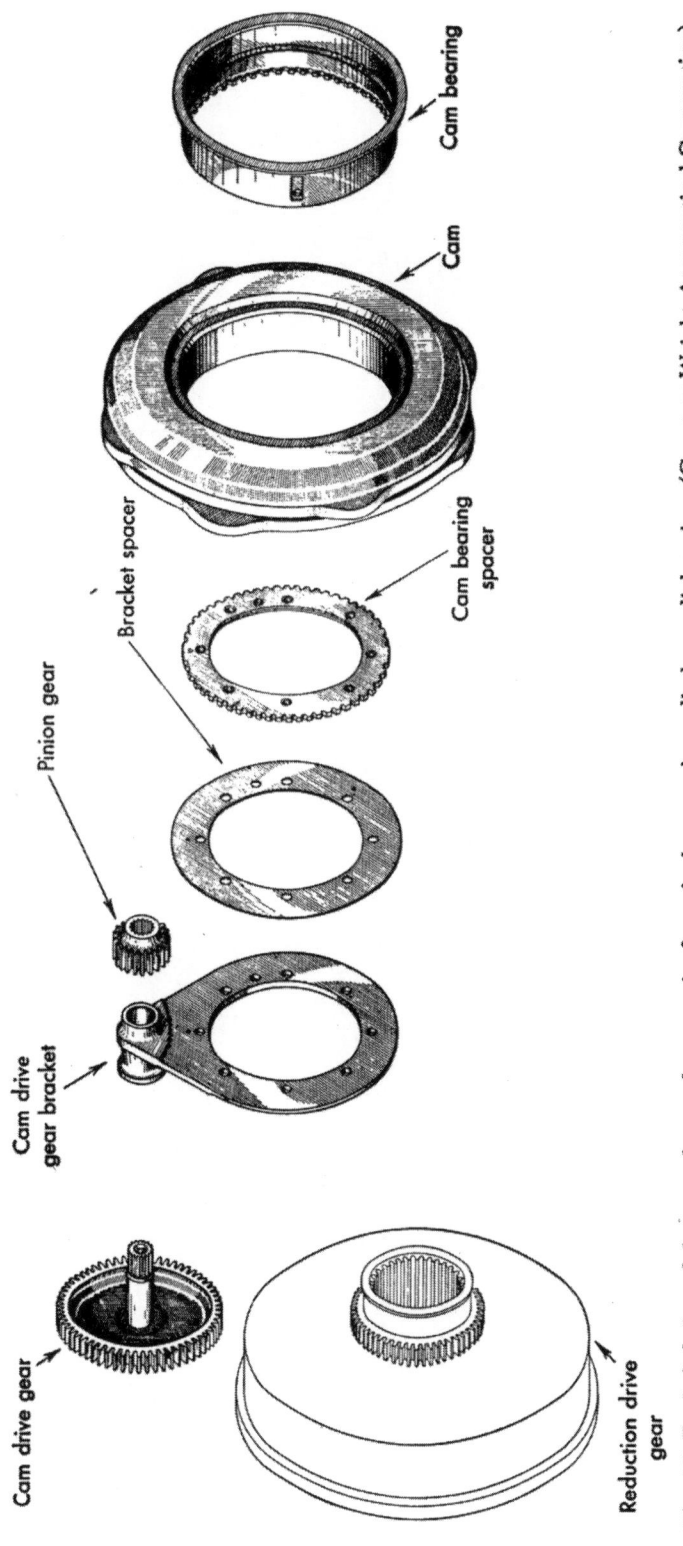

Fig. 85. Exploded view of the cam plate and accessories for a single-row, nine-cylinder, radial engine. (Courtesy Wright Aeronautical Corporation)

95

AIRCRAFT ENGINES

Since each valve must open and close once for each two revolutions of the crankshaft, the camshaft rotates at one half the crankshaft speed. On radial engines the cam mechanism is in the form of a plate that is

Fig. 86. A cutaway view of a radial engine showing the valve operating mechanism. (Courtesy Jacobs Aircraft Engine Company)

rotated at the proper speed. This speed varies with the number of cylinders and their arrangement. The cams are in the form of lobes arranged around the outer edge of the cam plate. These lobes are part of the cam plate.

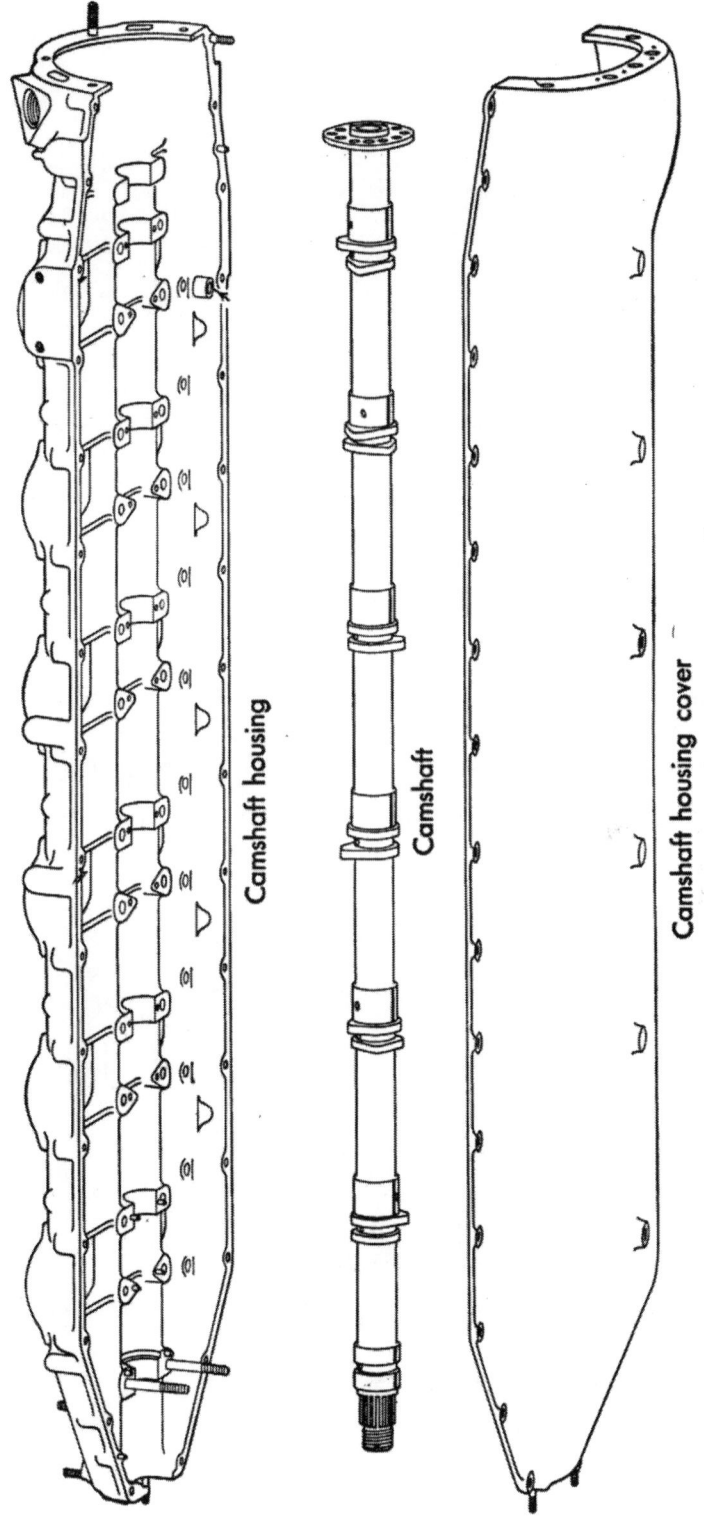

Fig. 87. Exploded view of the camshaft and camshaft housing for an inverted in-line engine. (Courtesy Ranger Aircraft Engines)

AIRCRAFT ENGINES

Table I shows the typical arrangement and speeds of cams for various radial engines.

TABLE I. CAM RINGS — RADIAL ENGINES

5 CYLINDERS		7 CYLINDERS		9 CYLINDERS		
Number of Lobes	Speed	Number of Lobes	Speed	Number of Lobes	Speed	Direction of Rotation
3	$1/6$	4	$1/8$	5	$1/10$	With crankshaft
2	$1/4$	3	$1/6$	4	$1/8$	Opposite crankshaft

It should be noted that in some engines the cam plate revolves in the same direction as the crankshaft, and in other engines it revolves in the opposite direction. It is necessary for the proper performance of the engine that the valves be opened and closed at exactly the right time. The camshaft is machined from steel forgings, and the cam lobes are usually hardened. Camshafts are driven by a gear train operated from the crankshaft.

Fig. 88. Cutaway view showing gear train from crankshaft which operates magneto drive shaft and the vertical drive to the camshaft. (Courtesy Ranger Aircraft Engines)

XI ELECTRICAL FUNDAMENTALS

In order to understand the operation of a modern ignition system, it is necessary to have a considerable understanding of electricity.

The ancients found that a certain stone, which they called a "lodestone," had the power to attract small particles of iron or small particles of the same kind of stone. The lodestone is the mineral magnetite, or magnetic iron ore. Thousands of years ago it was discovered that a piece of magnetite or magnetized steel, when suspended by a thread or balanced on a pivot, arranged itself in a north and south direction. This is a simple compass. The end of the magnet that pointed toward the north was called the north pole and the other end was called the south pole.

It was soon discovered that north poles repelled each other while north and south poles attracted each other. It was known by the ancients that a cat's fur, when rubbed, gave off sparks, and that amber, when rubbed, became charged with electricity. These early experimenters found that there were two kinds of electric charge. They also found that two charges of the same kind repelled each other, while charges of the opposite kind attracted each other.

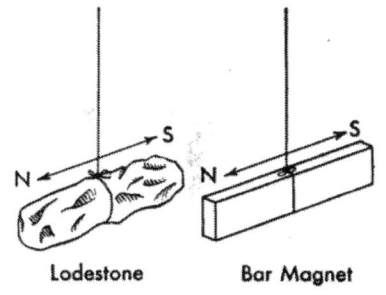

Fig. 89. A lodestone and a bar magnet suspended to form simple compasses.

Early in the history of electricity these different kinds of charges became known as "positive" and "negative." It was found possible to charge a body with enough electricity to make a spark jump. This kind of electricity does not flow steadily along a conductor and is known as "static electricity." This is the kind of electricity that is generated by stroking fur or scuffing over a wool rug. It is possible to light a gas jet by turning on the jet, scuffing the feet over a dry woolen rug, and touching the tip of the jet with a finger. A spark jumps to the metal part

AIRCRAFT ENGINES

of the jet and ignites the gas. This corresponds to igniting the air-fuel mixture in an engine cylinder by means of an electric spark.

Various types of static electric machines were developed, and experiments made to find uses for this kind of electricity.

The electric magnet, when placed under a piece of glass or paper, will attract particles of iron filings which arrange themselves in definite

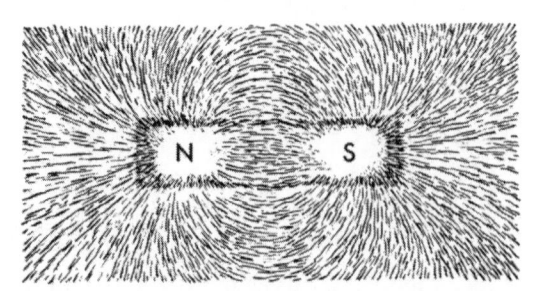

Fig. 90. The arrangement of iron filings to show the magnetic field of a bar magnet.

patterns. It is apparent from the arrangement of the iron filings that they tend to arrange themselves along lines that seem to connect the two ends of the magnet and pass through space around the magnet. These lines are magnetic lines of force, and the space through which they pass is an electric or magnetic field.

The earth acts as a great magnet and is surrounded by a magnetic field. A magnetic field exerts an effect on magnetic substances that may be brought into the field. A piece of iron that in itself is not a magnet, when brought into a magnetic field, becomes a magnet. This is known as "induced magnetism." This principle plays an important part in the ignition system of an engine.

Certain substances will carry an electric current, while others will not. Those substances that carry an electric current, or allow the flow of electricity through them, are conductors. Those substances which will not permit the flow of an electric current through them are known as nonconductors.

All metals are conductors of electricity, while such substances as glass, rubber, and plastic are nonconductors. Not all conductors of electricity are magnetic. A magnetic substance is one that is attracted by a magnet. Iron, nickel, and cobalt show magnetic properties.

The modern theory of electricity is based upon the theory of atoms. Every substance is made up of atoms or molecules which in turn are made up of even smaller particles. These smaller particles are positive

ELECTRICAL FUNDAMENTALS

and negative particles of electricity. The positive particle is called the "proton," and the negative particle is called the "electron." The electron weighs only about $\frac{1}{1840}$ of the proton. The hydrogen atom, which is the lightest of atoms, is thought to be made up of one proton and one electron. It is possible to remove some of its electrons from a body. When this is done, the body is said to have a positive charge. If electrons are added to the body, the body is said to be negatively charged. That is, it has an excess number of electrons. When a body is positively charged, it will attract a negatively charged body. If the negatively charged body is brought into contact with the positively charged body, electrons will move from the negatively charged body to the positively charged body in such quantities as to bring about a balance between the positive and negative particles. This movement of electrons is the basis of an electric current. If the bodies are not brought into contact with each other but are connected with a conductor, the electrons will flow along the conductor, producing an electric current. It can therefore be stated that electric current consists of a flow of electrons along a conductor.

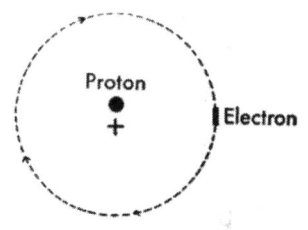

Fig. 91. Diagram to show the theoretical makeup of the hydrogen atom.

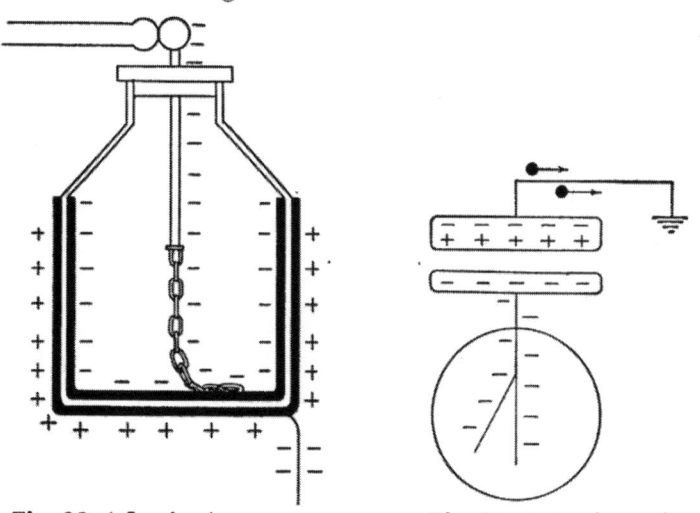

Fig. 92. A Leyden jar.

Fig. 93. A simple condenser.

A Leyden jar is made by taking a glass jar and coating part of the inside and outside with an electric conductor, such as tin foil. If one side is connected to the ground, a considerable charge of electricity may

101

AIRCRAFT ENGINES

be built up by bringing the other layer of tin foil into contact with a highly charged body or a source of electric current. This arrangement is a condenser. Each side, when charged, induces an opposite charge on the other layer of tin foil. A condenser is used in ignition systems to act as a temporary storage reservoir for a quantity of electricity.

Electric current has been compared with a water system. If two tanks on a level floor are connected at the bottom with a small pipe and one tank is filled with water while the other is empty, water tends to flow from the full tank to the empty tank. This tendency to flow can be measured by the pressure of the water in the connecting pipe. The amount of pressure will vary depending upon the depth of the water in the full tank. In water, this pressure is measured in pounds per square inch.

If two bodies having an opposite charge of electricity are connected with a conductor, there is a tendency for the electrons on the negatively

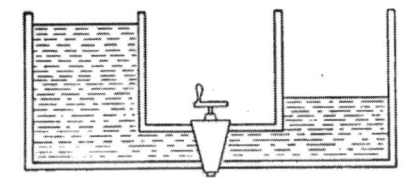

Fig. 94. Water tends to flow to the tank having the lowest water level.

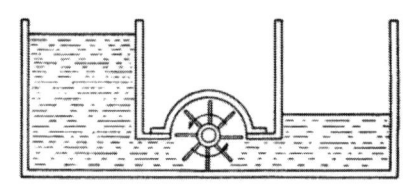

Fig. 95. Water flowing from one tank to the other.

charged body to flow to the positively charged body. This tendency to flow develops an electrical pressure. This pressure depends upon the amount of difference between the charges on the two bodies. This difference in electrical pressure is called "difference in potential," and the electric pressure is measured in volts. A volt is the unit by which electrical pressure, or difference in potential, is measured.

The rate at which the water will flow from one tank to the other depends upon the size and length of the connecting pipe. A small, long pipe offers more resistance to the flow of the water than does a large short pipe. A small, long wire offers more resistance to the flow of electric current than does a large wire of the same material. Electrical resistance is measured in ohms. An ohm, therefore, is the unit used in measuring electrical resistance.

If the pipe between the tanks is of such size that 3 gal. of water per minute flow from one tank to the other, the rate of flow is measured in units of gallons per minute. The rate of flow of electricity along a con-

ELECTRICAL FUNDAMENTALS

ductor is measured in amperes. The ampere, in measuring the flow of electricity, corresponds to the gallon in measuring the flow of water and does not tell how much water flows but only the rate of flow.

The coulomb is the unit used to measure the quantity of electricity. The coulomb does not measure the rate of flow, but measures the actual amount of electricity that has flowed over the conductor. The standard ohm is that amount of resistance offered to the flow of electric current by a column of mercury 106.3 cm. long and having a cross section of 1 sq. mm. One thousand feet of No. 10 copper wire has a resistance of almost exactly 1 ohm. The coulomb is that amount of electricity which will deposit 0.001118 g. of silver from a solution in 1 sec.

There is a close relationship between the ohm, volt, ampere, and coulomb. With a difference in potential of 1 v. and a resistance of 1 ohm, the rate of flow will be 1 amp. A current of 1 amp. will carry 1 coulomb of electricity past any point in the conductor each second. In other words, a coulomb is the amount of electricity that passes a given point in a wire in 1 sec. when 1 amp. of current is flowing.

Difference in potential, or voltage, is measured by means of a voltmeter. The rate at which the electric current is flowing is measured in amperes by an ammeter.

The power generated by the water flowing from one tank to the other is its ability to do work. The standard unit in the English system for rate of doing work is horsepower. One horsepower is that amount of power that develops 33,000 ft.-lb. per min., or 550 ft.-lb. per sec. A foot-pound is the amount of work done when 1 lb. is lifted 1 ft. against the pull of gravity. Horsepower is the rate of doing work and is not the amount of work done.

Electricity flowing along a wire has the ability to do work. This ability is measured in watts. A watt is that amount of power developed by 1 amp. flowing with a difference in potential of 1 v. The total number of watts may be determined by

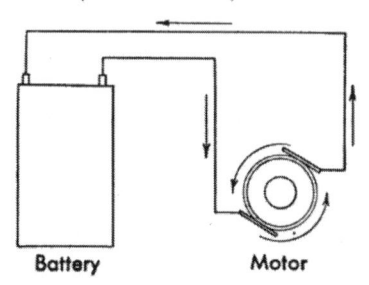

Fig. 96. A battery may be used to turn an electric motor.

multiplying the amperes flowing by the number of volts difference in potential. One horsepower is equal to 746 w. If power is used at the rate of 1 w. for 1 hr., 1 whr. of current has been used. This is the unit by which electricity is sold. A heating unit requiring 500 w. will use

500 whr. per hr. A kilowatt hour is 1000 whr. House current is commonly figured in kilowatt hours.

An electric circuit is made up of conductors connected in such a way that an electric current can flow through the circuit. If there is any point in an electric circuit that the current cannot pass, such as an open

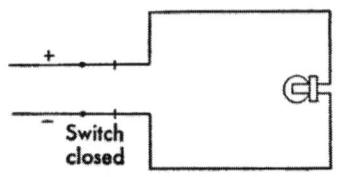

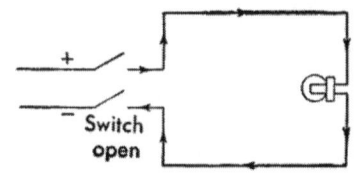

Fig. 97. Closed electrical circuit.

Fig. 98. Open electrical circuit.

switch, the circuit is broken. An electric circuit must always have one terminal or connection that is positively charged and one terminal or connection that is negatively charged, in order that current will flow through the circuit. If there is more than one path between the terminals over which the current can flow it is known as a "parallel circuit."

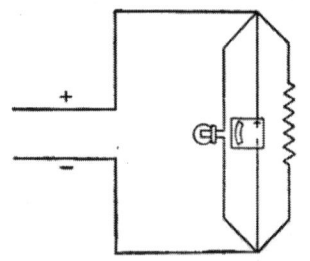

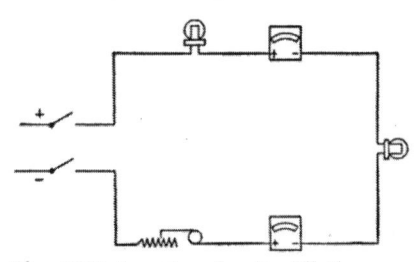

Fig. 99. A parallel circuit. The current has three paths it may flow.

Fig. 100. A series circuit. All the current must flow through all parts of the circuit.

In a series circuit, all of the parts of the circuit are so arranged that all of the current that flows through the circuit must pass through each of these parts in turn.

The first controlled electric currents were generated by means of batteries. If a dilute solution of acid, such as sulphuric acid, is placed in a glass and a strip of copper and a strip of zinc are placed in the solution in such a manner that they are not in contact, a difference in electrical potential will be built up on the strips. If these strips are connected by a conductor, an electric current will flow from one strip to the other. The tendency for an electric current to flow is called "difference in potential" and is measured in volts. The copper strip becomes charged positively and the zinc strip becomes charged negatively. This action is continuous, and there will be a continual flow of electrons

ELECTRICAL FUNDAMENTALS

from the zinc strip to the copper strip. It was formerly believed that the direction of flow of electric current was from the positive to the negative. The positive connection is the anode, and the negative connection is the cathode. This arrangement of acid and strips of metal is known as a "wet cell."

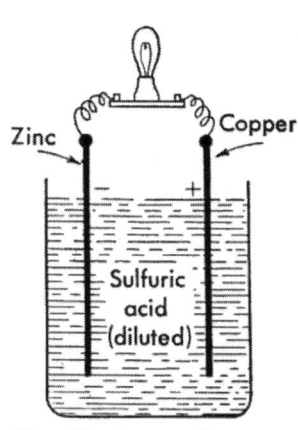

Fig. 101. A wet cell.

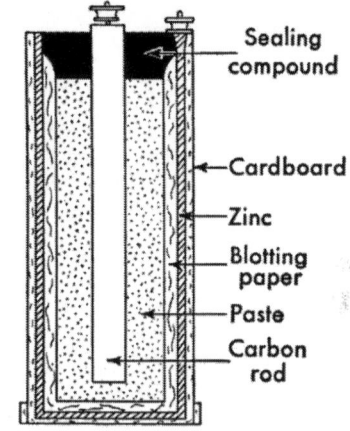

Fig. 102. A dry cell.

A dry cell, a common example of which is the ordinary flashlight battery, is made up of a zinc container in which is inserted a carbon rod. The space between the rod and the zinc container is filled with a paste made of manganese dioxide and granulated carbon. This paste is saturated with a solution of ammonium chloride (sal ammoniac) and zinc choride. The top of the cell is sealed with a pitchlike compound, and the whole is placed in a cardboard container. One terminal is fastened to the zinc container, and the other terminal is fastened to the carbon rod in the center. The carbon rod in this case is the anode, and the zinc container is the cathode.

A storage battery is used to store up electricity by chemical means. The dry cell, the wet cell, and the

Fig. 103. An aircraft storage battery. (Courtesy Delco-Remy Division, General Motors Corporation)

storage battery produce electricity by chemical means. In the storage battery, the chemical reaction is forced in one direction by charging with an electric current. When the positive and negative terminals are connected by conductors, the chemical reaction reverses itself, giving

105

off the electric current. A storage battery is made up of alternate plates of lead and lead oxide, separated by insulators and immersed in an acidproof box containing a sulphuric acid solution. The chemical equation for the storage battery is:

$$\underset{\text{Discharge} \longrightarrow}{\overset{\longleftarrow \text{Charge}}{PbO_2 + Pb + 2H_2SO_4 = 2PbSO_4 + 2H_2O}} + \text{electrical energy}$$

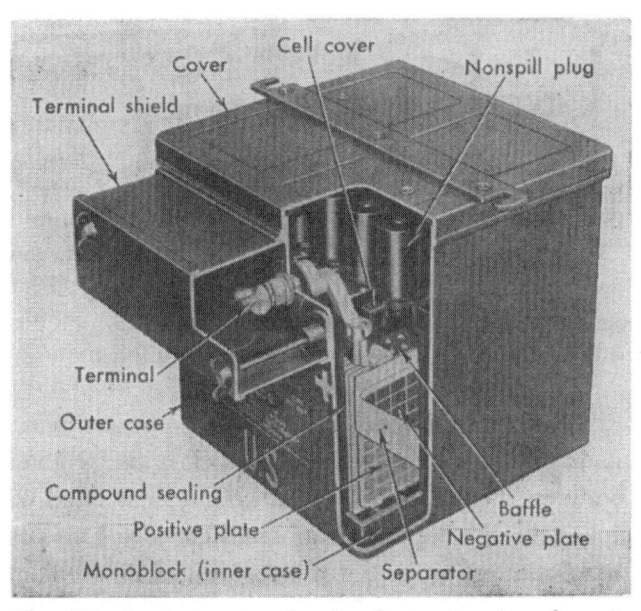

Fig. 104. A cutaway view showing the construction of an aircraft storage battery. (Courtesy Delco-Remy Division, General Motors Corporation)

During charging, the acid solution becomes more concentrated. Concentrated acid is much heavier than water, and the state of charge in the battery may be determined by measuring the specific gravity of the liquid.

When a conductor, such as a copper wire, is moved in an electric field in such a manner that magnetic lines of force are cut, an electric current is set up in the conductor. If the wire is moved through the magnetic field in the opposite direction, the current set up in the wire flows in the opposite direction. If the ends of the wire are connected, while the wire is moved rapidly back and forth through the magnetic field, an alternating current of electricity is set up in the circuit. This is the fundamental principle of an alternating current generator, or alternator. If an arrangement is made whereby the ends of the circuit

ELECTRICAL FUNDAMENTALS

are connected with the opposite ends of the moving wire each time the wire changes its direction, the current in the circuit will flow in one direction only, and a direct current is generated. This type of generator

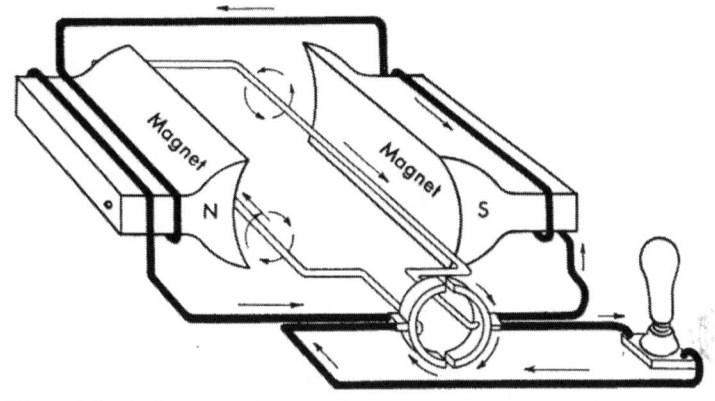

Fig. 105. A diagram showing how a conductor may be rotated through an electric field to produce an electric current. This is a simple shunt-wound generator.

is known as a direct-current generator. The device for connecting the opposite ends of the moving wire to produce a direct current is the commutator. A d-c generator has a commutator, while an a-c generator does not.

Fig. 106. An aircraft direct-current generator. (Courtesy Delco-Remy Division, General Motors Corporation)

In a generator, instead of a single wire moving through the magnetic field, many turns of wire are wrapped around a soft iron core that is revolved in a magnetic field between the poles of magnets. This revolving coil of wire is the armature, and the magnets are the field magnets. This type of arrangement, when used for ignition purposes, is the magneto. In the magneto the magnets revolve and the coils of wire remain

AIRCRAFT ENGINES

stationary. Whenever an electric current flows through a conductor, a magnetic field extending out into space is set up around the conductor. The lines of force build outward as the current increases and die down when the current decreases. If a wire carrying an alternating current of electricity is placed close to another wire, the second wire will have set up in it an alternating electric current that is opposite in direction to the alternating current in the first wire. This second current is an induced current. If the first wire carrying the alternating current is

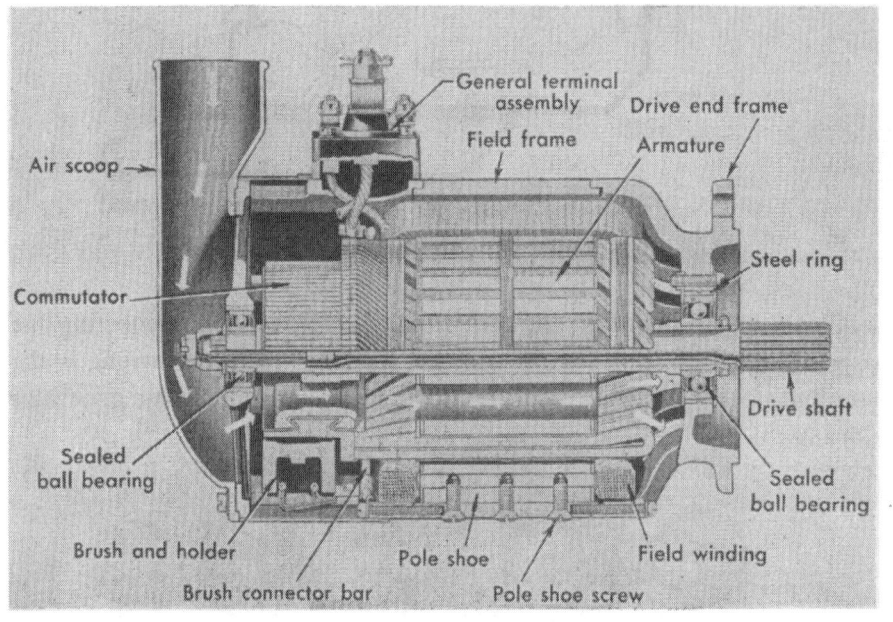

Fig. 107. A cutaway view of an aircraft generator showing construction and parts. (Courtesy Delco-Remy Division, General Motors Corporation)

wrapped around a soft iron core and the second wire is wrapped around the same core, but is insulated from the first wire, the voltage of the induced current in the second wire is in proportion to the number of turns of each wire around the core. This device is a transformer or induction coil.

Assume that the primary coil has 100 turns and carries a 10-v. alternating current, and the secondary coil consists of 10,000 turns. A 1000-v. current will be induced in the secondary coil. These coils may also be called primary and secondary windings. The windings in the two coils being in the ratio of 100 to 1, the voltage of the currents in the two coils is in the same ratio.

ELECTRICAL FUNDAMENTALS

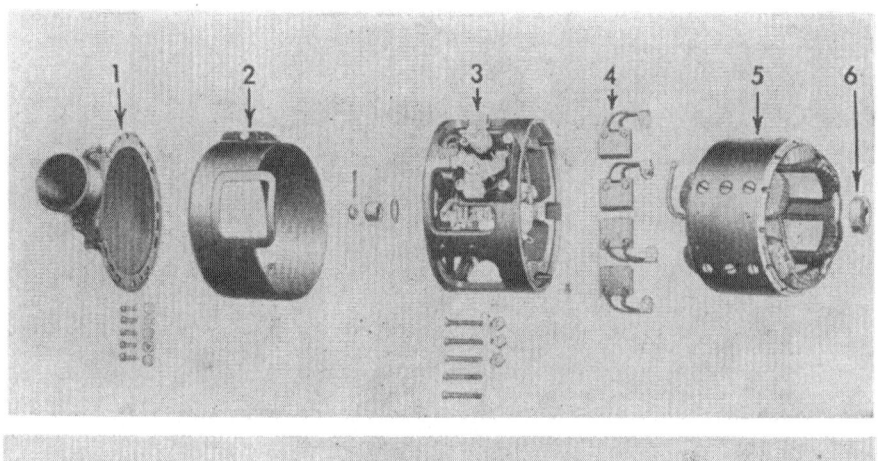

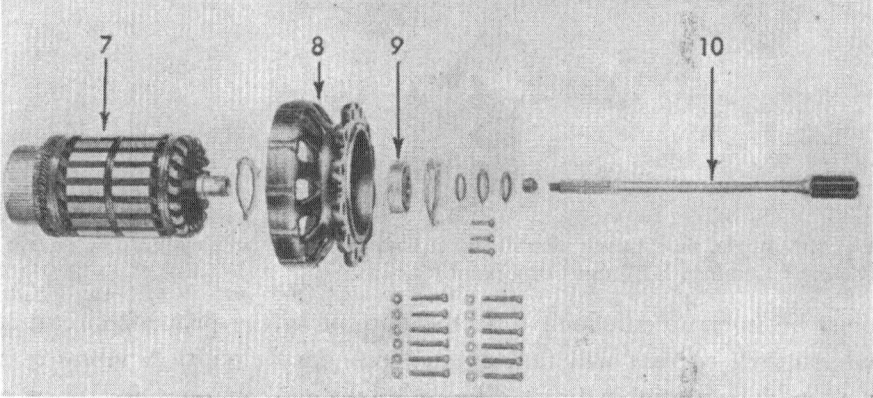

Fig. 108. An exploded view of an aircraft generator. (1) Air scoops; (2) cover band; (3) commutator end frame assembly; (4) brushes; (5) field frame assembly; (6) ball bearing; (7) armature; (8) drive end frame; (9) ball bearing; (10) drive shaft. (Courtesy Delco-Remy Division, General Motors Corporation)

If a direct current flows through the primary circuit, current will flow in the secondary circuit only during the time that the current in the primary circuit is building up or dying down. In order to obtain the effect of an alternating current, the direct current in the primary coil

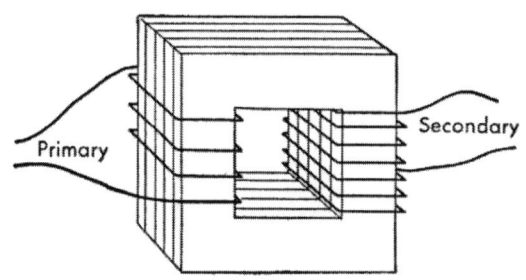

Fig. 109. A diagram of a simple transformer.

109

AIRCRAFT ENGINES

Fig. 110. A light, six-cylinder, aircraft engine having two magnetos mounted at the rear. (Courtesy Continental Motors Corporation)

must be interrupted. Each time the current in the primary circuit is interrupted, current will flow in the secondary circuit. A vibrator is used to interrupt the direct current in the primary circuit. When a vibrator is introduced into the primary circuit, it has the effect of

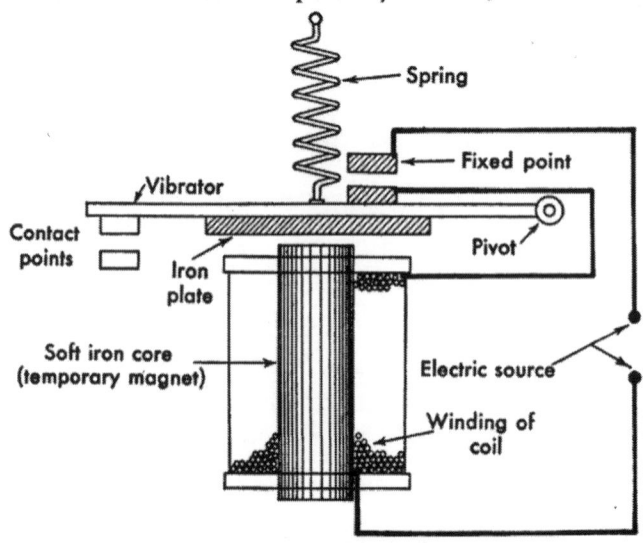

Fig. 111. A diagram to show the construction of a vibrator.

ELECTRICAL FUNDAMENTALS

opening and closing a switch many times a second. The high-voltage current generated in the secondary circuit of the induction coil is the current that is led to the spark plugs.

High-voltage current has the ability to jump across the gap between the electrodes of the spark plugs. This gap may vary from 0.010 to 0.020 in. Because air is a nonconductor of electricity, a high voltage of approximately 20,000 v. is required to jump this gap. The voltage is high but the amperage is low, being only a small fraction of 1 amp.

XII IGNITION SYSTEMS

Good engine performance depends on the proper operation of the ignition system. The entire problem of the ignition system is to produce an electric spark which will ignite the air-fuel mixture in the cylinder at the proper time.

Early ignition systems were of the make-and-break type. This type consisted of a pair of contact points placed within the combustion chamber of the engine. These points were part of a circuit carrying a direct current and they remained open during the compression period. The points closed and opened at the time a spark was desired in the combustion chamber. This spark consisted of a small electric arc which formed when the points were pulled apart. This type of spark worked quite successfully on low-speed engines, but was not effective for high-speed engines which required a short, intense spark accurately timed.

Ignition systems fall into two general classes: the battery system, such as that used on most automobiles today, and the magneto system, which is most commonly used on aircraft engines. Most early automobiles used the magneto ignition system, but now the battery system is more commonly used.

The battery ignition system has the following advantages:

1. A strong spark is obtained in the engine combustion chamber even at low engine speed, which makes starting less difficult.

2. If the aircraft is already equipped with a battery to furnish current for other equipment such as landing and navigation lights, the weight of the two magnetos is saved.

3. There are fewer moving parts in the battery system and less chance for mechanical failure.

4. The battery generating equipment furnishes a convenient supply of electric current for instrument lights and other electrical equipment in the aircraft.

IGNITION SYSTEMS

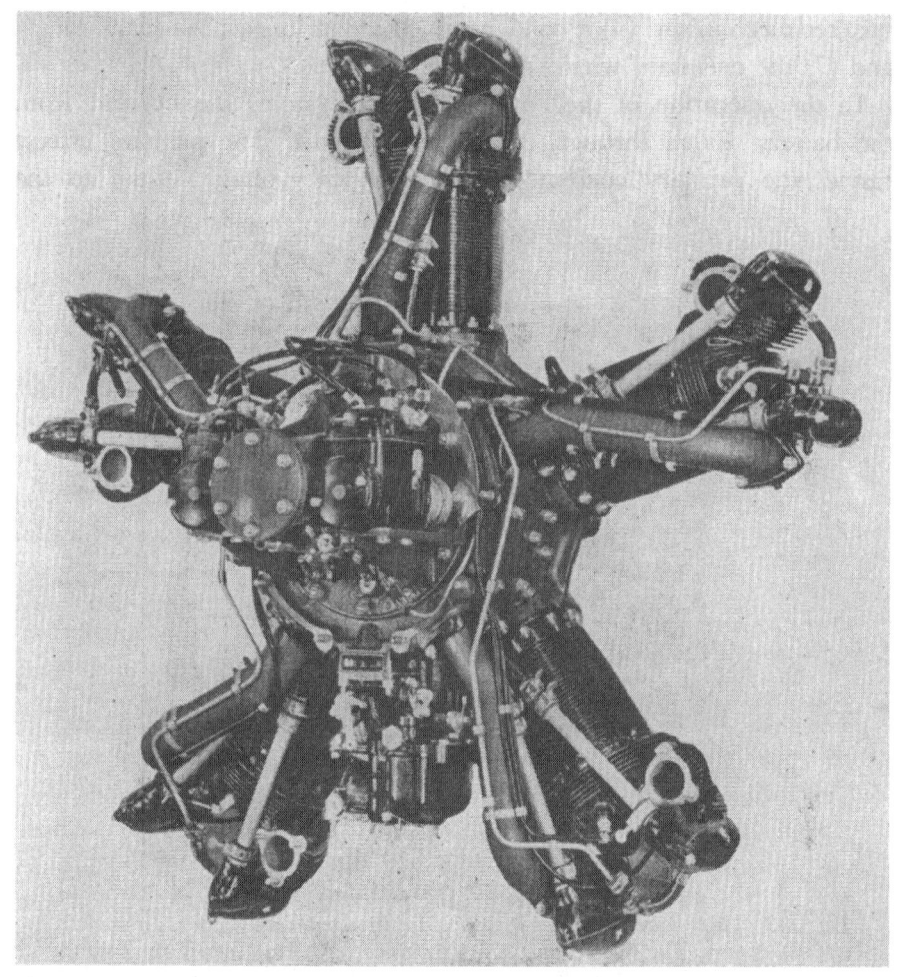

Fig. 112. Three-fourths rear view of a five-cylinder radial aircraft engine, showing ignition wiring. (Courtesy Kinner Motors Inc.)

The magneto system offers the following advantages:

1. When magnetos are used, a saving of weight is accomplished as a storage battery is not needed.
2. There is less danger of fire when magnetos are used.
3. There are fewer external connections which might become loose.
4. The magneto system occupies less space and has less wiring.
5. There is no danger of ignition failure due to the battery becoming discharged.

The battery ignition system consists of (1) the battery, which is the source of electrical current; (2) an induction or high-tension coil; (3) the breaker mechanism; (4) a condenser; (5) a distributor; (6) spark plugs; and (7) the necessary wiring and switches.

In the operation of the battery ignition system, the current from the battery is led through the primary circuit. The primary circuit carries the primary current, which is of low voltage, equal to the

Fig. 113. Three-fourths left rear view of an inverted, six-cylinder, air-cooled, aircraft engine. The two magnetos are mounted at the top rear of the engine. The hollow magneto drive shaft carries oil to the main bearings. (Courtesy Ranger Aircraft Engines)

voltage of the battery. The voltages of aircraft batteries are 6, 12, or 24 v. The primary circuit consists of the battery lead to the induction coil; the primary winding in the induction coil; and the leads to the breaker mechanism, one side of which is grounded to the engine as is one terminal of the battery. When the breaker points are closed, the current flows from one terminal of the battery through the primary windings of the coil, through the breaker mechanism, and back through the ground to the battery. The condenser is connected in parallel with the breaker points. With the exception of a switch which opens and closes the circuit, this completes the primary circuit. The main switch opens and closes the primary circuit.

IGNITION SYSTEMS

The secondary circuit of the battery ignition system consists of the secondary windings in the induction coil; the lead from one terminal of the induction coil to the distributor and to the rotor within the distributor head; and wires leading from the distributor head to the spark plugs. The current jumps the gap in the spark plug and flows back

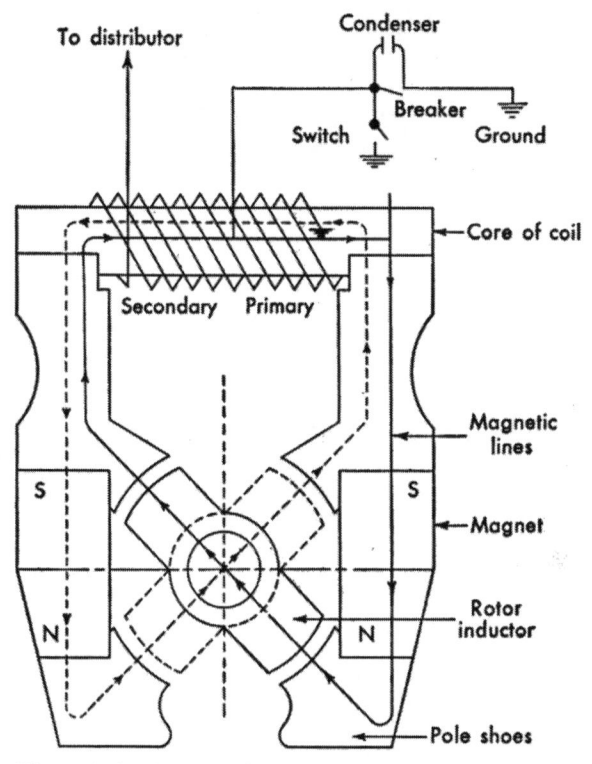

Fig. 114. A schematic drawing showing the fundamental operations of an aircraft magneto.

through the engine itself to the other terminal of the secondary winding which is grounded to the engine structure.

The storage battery furnishes a low-voltage direct current. As the current from the battery flows through the primary winding of the induction coil, a magnetic field which expands outward into space is set up around the primary winding, cutting the secondary winding. This field remains constant when there is no change in the amount of current flowing through the primary winding. As the points in the breaker mechanism close, the current starts to flow through the primary circuit, and the electric field is built up. This build-up is so gradual that the

115

AIRCRAFT ENGINES

current generated in the secondary winding is not intense enough to jump the spark gap in the spark plugs. At the instant a spark is desired, the points are rapidly opened, causing the sudden collapse of the magnetic field. This magnetic field cuts the secondary winding and induces a high voltage, some 20,000 to 25,000 v., in the secondary circuit.

Fig. 115. A modern aircraft-engine magneto. (Courtesy Eisemann Corporation)

When an electric current is suddenly interrupted, self-induction is set up in the circuit, which tends to keep the current flowing. This self-induction would cause an arcing across the points were it not for the condenser which is connected in parallel across the points. The surge of electric current, due to self-induction, is absorbed in charging the condenser. This prevents arcing across the points. Arcing would cause burning of the points, which would decrease the intensity of the current generated in the secondary circuit.

As the self-induced current in the primary circuit dies down, the condenser discharges back into the primary circuit. The points in the breaker mechanism are opened and closed by means of a cam arrangement on the rotor shaft. Attached to the top of the rotor shaft in the distributor head is the rotor connected by means of a sliding contact to the

IGNITION SYSTEMS

lead from the high tension or secondary winding of the coil. Each spark plug lead is connected to a metal block in the distributor head. The rotor has a metal point which clears the metal blocks in the distributor head by a few thousandths of an inch. This point is opposite the proper block connected with the lead to the spark plug at the time the spark is

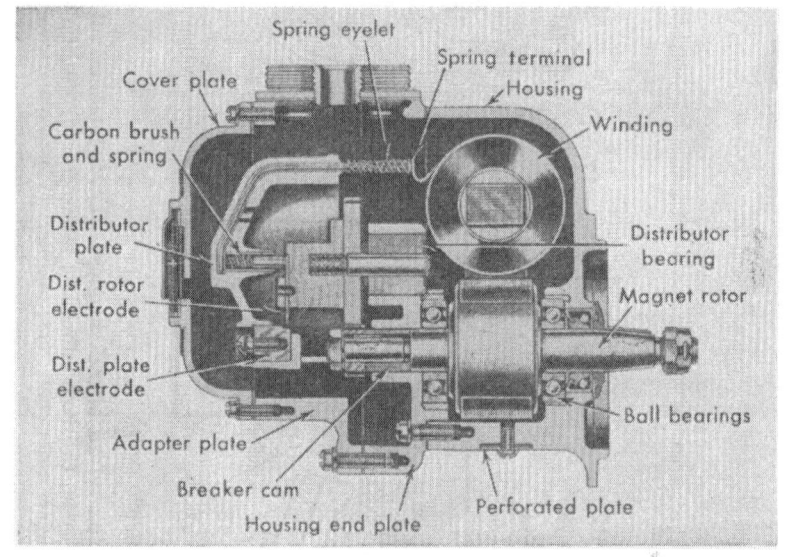

Fig. 116. A cutaway view showing construction of a magneto. (Courtesy Eisemann Corporation)

desired in that plug. The high-voltage current set up in the secondary circuit jumps across the gap between the rotor and the metal block and is carried to the spark plug, where it jumps the spark plug gap, producing the spark in the combustion chamber. The current then travels back through the ground to the terminal of the high tension coil, which is grounded to the engine structure. There is usually a safety gap installed in the high-tension coil to protect the high-tension winding from overheating or from being damaged by a short circuit, or in case a wire becomes loose and the secondary circuit is opened.

The magneto ignition system operates in a manner similar to the operation of the battery system, with the exception that a magneto is the source of current instead of a storage battery. The magneto furnishes a low-tension alternating current which flows through the primary circuit in a manner similar to the primary circuit of the battery ignition system. The modern aircraft magneto has included in the magneto itself the high-tension coil, the condenser, the safety gap, the distributor head,

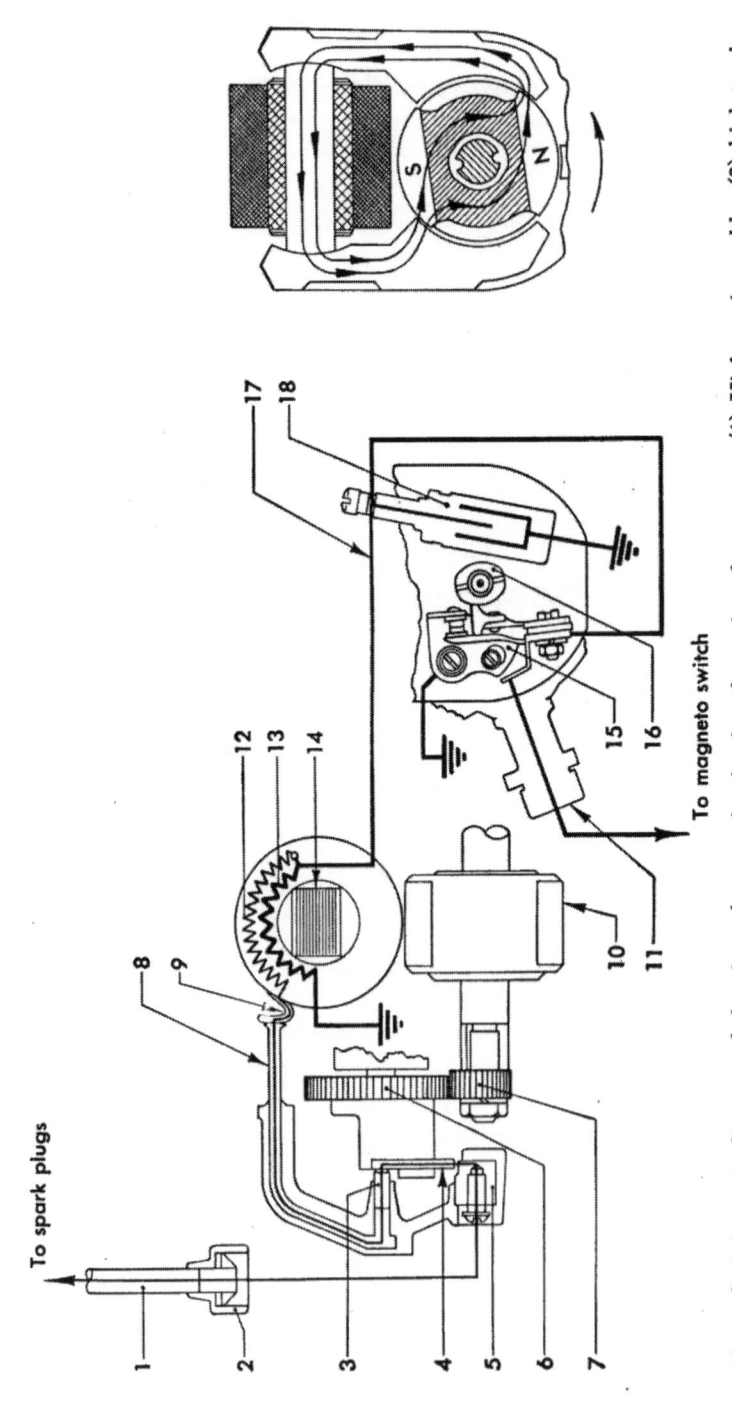

Fig. 117. Schematic diagram of electric and magnetic circuits of an aircraft magneto. (1) High-tension cable; (2) high-tension terminal; (3) carbon brush and spring; (4) distributor electrode; (5) distributor-plate electrode; (6) distributor gear; (7) pinion gear; (8) spring eyelet conductor; (9) spring clip; (10) magnet rotor shaft; (11) adapter; (12) secondary winding; (13) primary winding; (14) coil core; (15) breaker assembly; (16) breaker cam; (17) winding lead condenser; (18) condenser. (Courtesy Eisemann Corporation)

118

IGNITION SYSTEMS

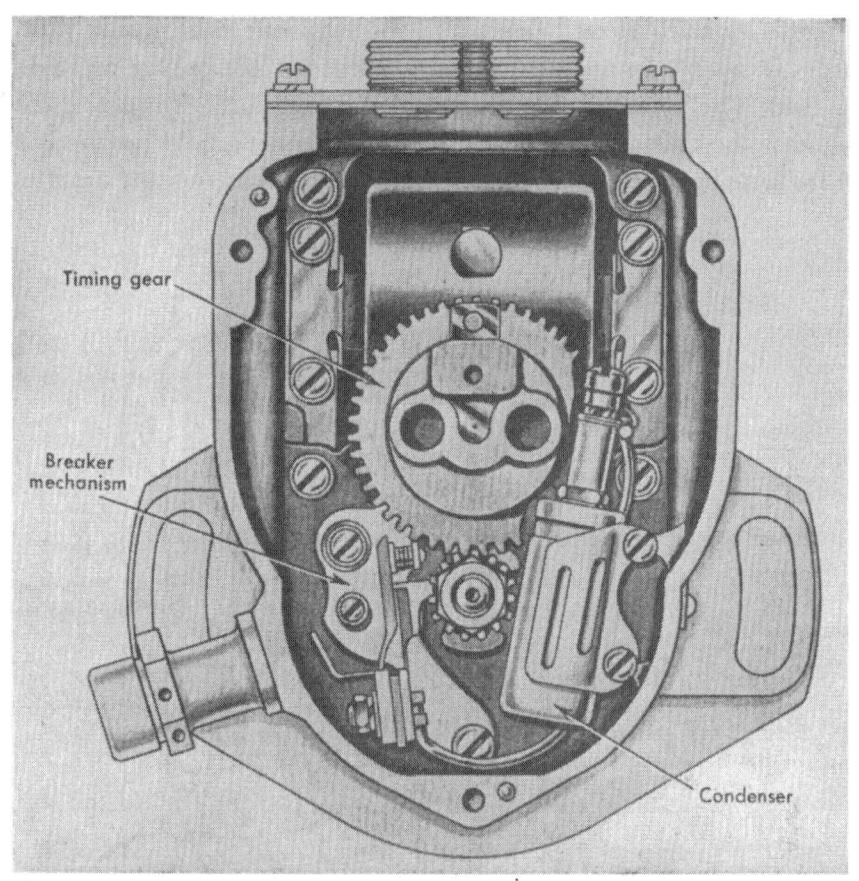

Fig. 118. A cutaway view showing condenser, breaker mechanism, and timing gears of an aircraft magneto. (Courtesy Eisemann Corporation)

the rotor, and the breaker mechanism. All parts of this system operate in the same manner as in the battery system.

In the magneto system the switch, when in the OFF position, closes the circuit, grounding the magneto so that no current will flow through the primary circuit. In case the ground wire becomes loose, the engine will continue to run even though the switch is placed in the OFF position. When this occurs, the pilot stops the engine by cutting off the fuel supply.

The current in the primary circuit of the magneto system is produced by electro-magnetic induction. Current may be produced by electro-magnetic induction in three ways:

1. By moving a conductor through a magnetic field.
2. By moving a magnetic field in such a manner that it cuts a stationary conductor.
3. By a fluctuating magnetic field cutting a stationary conductor.

AIRCRAFT ENGINES

In the ordinary electric generator, the magnetic field, which is produced by means of stationary magnets, is cut by coils of wire rotated in this field. The rotating coils are called the armature. In most aircraft magnetos the coils are stationary, and the magnetic field produced by the permanent magnets is rotated past the coils. The rotating armature

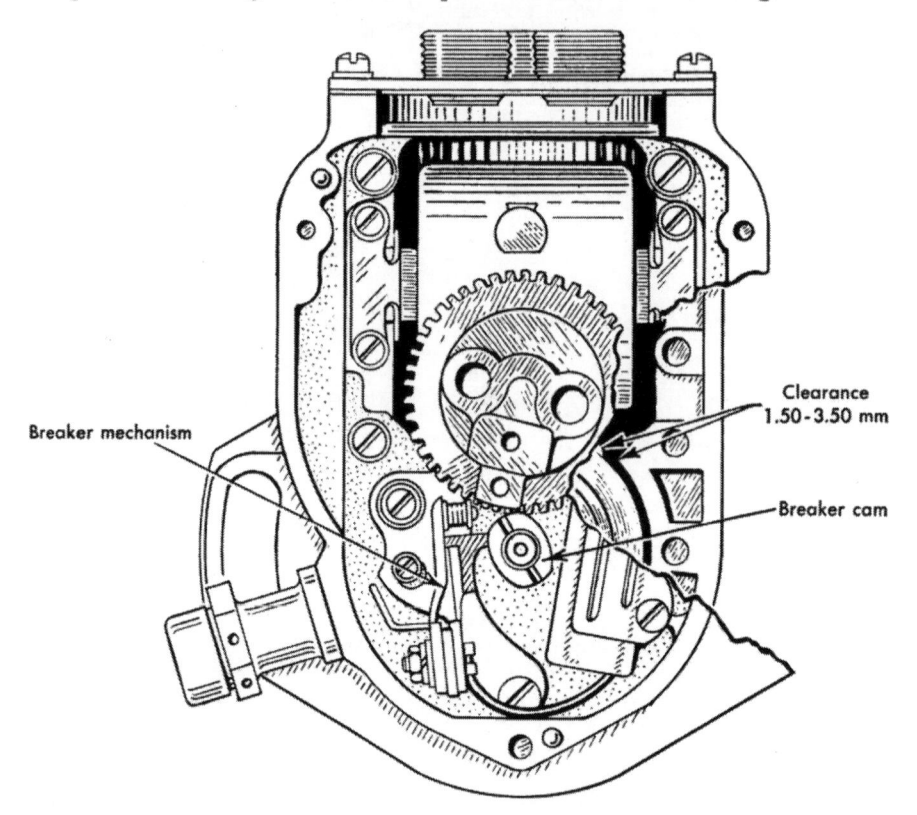

Fig. 119. Drawing of a magneto showing breaker cams, breaker mechanism, and clearance between armature and field coils. (Courtesy Eisemann Corporation)

is made up of the permanent magnets which rotate and carry the field past the coils. This armature may be made up of two, four, or eight poles. As the poles of the magnet pass a coil, a current is set up in one direction in the coil and, as the other poles of the magnet pass the coil, the current dies down to zero and then is set up in the opposite direction in the coil. This action produces an alternating, low-voltage current. This current is interrupted by the opening and closing of the breaker points. This causes the induction in the secondary windings of the coil of the high voltage current required for the spark.

Due to the fact that when the magneto is rotating at low speeds a

IGNITION SYSTEMS

weak spark is generated, a booster arrangement is usually included in the magneto system. This system consists of an auxiliary high-tension coil and a vibrator. The booster switch may consist of a push button or may be connected to the starter switch and it operates only while the engine is being started. The booster segments in the magneto are usually a short distance behind the main segments in order to retard the spark, thus preventing kickback while the engine is being started.

Fig. 120. An aircraft-engine magneto. (Courtesy Scintilla Magneto Division, Bendix Aviation Corporation)

Spark plugs are of a number of different types although they all function on the same principle. A spark plug consists of a threaded metal shell that screws into the spark-plug bushing in the cylinder head. The opening in the bottom of the shell has from one to four small electrodes fastened around its circumference. These electrodes project inward toward the center of the opening. Inserted into this shell is the spark-plug core. This core consists of insulating material surrounding the main electrode of the spark plug. This electrode projects into the opening between the electrodes fastened to the spark-plug shell. The space between the center electrode and the electrodes on the shell is called the

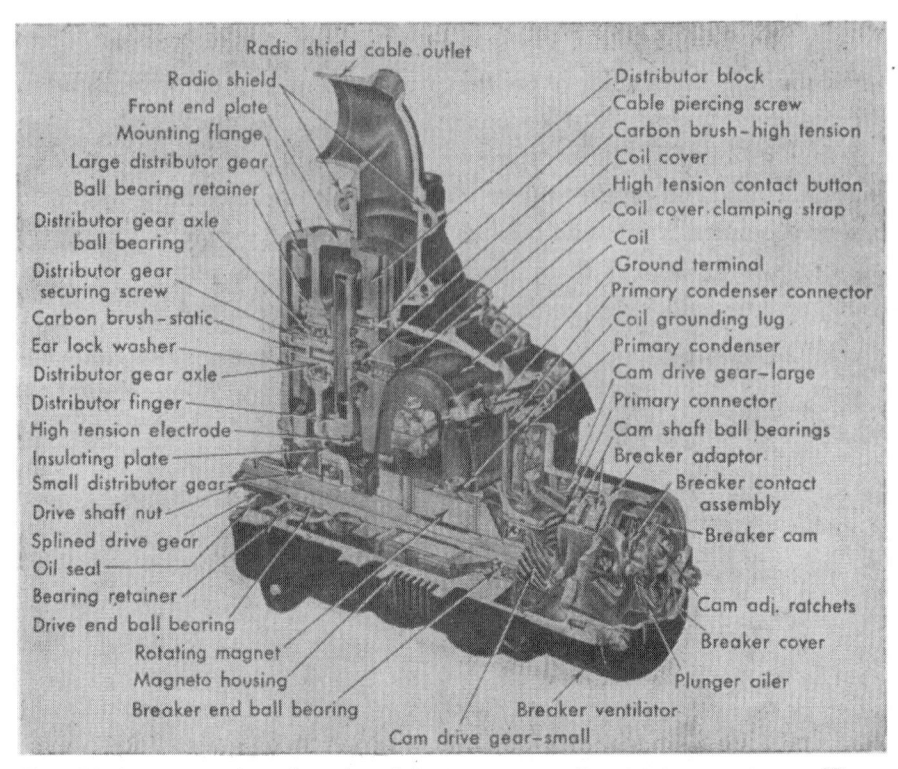

Fig. 121. A cutaway view of an aircraft-engine magneto showing its many parts. (Courtesy Scintilla Magneto Division, Bendix Aviation Corporation)

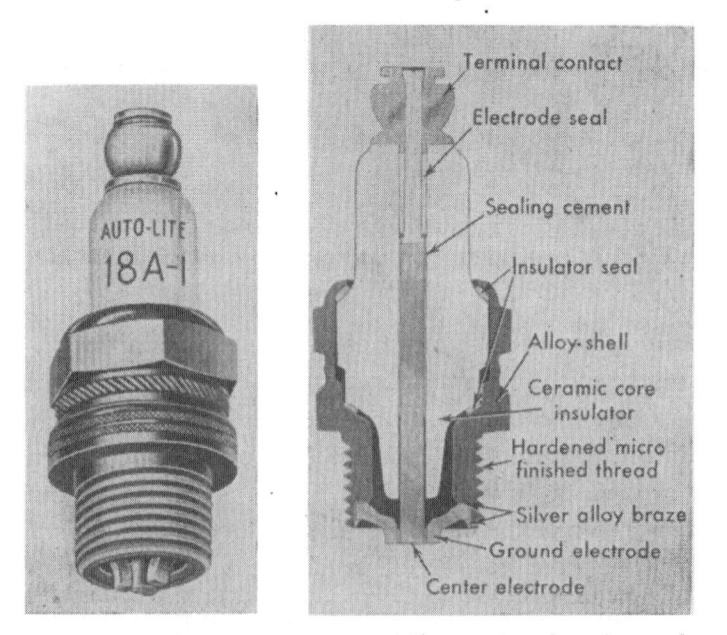

Fig. 122. Short-reach, unshielded, "cold"-type, aircraft-engine spark plug. (Courtesy The Electric Auto-Lite Company)

122

IGNITION SYSTEMS

spark gap. This gap is usually about 0.015 in., but should agree with the manufacturer's specification. The spark plug is assembled before it is placed in the engine. The high-tension wire carrying the current to the plug is connected to the main electrode which extends through the insulating core. The insulating material used in aircraft spark plugs is either mica or a special type of porcelain. Three types of spark plugs are used in various types of engines. Spark plugs are classified as hot, medium,

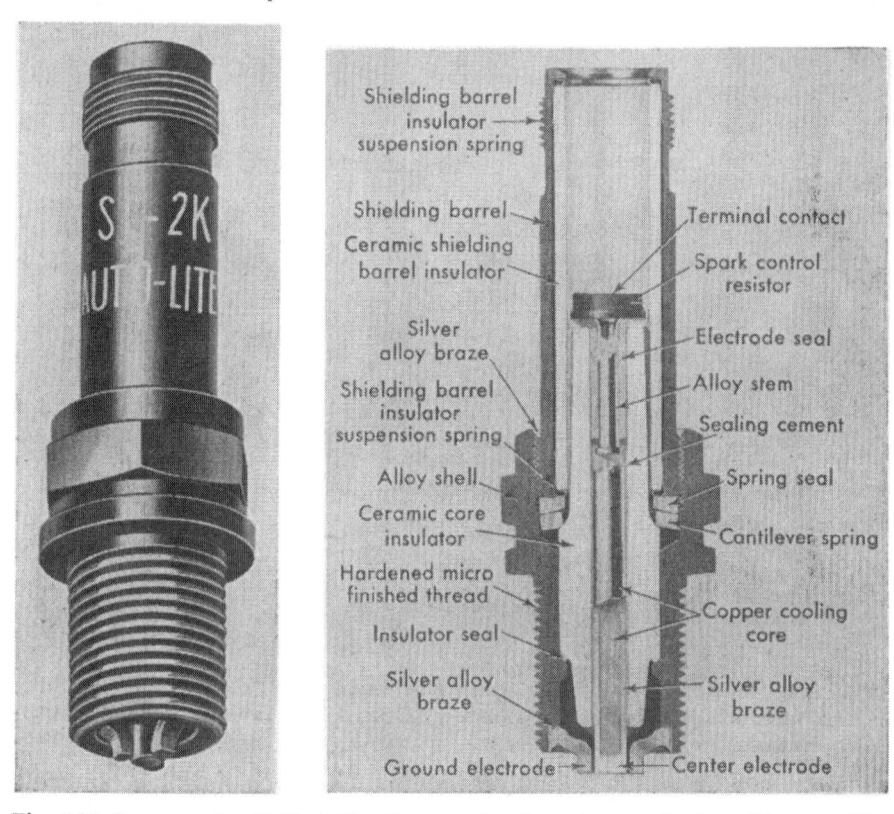

Fig. 123. Long-reach, shielded, "hot"-type, aircraft-engine spark plug. (Courtesy The Electric Auto-Lite Company)

or cold. The hot plugs are usually longer than the cold plugs and protrude farther into the combustion chamber. These plugs do not cool as rapidly as the cold plug. Hot plugs are more generally used on liquid-cooled engines. On air-cooled engines, arrangements are sometimes made to cool the plug itself. One type of plug has cooling fins as a part of the insulator. The contact between the center electrode and the high-tension wire leading to the plug must be such that it will remain secure at all times. The fastening may be made by means of a knurled nut, a spring

123

AIRCRAFT ENGINES

pressure plate, or a spring clip. Whatever the type of fastening, it must be such that it will not work loose because of the vibration of the engine.

The Civil Aeronautics Administration requires that all aircraft engines be equipped with dual ignition. This means that there are two spark plugs in each cylinder combustion chamber, each connected with

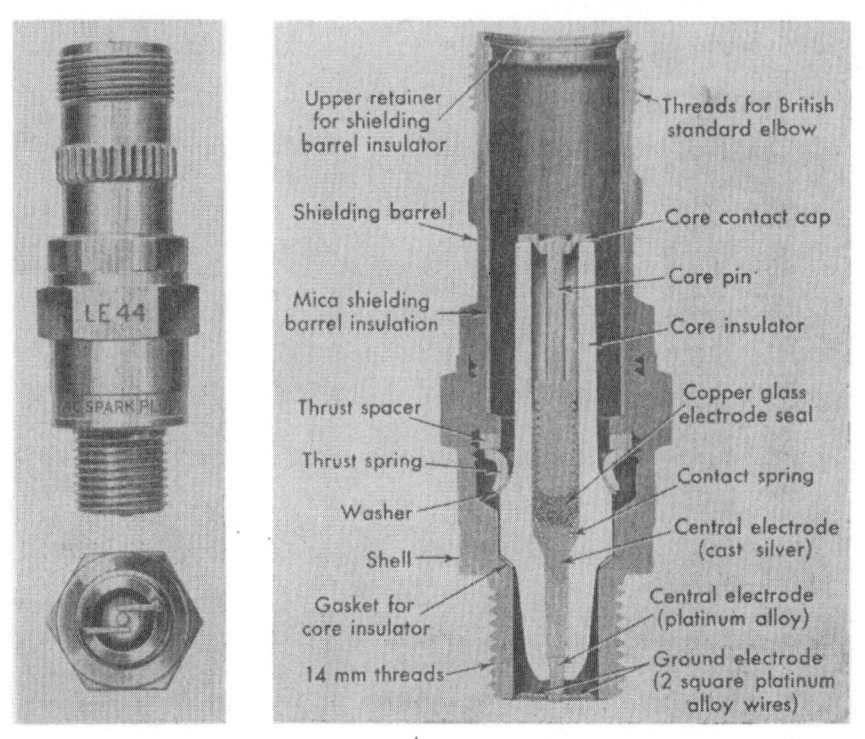

Fig. 124. Shielded aircraft-engine spark plug having two "shell" electrodes. The "shell" electrodes are of platinum. (Courtesy AC Spark Plug Division, General Motors Corporation)

a separate ignition system. These engines have two completely separate ignition systems. These separate sources of ignition may consist of two separate magnetos, two separate battery ignition systems, or one magneto for one set of plugs and a battery ignition for the other set. This dual ignition arrangement is primarily a safety factor, but is also a method by which engine power output is increased. An aircraft engine will operate on either set of plugs. When either set of plugs is cut out, such as by turning off one magneto, the engine continues to operate but loses from 50 to 100 r.p.m. Using only one set of plugs may cause detonation to take place, particularly when the engine is operating under full power.

IGNITION SYSTEMS

Dual ignition ignites the fuel charge in two places at about the same time, usually on opposite sides of the combustion chamber. This leads to more rapid burning of the charges and a more nearly equal distribution of the forces developed. Many engines have an arrangement whereby a battery system may be connected with one set of plugs during the time

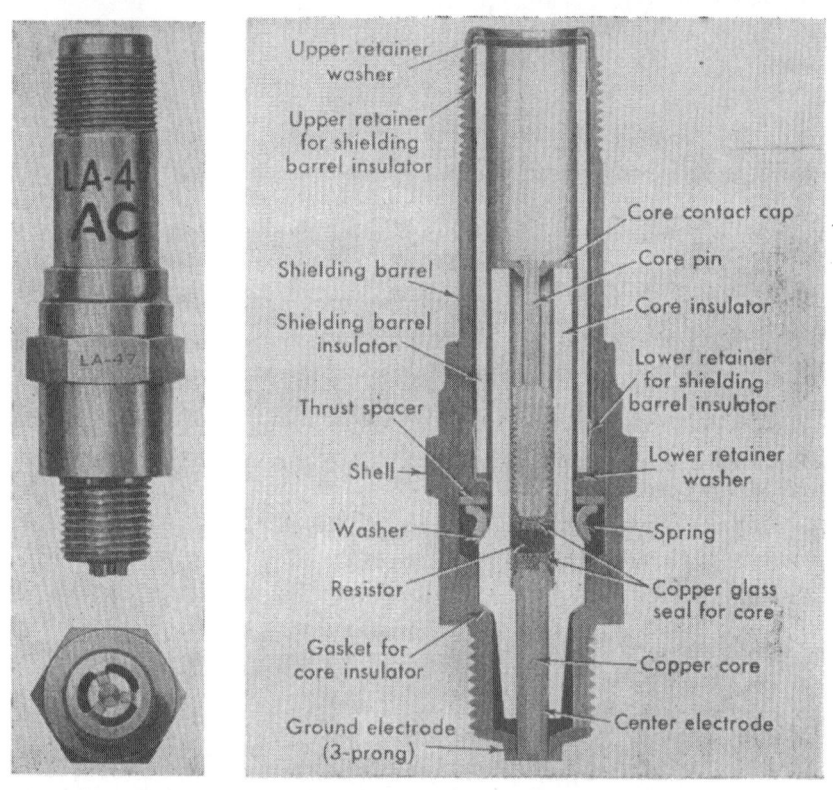

Fig. 125. Shielded aircraft-engine spark plug having three "shell" electrodes. (Courtesy AC Spark Plug Division, General Motors Corporation)

the engine is being started. This assists in starting the engine, as a hot spark is developed at low engine revolutions per minute.

When a battery is used in an aircraft, it is necessary to have some arrangement whereby the battery will charge while the aircraft is in operation. Usually the battery receives a continuous charge whenever the engine is being operated at ordinary speeds. Some of the older aircraft, as well as some of the present-day light aircraft equipped with magneto ignition systems, have a battery installed to operate the lights or radio equipment. When no provision is made for charging the battery while in flight, the battery is often found to be run down or inoperative at critical times.

125

AIRCRAFT ENGINES

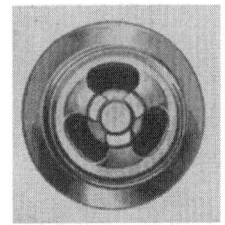

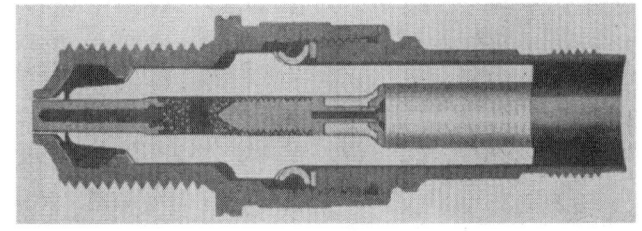

Fig. 126. High-altitude aircraft-engine spark plug. (Courtesy AC Spark Plug Division, General Motors Corporation)

Fig. 127. An aircraft engine-driven generator. (Courtesy Delco-Remy Division, General Motors Corporation)

The battery-charging system consists of a d-c generator, usually of the shunt-wound type. In this type of generator, the field circuit, or winding, is connected across the armature circuit. This arrangement allows a variation in current output. The amount of current going into the battery depends upon the amount of current flowing through the field circuit. When the battery is low or when a considerable amount of

IGNITION SYSTEMS

current is used as in starting, the charging rate will be high. As the battery becomes nearly charged, the charging rate becomes approximately zero. The generator is operated by gearing from the crankshaft of the engine.

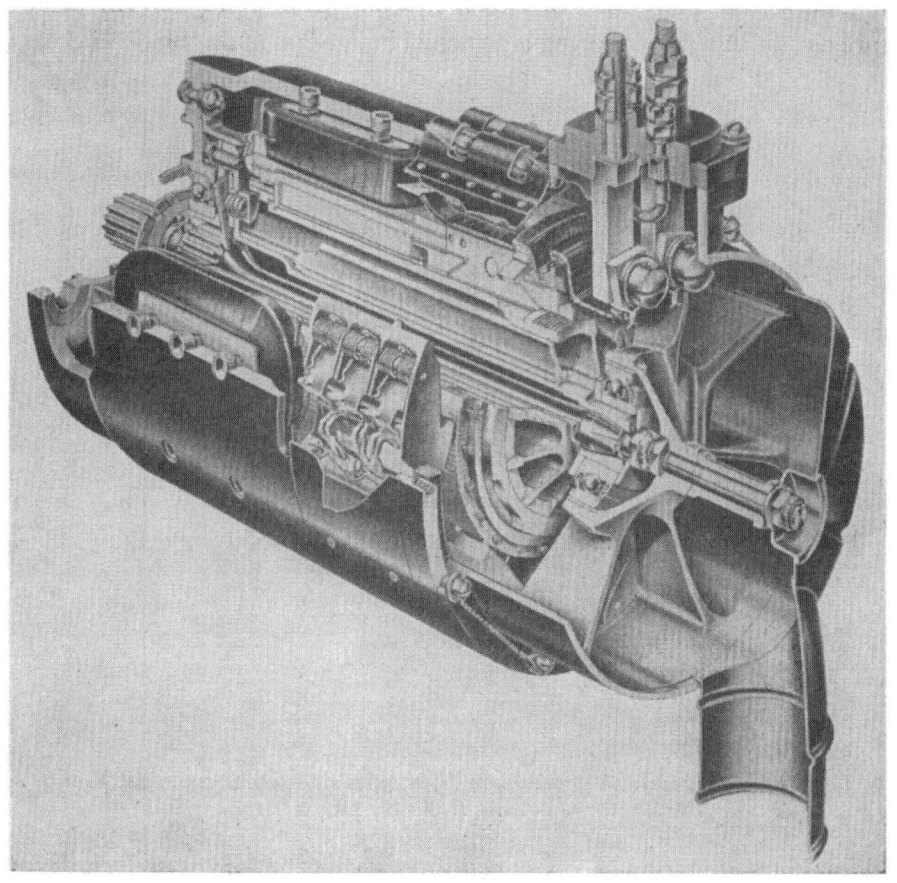

Fig. 128. A cutaway view of an engine-driven generator assembly. (Courtesy Jack & Heintz, Inc.)

The battery charging system usually consists of (1) a generator, (2) a voltmeter, (3) a field switch, (4) a voltage regulator, (5) a current limiter, (6) a reverse current cutout, (7) an ammeter, (8) a master switch, and (9) a battery. The voltage regulator, current limitator, and reverse current cutout are usually mounted together in a single unit called the control panel. The field switch may be mounted on the control panel if it is installed near the pilot in the cockpit or on the instrument panel if the control panel is mounted elsewhere.

The **voltage** regulator consists of a core around which is wound a coil

AIRCRAFT ENGINES

of fine insulated wire which is connected across the armature circuit. This means that the winding will be affected by the voltage in the circuit. A flat blade is mounted close to one end of the core. This blade is pivoted and has a pair of contact points at one end in such a manner that they may be opened or closed by changes in voltage in the armature circuit. A suitable spring is attached to one end of this blade and also

Fig. 129. Armature and commutator from an engine-driven generator. Commutator being checked for roundness. (Courtesy Jack & Heintz, Inc.)

to an adjusting screw. A resistance is connected to the voltage regulator in such a manner that the current flowing to the generator field will pass through the resistance when the points are opened. This reduces the flow of current to the field and regulates the output of the generator. When the voltage reaches a maximum force, the core will be magnetized, opening the points, thus causing the current to the field winding to be reduced. This holds the output of the generator to the desired limit. If the armature circuit is overloaded, the voltage will not be high enough to sustain the points and they will remain closed, thus permitting full current flow to the field and developing the maximum output of the generator. If the circuit is too highly loaded, the voltage regulator will not function and the generator may be burned out.

IGNITION SYSTEMS

The current limitator is used to protect the generator in this case. The current limitator consists of a core with its winding, the pivoted

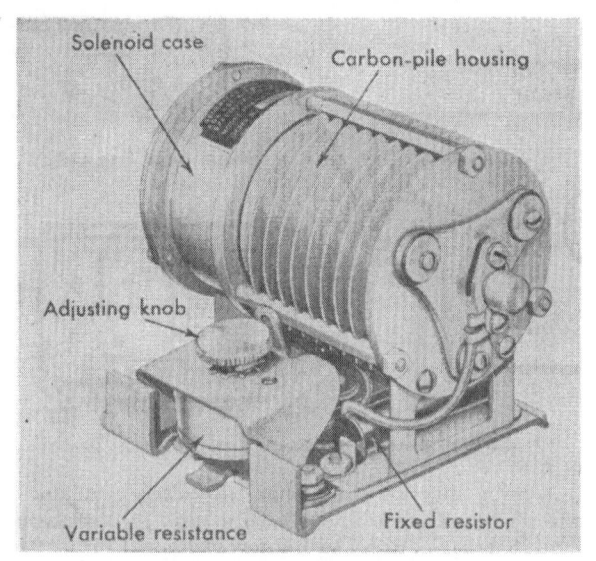

Fig. 130. A carbon-pile voltage regulator. (Courtesy Delco-Remy Division, General Motors Corporation)

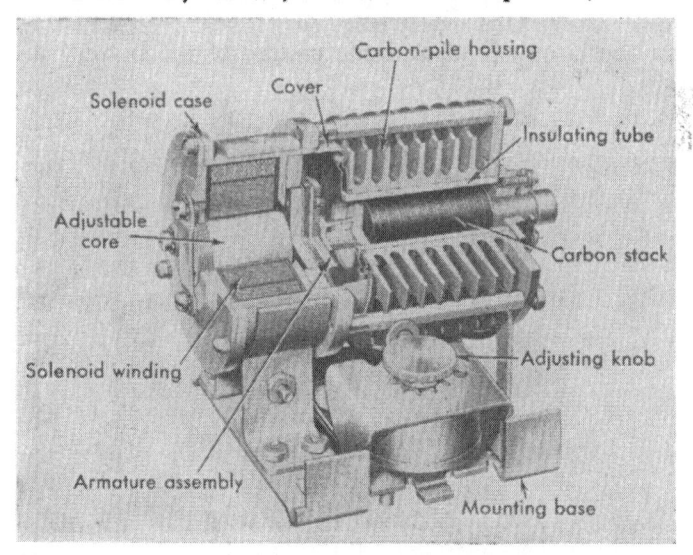

Fig. 131. A cutaway view of a carbon-pile voltage regulator. (Courtesy Delco-Remy Division, General Motors Corporation)

blade, the points, and spring, and is quite similar to the voltage regulator. The coil, which is wound around the core, is connected in series with one side of the armature circuit. This coil is affected by the amperage or rate of flow of current in the armature circuit. The resistor and

AIRCRAFT ENGINES

point circuits of the voltage regulator and the current limitator are connected to the field winding of the generator in such a manner that

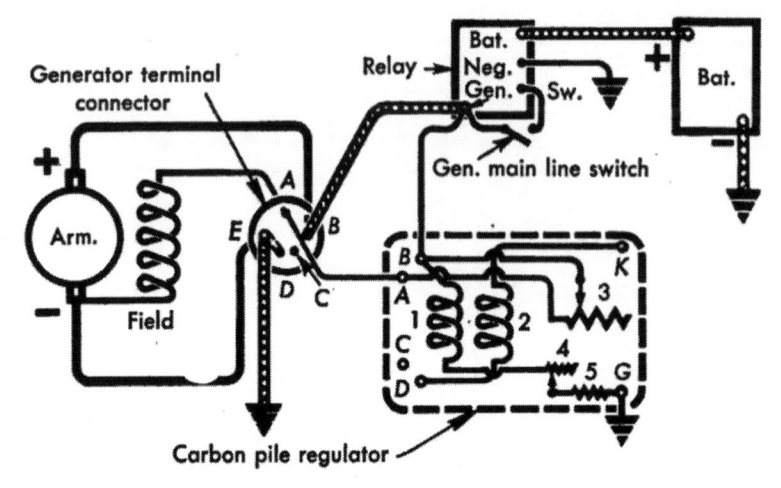

Fig. 132. A wiring diagram of a carbon-pile voltage regulator. (1) Regulating coil; (2) equalizing coil; (3) carbon pile; (4) variable resistance; (5) fixed resistance. (Courtesy Delco-Remy Division, General Motors Corporation)

either one may control the flow of field current and output of the generator. When the circuit load and the charge of the battery are normal,

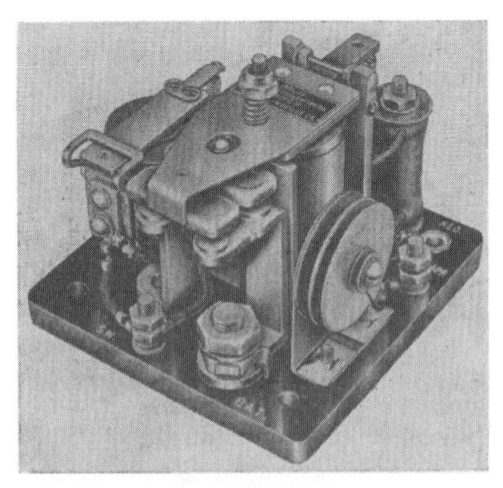

Fig. 133. A relay current control. (Courtesy The Leece-Neville Company)

the voltage regulator will regulate the output of the generator by alternately opening and closing the points at a rate of approximately 45 to 90 times per second. The resistor is cut in and out of the field circuit in

IGNITION SYSTEMS

order to maintain approximately 14.5 v. for a 12 v. battery. If the circuit is overloaded, there will be a voltage drop, causing an increase in the flow of current to the field winding. This, in turn, increases the amperage or flow of current from the generator. The spring control is so adjusted that when the output reaches the maximum for which the generator is designed, the current limitator points will open, decreasing the flow of current from the generator. This device functions only when necessary to limit the output and protect the generator from being burned out.

Fig. 134. Reverse-current cutout mounted on a control panel with a current regulator. (Courtesy The Leece-Neville Company)

The reverse-current cutout consists of a wire-wound iron core on one end of which is mounted a blade and contact points. A spring and adjusting screw assembly is arranged in such a manner as to hold the point open when the system is not functioning. This device prevents the battery current from flowing back through the generator when the engine is stopped.

The field switch provides a means of preventing the flow of current to the field in case the electric load is very light and there is no need for current to be sent to the battery.

The voltmeter and ammeter are installed in the circuit to inform the pilot of the voltage of the current being generated and the amount of current flowing.

The master switch, which should be fused, furnishes a means of opening the circuit when the aircraft is stored or is not in use. This switch and fuse eliminate a possible fire hazard.

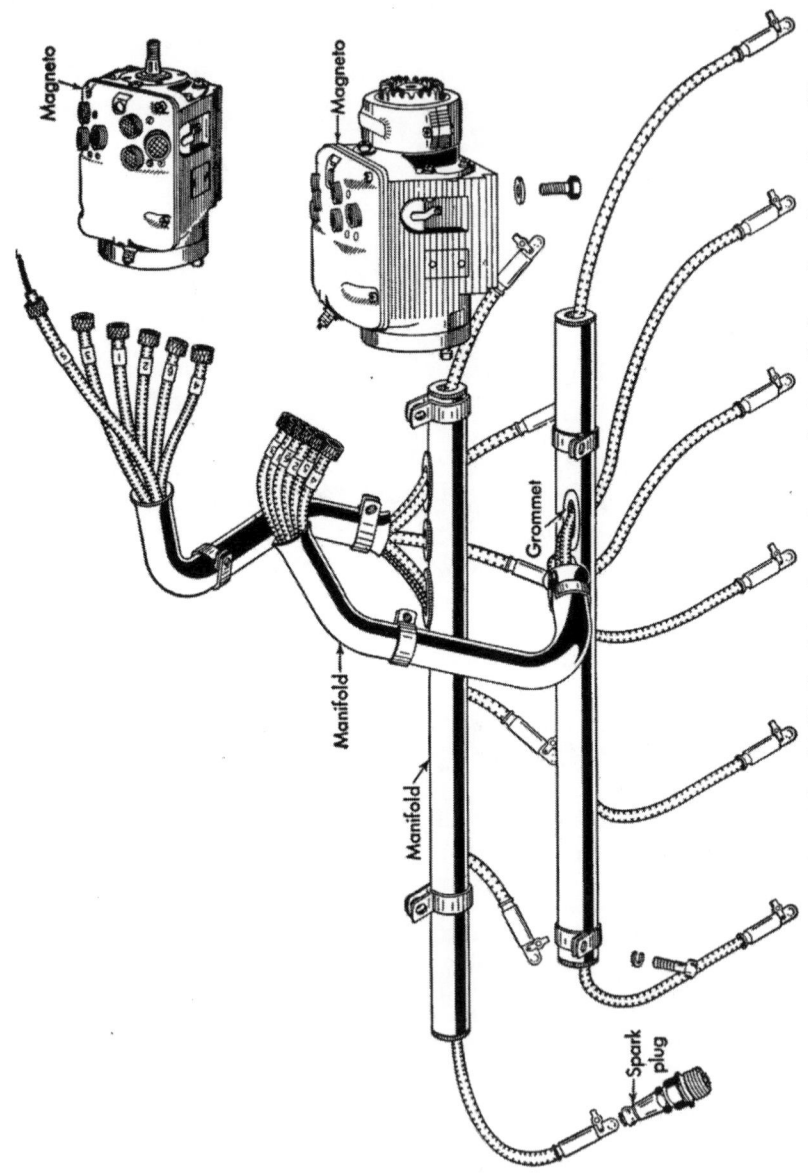

Fig. 135. Magnetos and standard (unshielded) wiring for an inverted, six-cylinder, aircraft engine. (Courtesy Ranger Aircraft Engines)

IGNITION SYSTEMS

The firing order of an engine has considerable effect upon vibration. As a rule, the cylinders of an engine do not fire consecutively, but the firing is spaced to produce the least amount of vibration and torsional strain. Radial engines have always been built with an odd number of cylinders in each row, with the exception of some of the new two-cycle aircraft engines now being produced. The firing order for a nine-cylinder

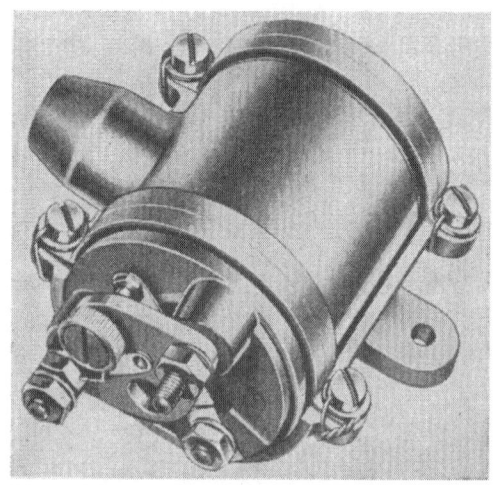

Fig. 136. An ignition booster coil. (Courtesy The Electric Auto-Lite Company)

engine is usually 1–3–5–7–9–2–4–6–8. One cylinder is skipped each time. In in-line engines, opposed engines, and multiple in-line engines, the firing order is also arranged to produce a minimum amount of vibration. For example, one small, six-cylinder, opposed engine has a firing order of 1–5–3–6–2–4. In one of the large, two-row, radial engines of fourteen cylinders, the back row is numbered 1–3–5–7–9–11–13, and the front row is numbered 2–4–6–8–10–12–14. The firing order is 1–10–5–14–9–4–13–8–3–12–7–2–11–6. Note that the firing skips from row to row of alternate firing.

The timing of the ignition of an engine consists of adjustments which determine the exact time that the spark occurs in the combustion chamber. The spark does not occur at the exact time that the piston reaches top dead center but somewhat in advance of this point. This adjustment, which is known as spark advance, may be as great as 45° of crankshaft rotation ahead of top dead center. If the spark occurs after top dead center, which is usually the case while starting the engine to prevent kickback, the spark is said to be retarded.

AIRCRAFT ENGINES

From the time the spark ignites the mixture, a definite time interval is required until the maximum temperature and pressure are reached in the combustion chamber. This period of time is affected by several factors such as (1) the speed of the engine, (2) the condition of the mixture in the combustion chamber, (3) the compression ratio, (4) the shape of the combustion chamber, (5) the spark plug location, (6) the quality of the fuel, and (7) the fuel-air ratio. A period of time equal to

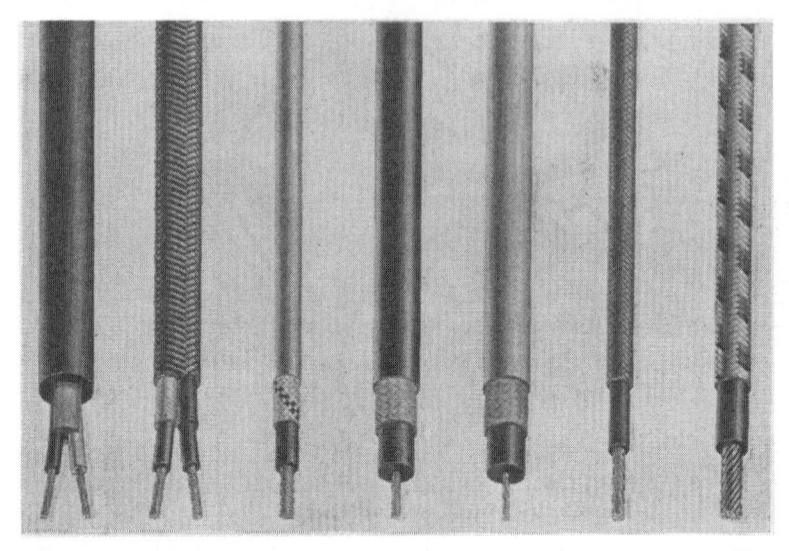

Fig. 137. Various types of aircraft-engine ignition, high-tension wires. The third from the right is called "steelduct." (Courtesy The Electric Auto-Lite Company)

about 0.03 sec. is allowed for the burning of the fuel. An engine operating at 2300 r.p.m. would have a spark advance of approximately 35° before top dead center. This advance would allow the maximum temperature and pressure to be developed in the combustion chamber at approximately the time the piston passed top dead center. A lean mixture burns more slowly than does a proper mixture and leads to overheating. With dual ignition, less spark advance would be needed than with a single spark plug. In staggered ignition timing, the exhaust-side spark plug, that is the one located nearest the exhaust valve, will fire from 4° to 8° of crankshaft rotation before the plug located near the intake valve. If the spark plugs are located at nearly right angles to the valve, the firing may be synchronized, that is, the two plugs fire at exactly the same time. In high-powered single and twin-row radial engines, because of the connecting-rod arrangement, it may be necessary

IGNITION SYSTEMS

to use compensated timing magnetos. The lobes on the breaker cam of these magnetos are spaced at unequal intervals to cause the magneto to fire unevenly but in proper relation to the variations in piston location caused by the elliptical travel of the knuckle pins. When timing this type

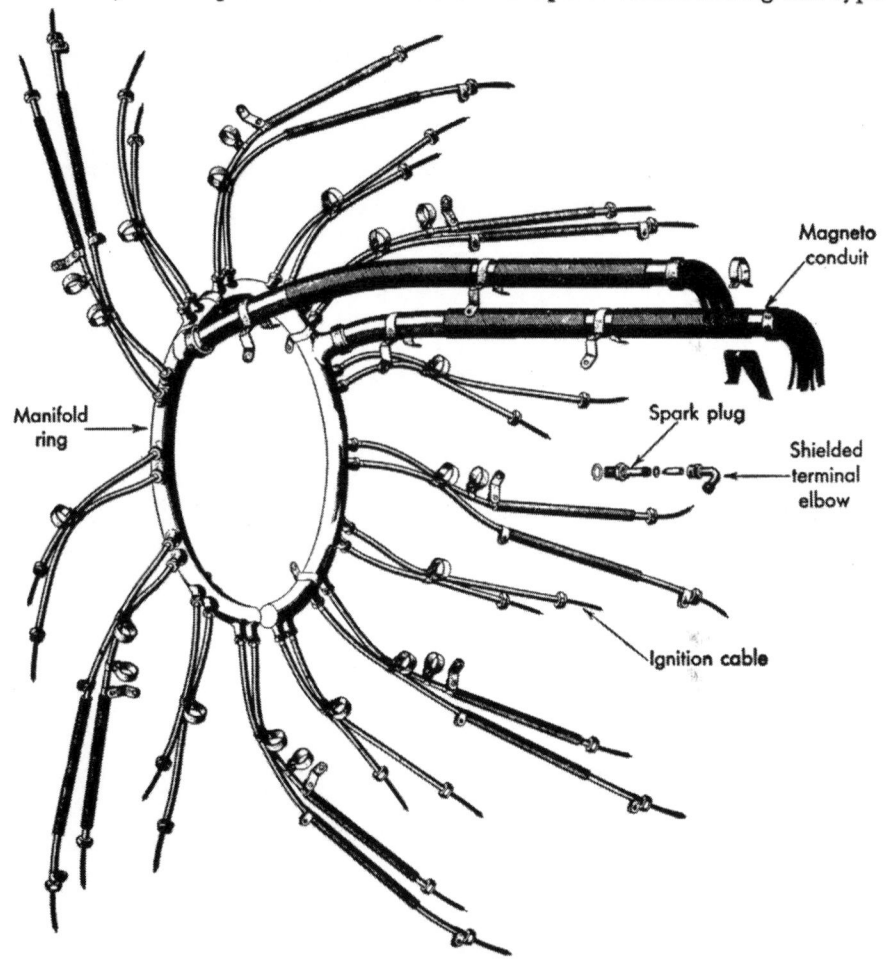

Fig. 138. A radio, shielded, ignition wiring system for a 14-cylinder, two-row, radial aircraft engine. (Courtesy Wright Aeronautical Corporation)

of magneto to the engine, the lobe marked with a zero must be the one used in opening the breaker points, and the magneto must be timed to the proper cylinder. When this is done, the unequally spaced lobes will be in the proper relation to the cylinders they are designed to fire.

The wires used to carry the high-tension current in the ignition system consist of several fine strands of wire enclosed in a thick rubber insulation. One end of each spark plug wire has a suitable terminal for

AIRCRAFT ENGINES

attaching to the spark plug, while the other end is attached securely to the metal block in the distributor head.

If the aircraft is equipped with radio, the entire ignition system must be shielded. This is done by enclosing all parts of the electrical system in proper metal shields. The wires are usually covered with a flexible, woven-metal conduit, and a shield is arranged to cover the spark plugs. Aircraft equipped with radio usually have the metal parts of the plane bonded, which means that all the metal parts of the aircraft are connected by means of suitable conductors that are properly grounded.

XIII FUEL AND FUEL SYSTEMS

A fuel is any substance that, when burned, produces heat energy. The internal-combustion engine is a heat engine. The power developed by an internal-combustion engine is obtained by burning a fuel with the oxygen in the air.

Fig. 139. This light airplane has the main fuel tank under the cowl in front of the pilot. It has a gravity fuel system, and extra tanks may be placed in the wings or in the center section. (Courtesy Piper Aircraft Corporation)

Fuels are of many types and may be either solids, liquids, or gases. Internal-combustion engines have been designed which make use of solid fuel in a finely divided form. Other internal-combustion engines operate on fuels in the form of gases which are mixed with air, forming the mixture burned in the cylinders.

The term gas is incorrect when applied to ordinary motor fuels. Most internal-combustion engines operate on liquid fuels. These fuels are vaporized or sprayed into the air in a finely divided state, forming the mixture that is burned in the cylinder.

Experiments have been conducted with many liquid fuels, such as the alcohols, benzol, and various petroleum products. Almost all inflammable liquids have been tried at various times, but the petroleum

AIRCRAFT ENGINES

product, gasoline, has been found to be the most satisfactory. However, the alcohols, ethyl and methyl, can be used quite successfully in internal-combustion engines. The heat of combustion of alcohol is only about 13,000 B.t.u. as compared to 18,000 B.t.u. for benzol, and 20,000 B.t.u. for gasoline. Ethyl alcohol is common grain alcohol. Methyl alcohol is the so-called wood alcohol. A mixture of ethyl and methyl alcohols forms the basic alcohol blend for use in internal-combustion engines. Alcohol

Fig. 140. This light bomber has the main fuel tanks located in the wings and is equipped with a pressure fuel system. (Courtesy The Glenn L. Martin Company)

will stand higher compression ratios without detonation than will gasoline. Since alcohol does not vaporize as readily as gasoline, it is difficult to start a cold engine on ordinary grain alcohol. To make starting easier when alcohol is used, some inflammable liquid, such as ether or benzine, that will vaporize readily at ordinary temperatures must be added. Pure alcohol tends to preignite when high temperatures develop in the engine. Alcohol does not mix readily with gasoline. Alcohol containing less than 5 per cent water is expensive to manufacture.

Three alcohol blends have been used quite successfully. One blend of 20 to 40 per cent benzine and the remainder alcohol gives good results. This blend has good antiknock properties. Another blend contains 43 per cent ether and 57 per cent alcohol and is known as Natalite. A third blend contains approximately 17 per cent benzol, 35 per cent gasoline, 40 per cent alcohol, and approximately 8 per cent ether or some other highly volatile substance. This blend is known as Alcogas.

The petroleum fuels all belong to the hydrocarbon group. Hydrocarbons are compounds made up of hydrogen and carbon only. Alcohol

FUEL AND FUEL SYSTEMS

contains, in addition to the hydrogen and carbon, considerable oxygen. Benzol belongs to the hydrocarbon group and is obtained from coal. Benzol may be used in engines with high compression ratios without detonating. Gasoline and naphtha, which are also hydrocarbons, have been blended successfully with benzol. The benzol added to these blends may vary from 20 to 60 per cent. The benzol reduces the detonating characteristics of light gasolines and naphthas.

Fig. 141. A light, air-cooled, opposed-type aircraft engine with the carburetor installed below the engine in front of the oil tank. (Courtesy Continental Motors Corporation)

Petroleum is the name given to the mineral oils. Petroleum, as it is obtained from the earth, contains many hydrocarbon compounds. These compounds are obtained from crude petroleum by fractional distillation. This is a process whereby the various parts or fractions of the crude petroleum are separated by making use of their different boiling points. The substances having the lowest boiling points, such as naphtha and light gasoline, distill off first. Heavier fractions, such as regular gasoline and distillates, have higher heat values. These heavier fractions are not as volatile as those that come off first; therefore, starting engines on these fuels is difficult.

139

Fig. 142. A rear view of a six-cylinder, inverted, air-cooled aircraft engine with carburetor installed at the left side of the cylinders. (Courtesy Ranger Aircraft Engines)

FUEL AND FUEL SYSTEMS

Standard grades of gasoline are made up of a large number of compounds, some of which vaporize at low temperatures, making starting easier. The heavy fractions add heat value to the fuel.

Some of the important characteristics of a fuel for an internal-combustion engine are volatility, vapor pressure, heat value, antiknock value, and purity. Volatility is the characteristic of a liquid that determines the readiness with which it changes from the liquid to the gaseous form at various temperatures. A highly volatile liquid changes to gas readily at low temperatures, while a liquid having a low volatility changes to gas only at high temperatures. Vapor pressure is the tendency of a liquid to change to the gaseous form at various temperatures. Vapor pressure is generally one of the specifications given for a fuel. Usually, the vapor pressure is specified to be less than 7 lb. per sq. in. at 100° F. Fuel, when vaporized, expands to about 200 times its liquid volume. If a fuel vaporizes too rapidly, or the vapor pressure is too high, vapor locks may develop. A vapor lock is caused by some part of the fuel changing to a gas and preventing the liquid fuel in the lines or carburetor from flowing freely.

The heat value of a fuel is the number of B.t.u.'s given off when a given quantity of the fuel is completely burned with oxygen. Definite grades of fuels have specified heat values.

Antiknock value is that ability of the fuel to be highly compressed without detonating. To lower the detonating properties and raise the antiknock value, a chemical substance such as tetra-ethyl-lead is added to the fuel. Gasolines containing this substance are usually called ethyl gasolines. Practically all gasoline used in aircraft engines has tetra-ethyl-lead added. Some corrosion may be caused by fuels containing this substance. However, the corrosion is not severe enough to cause any appreciable damage. The antiknock value of a fuel is usually given by its octane rating.

The technical definition of octane rating is "the percentage of iso-octane in a mixture of iso-octane and normal heptane required to match the performance of the fuel being tested in a special test engine under controlled conditions." Ordinarily, the octane rating is given simply to show the antiknock characteristics of a fuel.

The compression ratio of an engine determines the octane rating of the fuel to be used. For example, if an engine has a compression ratio of from 5:1 to 6:1 and is not supercharged, it uses a fuel having an octane rating of from 65 to 75. The same engine, if supercharged, might have

AIRCRAFT ENGINES

the compression ratio increased to 6:1 or 7:1, and the octane rating would be raised to 75 or 80. Supercharged engines may require octane ratings up to 100. An engine having a compression ratio of 7.5:1 and a supercharger ratio of from 8:1 to 12:1 would require a fuel having an

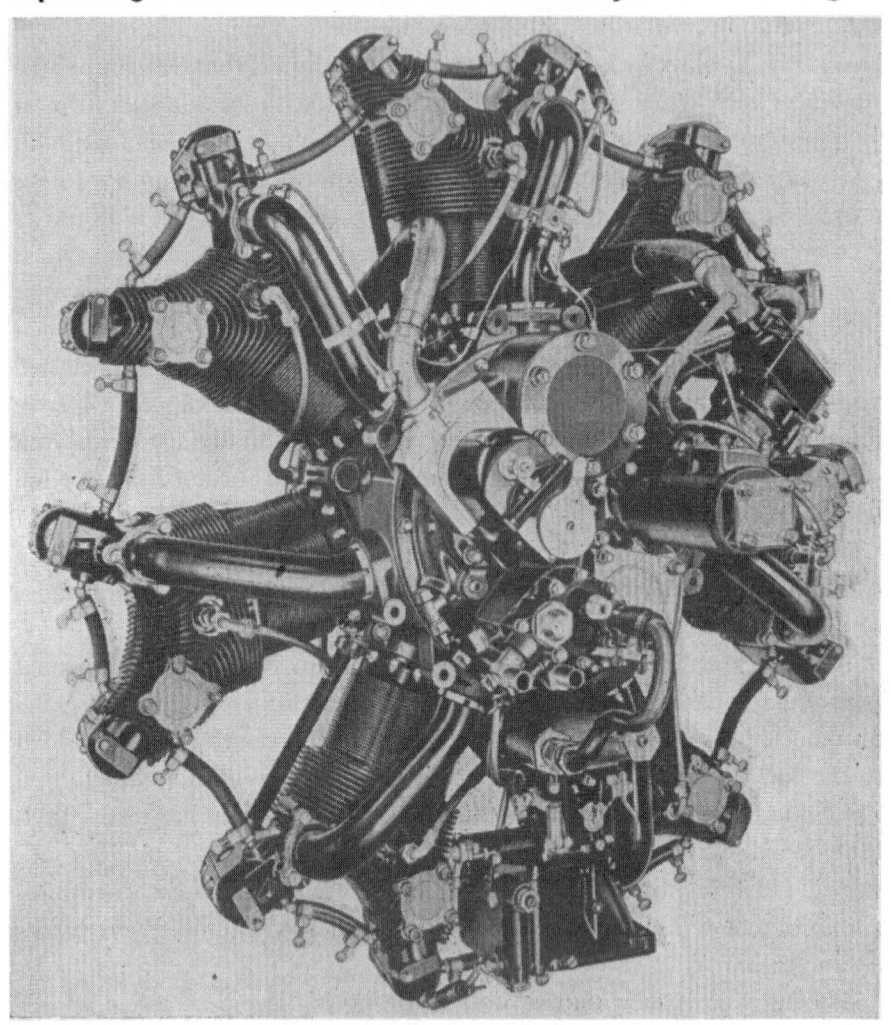

Fig. 143. A rear view of a seven-cylinder, air-cooled, radial aircraft engine. The carburetor is installed between the two bottom cylinders. (Courtesy Jacobs Aircraft Engine Company)

octane rating of from 87 to 100. A satisfactory fuel is one that vaporizes readily enough to make starting easy, and yet has sufficient heat value to enable the engine to develop its full-rated horsepower. The fuel should be high enough in antiknock rating so that detonation will not occur under full throttle and maximum engine temperatures.

FUEL AND FUEL SYSTEMS

Regardless of the amount of fuel used, power depends upon the air-fuel mixture. The maximum amount of power from the fuel is obtained when the mixture consists of about 12 to 14 parts of air to 1 part of fuel. Each pound of air requires from 0.072 to 0.083 lb. of fuel. When the amount of fuel furnished is less than 1 part of fuel by weight to 14 parts of air by weight, the mixture is lean. When more than 1 part of fuel is used to 12 parts of air by weight, the mixture is rich. When more fuel

Fig. 144. Air scoops are a part of an engine assembly and assist in supplying the air to be mixed with the fuel. (Courtesy Douglas Aircraft Company, Inc.)

is used, the mixture is full-rich or over-rich. When less than 1 part of fuel to approximately 15 parts of air is used, the mixture is over-lean. This type of mixture may be used in ratios as low as 1 part of fuel to 16 parts of air, when maximum fuel economy is desired. This mixture might be used at cruising speed with a low throttle setting.

When the throttle is in the idling position, not enough air is admitted to vaporize the fuel completely. The idling mixture is quite rich and tends to burn more slowly than when the correct mixture is used at cruising or full throttle.

The atmosphere, regardless of the altitude, always contains approximately 21 per cent, by volume, of oxygen, 78 per cent nitrogen, and approximately 1 per cent of other gases. When flying at high altitudes, the supply of fuel is decreased by "leaning" the mixture. Actually the air-fuel ratio remains the same, but the amount of fuel that is supplied

AIRCRAFT ENGINES

to the engine on each intake stroke is decreased. At high altitudes, in a non-supercharged engine, a cylinder full of air does not contain as much air by weight as it does at low altitudes.

Fuel systems are designed to supply a sufficient amount of fuel to the engine and carry the reserve fuel necessary for extended flights. In addition to the function of supplying fuel to the engine, the fuel system is designed to clean the fuel, removing any impurities. A satisfactory fuel system must be designed to operate under all conditions of altitude

Fig. 145. A fuel supply pump. (Courtesy Candler-Hill Corporation)

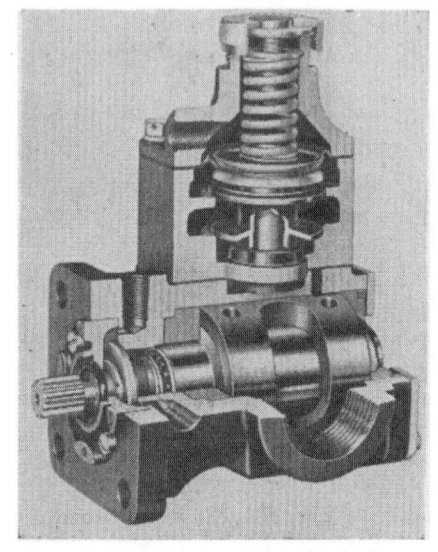

Fig. 146. A cutaway view of a fuel supply pump. This pump is of the vane type. (Courtesy Candler-Hill Corporation)

and the various positions of the aircraft. Most early aircraft simply had a tank containing the fuel, which was fed to the carburetor by gravity. This system is still used in light aircraft having a high wing in which the fuel tank may be placed. The fuel tank may be installed in front of the pilot in the upper part of the fuselage. When it is impossible to place the fuel tank high enough to ensure a steady flow of fuel to the carburetor by gravity, a pressure system must be installed. In some of the earlier aircraft, pressure was obtained by pumping air into the fuel tank. In this system it was necessary that the tank be airtight. Most pressure systems at present incorporate a pump that supplies fuel under 3 to 7 lb. per sq. in. pressure to the carburetor. The fuel is supplied from a vented fuel tank, and the pressure system is equipped with a by-pass valve that regulates the pressure to the carburetor. A typical gravity

FUEL AND FUEL SYSTEMS

system consists of a vented fuel tank; shut-off cock; strainers; and a primer. The tank must be located high enough above the carburetor to ensure a steady flow of fuel. In this system the fuel usually flows downward from the fuel tank to a strainer and sediment trap located below the carburetor.

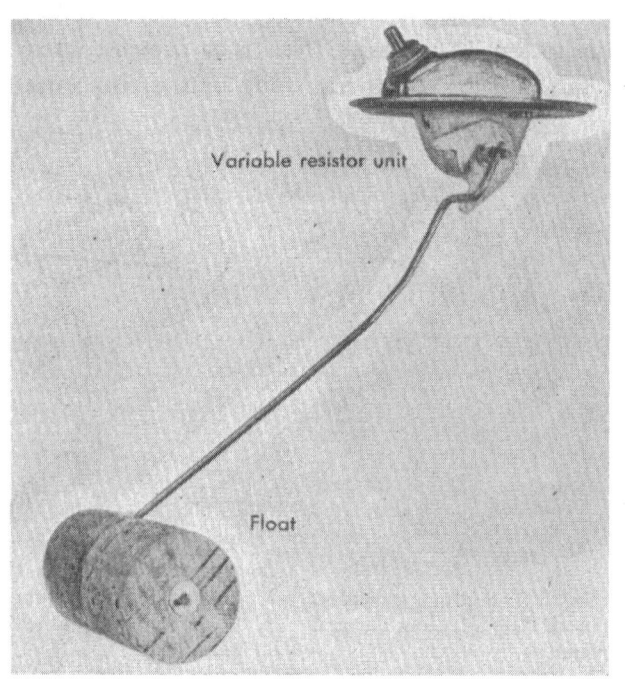

Fig. 147. An electric fuel-gauge unit which is installed in the fuel tank. (Courtesy AC Spark Plug Division, General Motors Corporation)

A simple pressure system consists of a vented fuel tank that may be located well below the carburetor level, a shut-off cock, a strainer and sediment trap, a power-driven pump including a by-pass relief valve, a fuel pressure gauge, a primer, and a wobble pump. A wobble pump is a hand pump to be used in case of failure of the power-driven pump. It is usually equipped with a by-pass relief valve. The fuel lines should be as short as possible and so arranged as to minimize danger from vapor locks. These lines should be fastened securely to prevent vibration. Whenever possible, sharp bends, elbows, and threaded connections should be avoided. Fuel lines should not be carried close to highly heated parts of the engine. The fuel line leading to the pump should be arranged so that it is kept full of fuel at all times. Vapor locks have a greater tendency to form between the pump and the fuel tank than

AIRCRAFT ENGINES

between the pump and the carburetor. When vapor forms in this line, it is impossible for the pump to lift the fuel from a point lower than the pump itself. This condition might lead to engine stoppage.

Fuel tanks are usually constructed of welded or riveted aluminum alloy. Fuel tanks should be arranged with outlets in such positions that, in normal flight, all fuel in the tank may be used. A sump should be arranged in the bottom of the tank. This is to prevent uncovering of the outlet when the fuel is sloshed from one part of the tank to another.

Fig. 148. An electric fuel-gauge indicator. (Courtesy AC Spark Plug Division, General Motors Corporation)

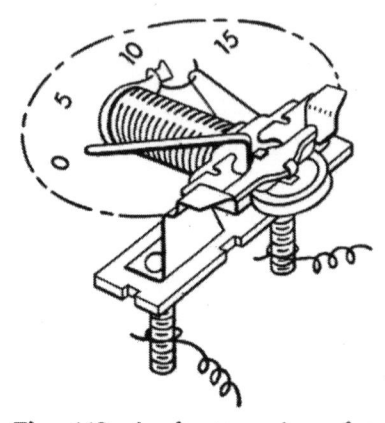

Fig. 149. A phantom view of a simple voltmeter such as is used as an electric fuel-gauge indicator. (Courtesy Wright Aeronautical Corporation)

Baffles are usually installed in the tank to prevent excessive surging. It is necessary that fuel tanks be vented both to allow air to replace the fuel drawn from the tank, and to allow expanding vapors to escape. The air and vapors in the space above the fuel expand and contract with changes in temperature.

Gauges are provided to inform the pilot of the amount of fuel available. These gauges may be of the direct-reading float type, or remote-indicating electrical type. In an electric, float-operated, rheostat gauge, a float and rheostat are placed in the fuel tank. The position of the float is determined by the fuel surface level causing a contact to slide over the rheostat. This operation changes the amount of electric current flowing through a gauge on the instrument panel. The amount of current flowing through the instrument moves a needle and thus indicates the fuel level.

A hydrostatic fuel gauge is operated by the difference in hydrostatic

FUEL AND FUEL SYSTEMS

pressure of the fuel in the tank. Most aircraft have more than one fuel tank, although in light aircraft the second tank may contain an emergency supply only. When more than one main tank is used, a suitable valve should be installed. This valve is usually of the selector type, making it possible to turn on each tank separately or all tanks at the same time.

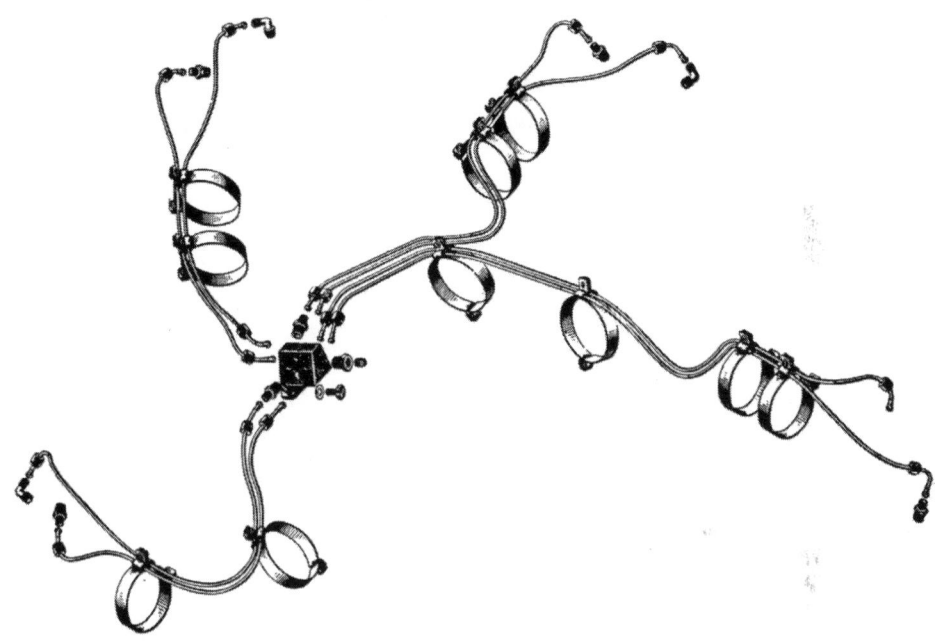

Fig. 150. The priming system of a 14-cylinder, twin-row, radial aircraft engine. (Courtesy Wright Aeronautical Corporation)

In a pressure fuel system, the pumps may be engine or electrically driven. These pumps may be of the vane or gear type. The gear type of pump is not as satisfactory as the vane type, as it does not prime as readily and there is considerable leakage through the pump when the gear teeth become worn. Vane-type pumps may have two or four vanes. This pump is dependable, light in weight, and compact.

Primers are usually installed in the fuel system for the purpose of furnishing a supply of fuel to the cylinders when starting the engine. A primer is a simple plunger pump and injects the fuel directly into the induction system.

The fuel-pressure gauge is usually of the Bourdon type. Some systems are equipped with a device that gives warning when the fuel pressure or fuel in any tank becomes low.

XIV CARBURETORS AND SUPERCHARGERS

A carburetor is a device used to mix the fuel with air in the proper proportion to furnish the correct air-fuel mixture to the cylinders. The carburetor also assists in vaporizing the fuel so that it enters the cylinder in a finely divided state. The modern carburetor is a precision instrument, providing accurate control of both the air flow and the amount of fuel.

The first attempts at carburetion consisted of such practices as bubbling air through gasoline, drawing a stream of air over sponges or wicks partly submerged in the gasoline, and drawing air through a spray of gasoline. Some attempts were made to heat the gasoline by passing the exhaust gases over the fuel pipe and then allowing the heated gasoline to spray into the air stream as it passed through the intake pipes to the cylinders.

Carburetors are divided, roughly, into two classes, the injector type and the float type. The float type of carburetor is used on most light aircraft. The fuel in this carburetor is maintained at a constant level by the float mechanism. The float chamber in the carburetor is supplied with fuel from the fuel tanks. A hollow float, consisting of an airtight metal cell, regulates the level of the fuel in the chamber by means of a needle valve. As the fuel level falls the float falls with it, opening the needle valve and allowing more fuel to flow into the chamber. Fuel may flow into the chamber by means of gravity, or may be supplied to the carburetor by means of a fuel pump under a pressure of approximately 3 to 5 lb. per sq. in.

A Venturi arrangement is used to increase the velocity of the air passing through the carburetor. This increased velocity creates a greater suction. A Venturi is a tube, part of which is smaller in cross section than the rest of the tube. It has been found by experiment that, when a fluid is passed through such a tube, the velocity of the fluid increases as it passes through the restricted area. The main spray nozzle or jet is located in the restricted part of the Venturi.

CARBURETORS AND SUPERCHARGERS

The amount of air passing through the carburetor is regulated by means of a butterfly valve connected with the throttle control. This is the throttle valve.

The construction of the carburetor may be such that the stream of air passes upward or downward through the carburetor. When the carburetor is so constructed that the air flow through the carburetor is

RIGHT   **LEFT**

Fig. 151. Right and left side views of a typical carburetor such as is used on fifty- to seventy-five horsepower aircraft engines. (Courtesy Marvel-Schebler Carbureter Division, Borg-Warner Corporation)

upward, it is known as an "updraft carburetor." If the construction is such that the air flow through the carburetor is downward, it is called a "downdraft carburetor."

In the passage leading to the spray nozzle is located the main metering jet which consists of a hole of definite size that regulates the flow of fuel. The main supply of fuel for the engine passes through the main metering jet, out through the spray nozzle into the stream of inrushing air. If no other arrangement were made, a solid stream of fuel would be drawn from the spray nozzle into the carburetor and, as the engine gained in speed, a larger and larger amount of pure fuel would be drawn into the air stream.

Therefore, an air bleed is provided to prevent an overrich mixture at high speed. This is simply an arrangement where, as the suction increases, a certain amount of air is allowed to mix with the fuel before it

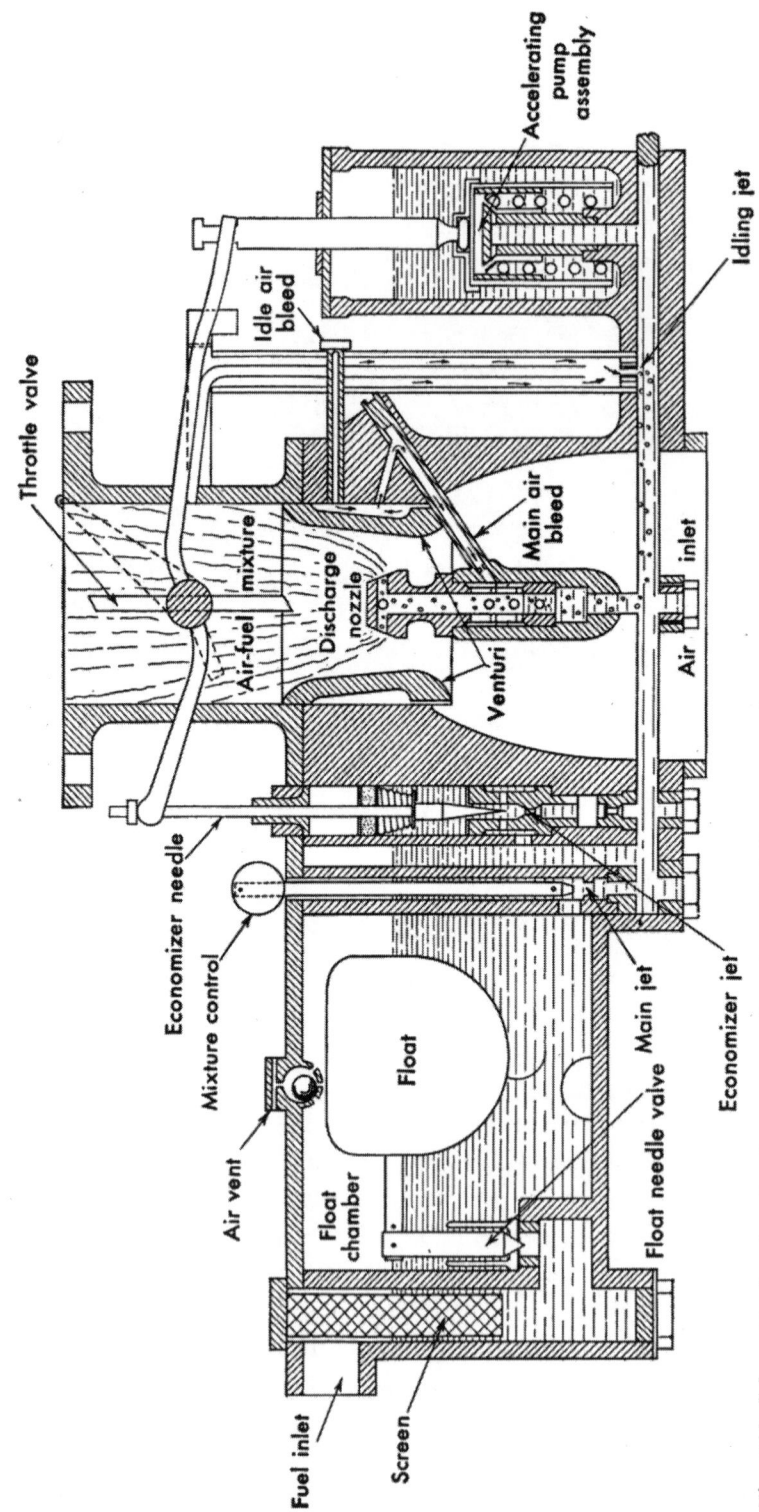

Fig. 152. Schematic drawing showing construction of a typical float-type carburetor.

CARBURETORS AND SUPERCHARGERS

enters the air stream. When the engine is operating at idling speed with the throttle closed, there is not enough suction to draw fuel through the main jet system to keep the engine in operation. The carburetor is equipped with an idling system which permits the necessary amount of fuel to be drawn into the intake manifold through an opening above the Venturi and throttle valve. The idling system may also have an air bleed to prevent too much fuel from being drawn into the manifold. Figure 152 shows the main parts of a simple float type of carburetor. The air supply

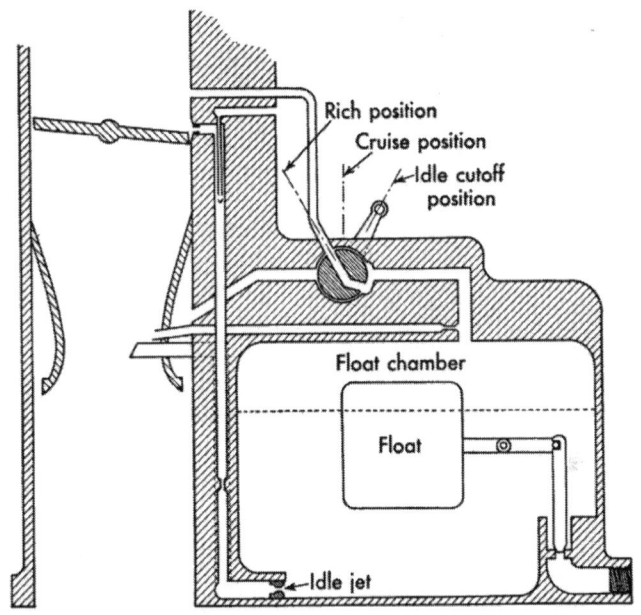

Fig. 153. A typical idler system and idle cutoff used on modern carburetors.

to the carburetor is through a large opening leading to the Venturi chamber. This opening may be exposed to the air stream, but is usually connected with an air scoop which produces a ram effect due to the velocity of the air stream. This ram effect builds up pressure and assists in increasing the amount of air-fuel mixture that is supplied to the cylinders.

The suction on the main spray nozzle depends upon the Venturi effect on the incoming air stream. The Venturi does not affect the action of the idling jet. With the throttle closed and the engine idling, the partial vacuum above the throttle valve draws the fuel in through the idling jet in sufficient quantities to supply the needs of the engine at this low speed. Aircraft engines idle at speeds slightly above 500 r.p.m. If the throttle is suddenly opened, the idling jet ceases to function and

AIRCRAFT ENGINES

there is an appreciable time before sufficient suction is caused by the Venturi to start the fuel flow through the main discharge nozzle. To prevent the engine's stopping due to a sudden opening of the throttle, an accelerating system is usually built into the carburetor. This may be in the form of an accelerating well or an accelerating pump. The accelerating well is a space surrounding the main discharge nozzle which is filled with gasoline during the time that the engine is idling.

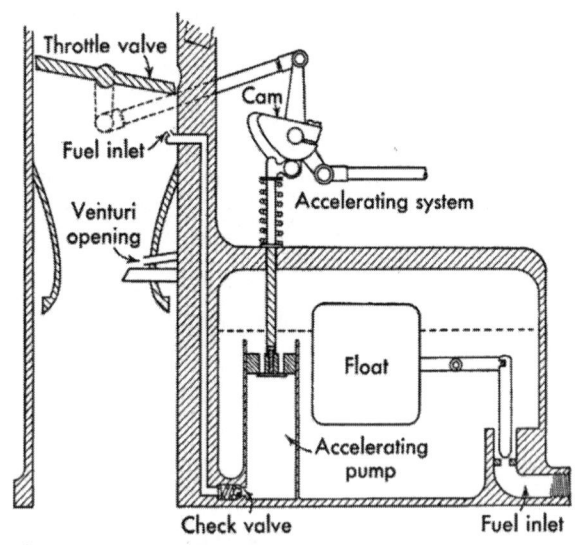

Fig. 154. A typical accelerating carburetor pump.

The sudden opening of the throttle causes the fuel to be sprayed out through the main nozzle, furnishing fuel for the sudden pick-up in speed. An accelerating pump is a plunger pump operated by the throttle control as it approaches the full-open position. This pump forces an extra supply of fuel out through the main discharge nozzle. When in the idling position, if the throttle is opened slowly, the idling jet continues to function for 200 to 300 revolutions, allowing time for the Venturi arrangement to start the flow of fuel through the discharge nozzle.

Many high-powered engines have an automatic accelerating device consisting of a spring bellows. When the engine is idling, the suction developed in the intake passage draws the bellows to one side, allowing additional fuel to accumulate on the other side of the bellows. When this suction is decreased by the opening of the throttle, the bellows springs back to its original position, forcing fuel into the intake passages. In starting an engine equipped with the plunger type of accelerating

CARBURETORS AND SUPERCHARGERS

system, additional fuel may be forced into the carburetor by pumping the throttle.

The simple float type of carburetor is used on most aircraft engines of low horsepower. In most engines ranging from approximately 150 to 450 hp., fuel is supplied to the carburetor by a fuel pump. Mixture con-

RIGHT    LEFT

REAR

Fig. 155. Right and left side and rear views of a typical up-draft carburetor used on 150- to 450-horsepower aircraft engines. (Courtesy Marvel-Schebler Carbureter Division, Borg-Warner Corporation)

trol systems and economizer systems are usually built into this type of carburetor. The fuel pump supplies the carburetor with fuel under constant pressure, regardless of the position of the aircraft. To take care of the decrease in air density at high altitudes, a mixture control cuts down the amount of fuel flowing through the carburetor, preventing an overrich mixture. One type of mixture control consists of a needle valve that controls part of the flow of fuel to the main metering jet. This needle is controlled by a lever in the cockpit. At low altitudes, this needle valve is wide open. The pilot may regulate the flow of fuel to the main metering jet by operating the mixture control at increased altitudes. When the mixture control is set at full-lean, the needle valve is completely

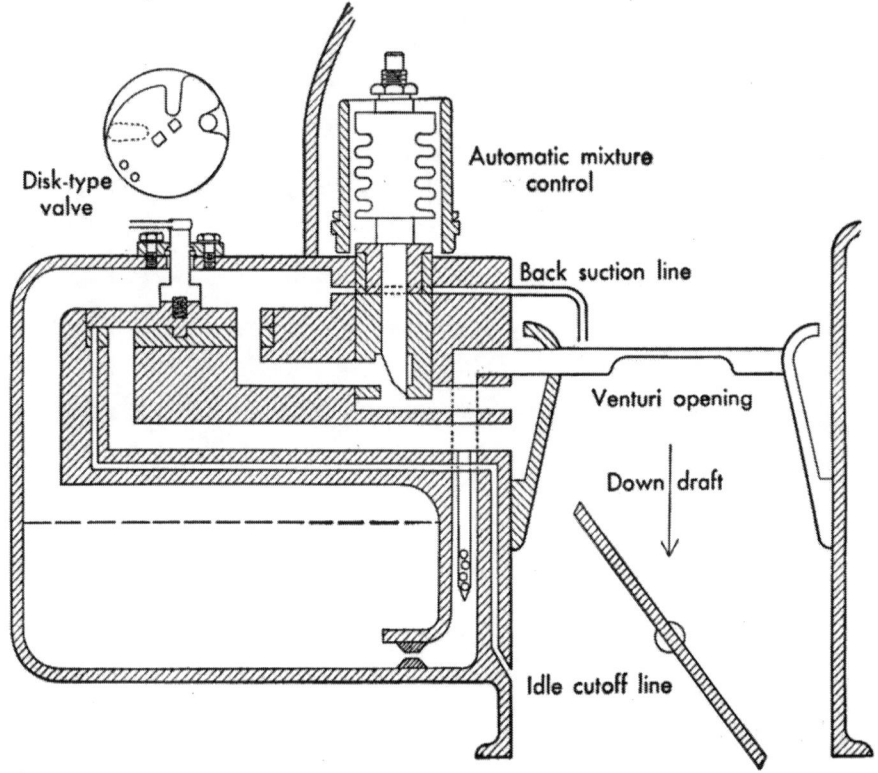

Fig. 156. An automatic mixture control.

closed. This type of control simply reduces the amount of fuel reaching the main metering jet. The back-suction type of mixture control depends upon decreased atmospheric pressure in the float chamber. When the mixture lever is set at full-lean, it opens a passageway from the float chamber to the area of reduced pressure caused by the Venturi. The decreased pressure in the Venturi area draws some of the air from the float chamber, decreasing the pressure within the chamber. This decreased pressure reduces the flow of fuel to the main metering jet.

On most high-altitude engines, the carburetor is equipped with an automatic mixture control. The back-suction valve in this type of carburetor is connected with a sealed bellows which expands with decreased atmospheric pressure. When the bellows expands, it opens the valve in a

CARBURETORS AND SUPERCHARGERS

passage in the carburetor, allowing back-suction to develop and decreasing the air pressure on the fuel supply in the float chamber. The bellows may also be arranged to close a valve, thereby cutting down the supply of fuel to the main metering jet.

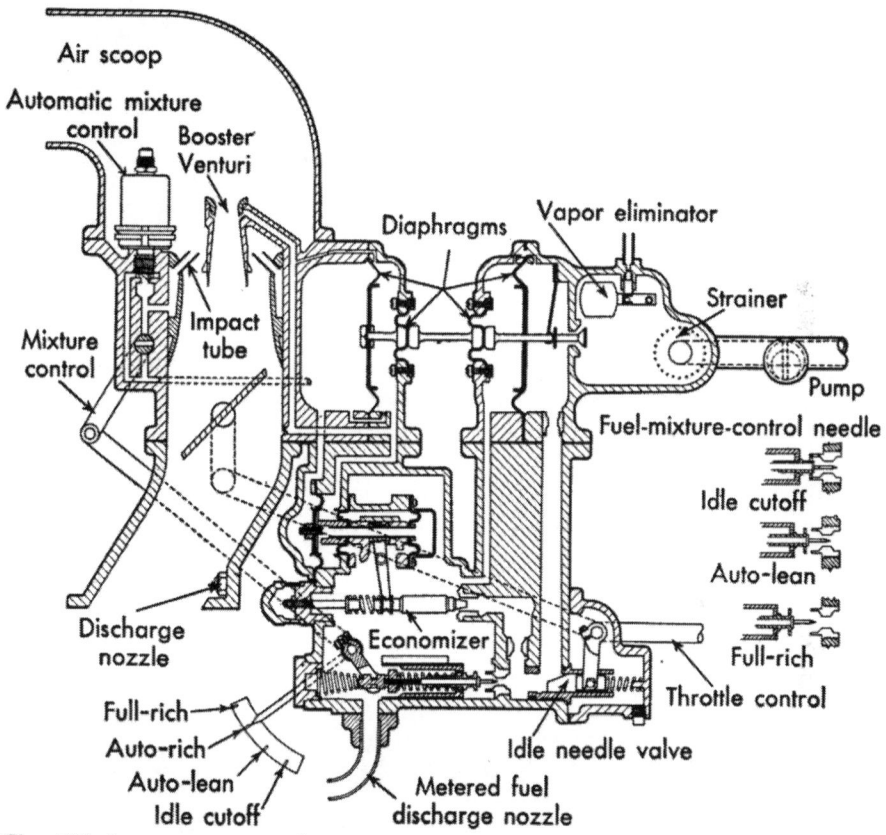

Fig. 157. A pressure-type carburetor.

When the engine is operated at full throttle, a rich mixture may be necessary to prevent the engine from overheating. An economizer system furnishes additional fuel when the throttle is in the fully opened position. The economizer does not cut down the supply of fuel, but protects the engine from overheating during take-off and climb.

Most carburetors are equipped with some type of heating arrangement. Carburetor heat is applied to prevent icing or to supply heated air when operating the engine under conditions of extreme cold.

On large engines an idle cutoff is installed. The idle cutoff closes the idling jet, forcing the engine to stop from lack of fuel. This prevents any chance of the engine's kicking back due to an air-fuel-mixture

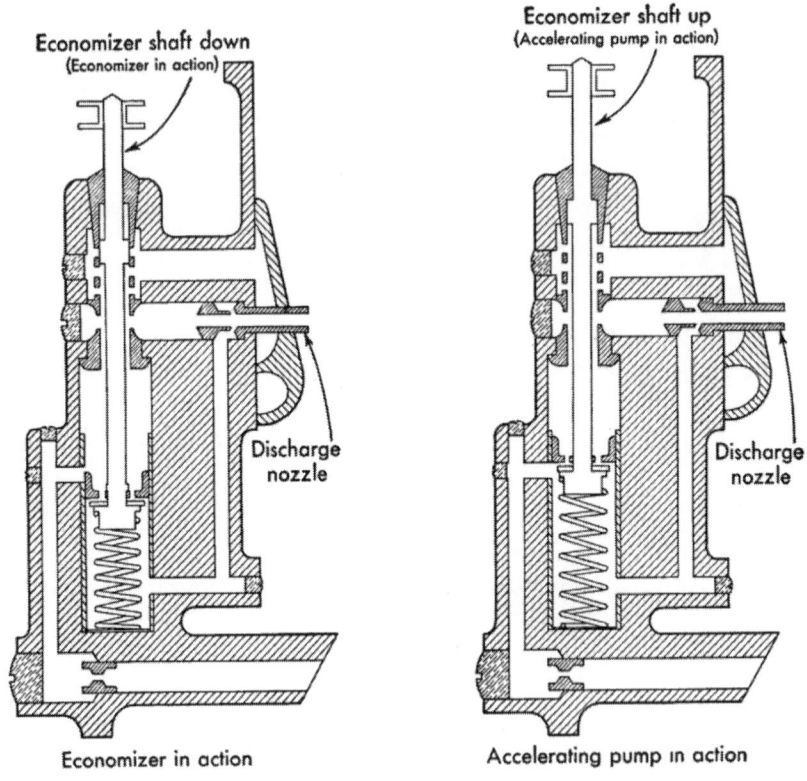

Fig. 158. An economizer and accelerating pump.

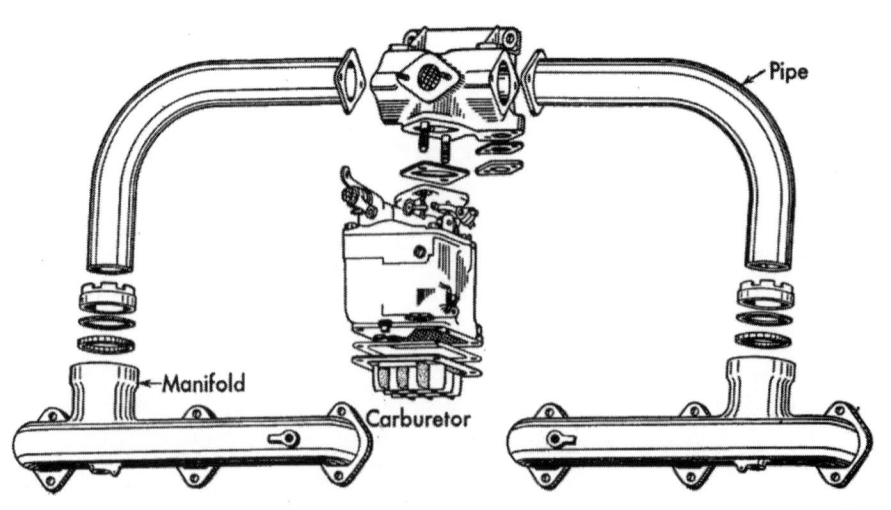

Fig. 159. Carburetor and intake manifold for a six-cylinder, inverted, in-line aircraft engine. (Courtesy Ranger Aircraft Engines)

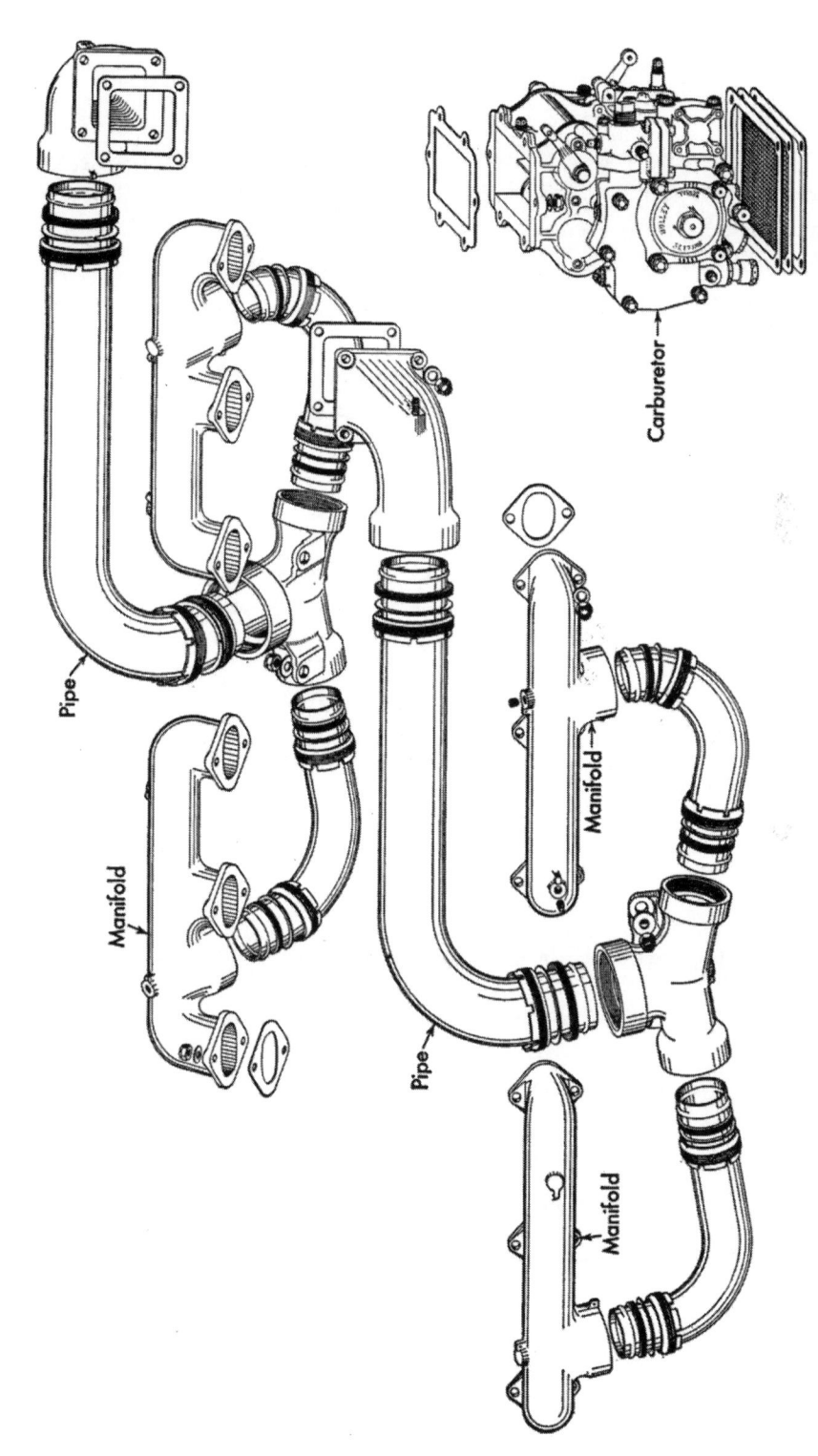

Fig. 160. Carburetor and intake manifold for a 12-cylinder, inverted, V-type aircraft engine. (Courtesy Ranger Aircraft Engines)

charge in the cylinder at the time the engine comes to a standstill. Even when idling, if the engine is stopped by operating the ignition switch, the charge in one or more cylinders may ignite because of the heated cylinder. This may cause a kickback. On engines equipped with lightweight propellers, this is not serious. However, on large engines equipped with propellers weighing as much as 800 lb. operated through reduction gears, a kickback may cause serious damage.

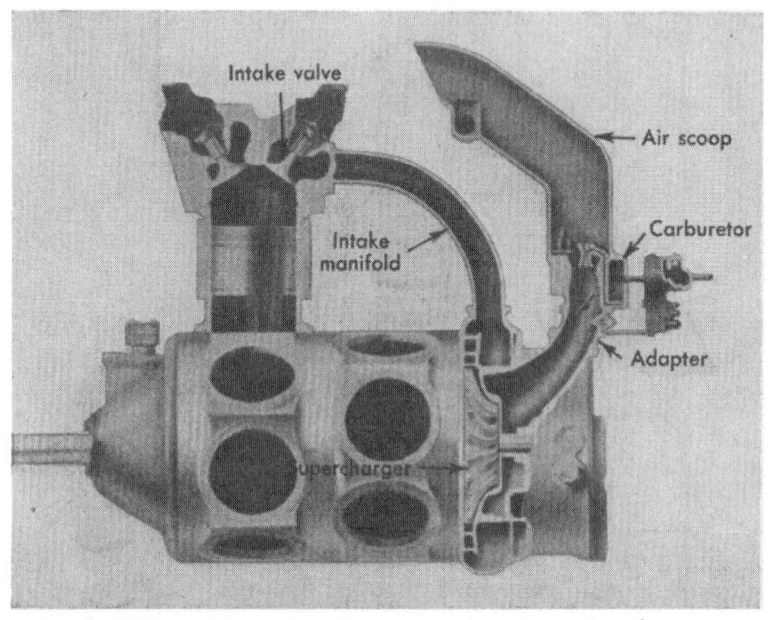

Fig. 161. Induction system including the supercharger for a radial aircraft engine. (Courtesy Wright Aeronautical Corporation)

The induction system of an engine consists of the intake pipes leading from the carburetor or supercharger to the engine, the carburetor, supercharger, and air scoops. The air-fuel mixture from the carburetor is carried to the cylinder by the intake manifold. This is a system of tubes arranged to supply as nearly as possible an equal amount of the air-fuel mixture to each cylinder. These pipes also act as evaporation chambers to assist in vaporization of the fuel particles on their way to the cylinder. The intake manifold is sometimes heated to assist in vaporizing the fuel. The final vaporization of the fuel takes place in the hot cylinder during the intake and compression strokes. Excessive manifold or carburetor heat should not be used. Only enough heat to vaporize the fuel is necessary. Any additional heat may cause detonation of the charge when it is compressed in the cylinder.

CARBURETORS AND SUPERCHARGERS

The intake manifold requires careful design in order that the air-fuel mixture may flow smoothly to the cylinders. The manifolds are designed to take advantage of the ram effect built up by the flow of the air-fuel mixture. The firing order of the engine is arranged in such a manner that, as nearly as possible, a steady stream of air-fuel mixture passes through the intake manifold. If the flow of air-fuel mixture were stopped

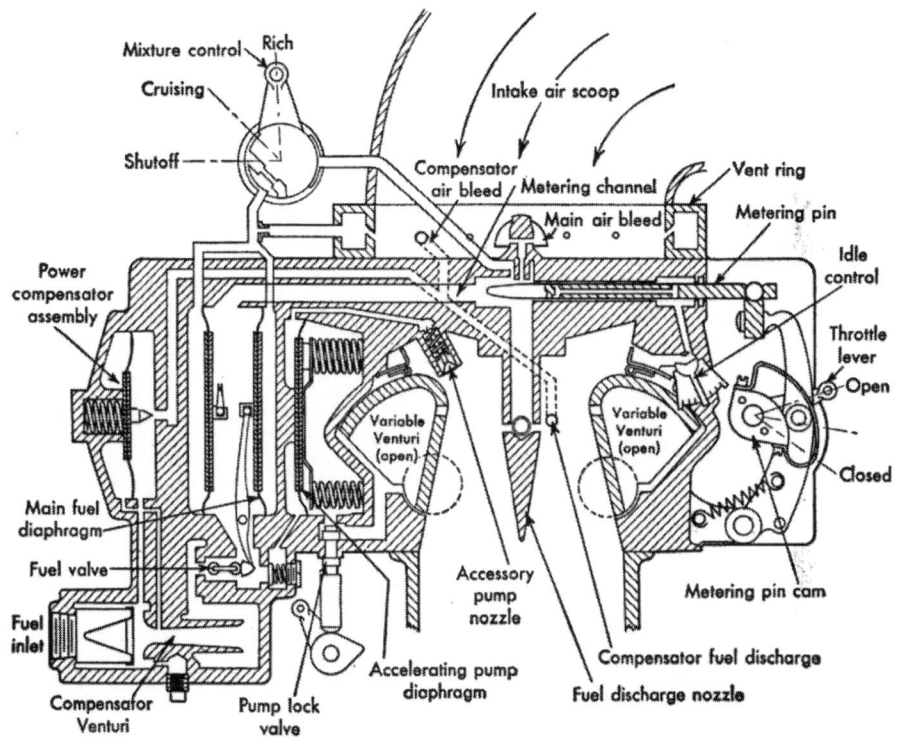

Fig. 162. A variable Venturi carburetor showing "full throttle" position.

completely, it would require some time to start the flow again. Since this mixture, like all gases, is elastic, the intake stroke in a cylinder might possibly be complete before the entire column of air-fuel mixture was in motion. This would result in a decreased charge in the cylinder and loss of power. When the column of air is in rapid motion past the intake pipe of one cylinder due to the intake of another cylinder, the ramming effect helps increase the amount of fuel forced into the first cylinder if the intake valve on this cylinder opens at approximately the same time as the valve on the second cylinder closes. The greater the number of cylinders drawing their supply of air-fuel mixture from a manifold, the more continuous is the flow in the manifold. An engine may

AIRCRAFT ENGINES

develop more power at cruising speed with an intake manifold having a diameter of 1¾ in. than it would equipped with a manifold having a diameter of 1½ or 2 in.

The float type of carburetor is affected by changes in the position of the aircraft. On aircraft engines having a wide range of performance, a carburetor is used in which the float arrangement has been replaced

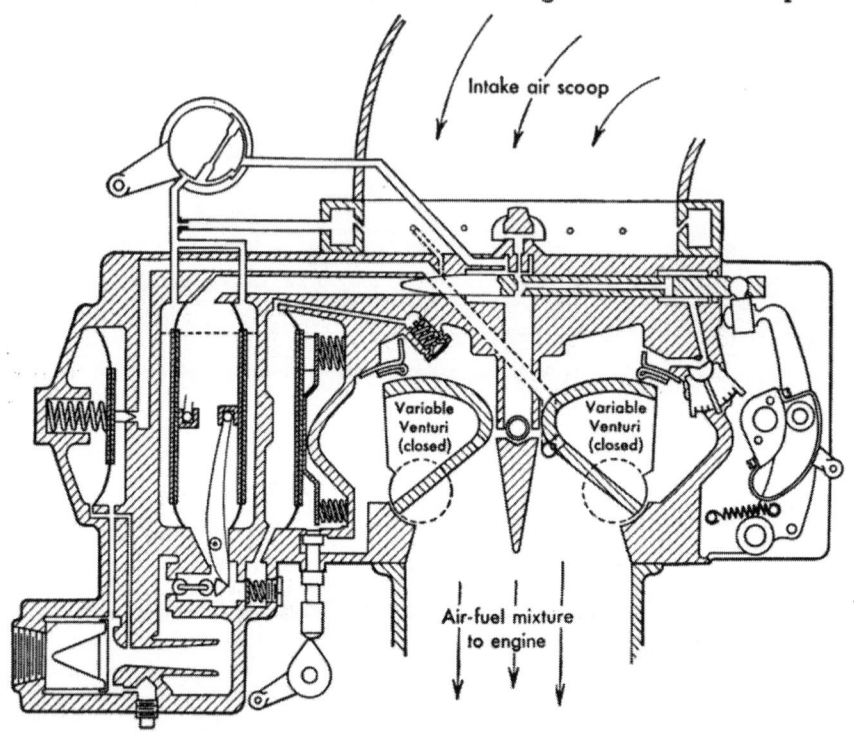

Fig. 163. A variable Venturi carburetor showing "idling" position.

by a diaphragm arrangement which maintains a full fuel chamber at all times. In the diaphragm type of carburetor, the fuel enters the air stream after the air has passed through the Venturi and throttle. The air-fuel mixture therefore does not flow over the carburetor parts.

In the standard float type of carburetor, the fuel is mixed with the air just as it passes through the Venturi. Evaporation of fuel lowers the temperature of the air stream. When this temperature is near the freezing point, ice may form. The diaphragm type of carburetor reduces to practically zero the chances of ice forming from this cause.

The main parts of a variable Venturi carburetor consist of a fuel metering needle, main air bleed, nozzle bar, and rotating Venturi parts. The rotating Venturi is in two sections connected by gears and operated by

CARBURETORS AND SUPERCHARGERS

the throttle connection. As the parts of the Venturi rotate, the opening between them opens and closes. These rotating parts act as a throttle valve. When the engine is not running, the atmospheric pressure within the fuel chamber is equal to the outside atmospheric pressure. The fuel level is approximately 1½ in. above the center of the diaphragm, and the fuel inlet valve is closed by the pressure exerted against

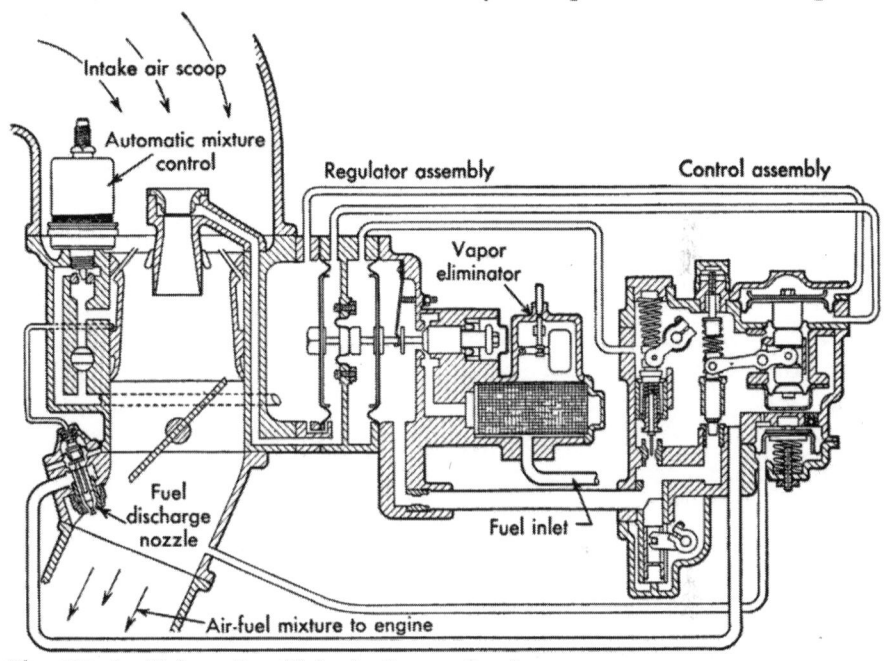

Fig. 164. An "injector" or "injection" type of carburetor.

the diaphragm by the fuel in the chamber. Fuel is furnished to the chamber under pressures of approximately 6 or 7 lb. per sq. in. When the engine is running, the incoming air creates a suction which draws fuel from the fuel chamber through the metering needle opening. An air bleed is provided in the passage between the fuel chamber and the metering needle. The metering needle is operated by a connection on the throttle shaft. A nozzle bar replaces the conventional discharge nozzle and is placed just below the Venturi parts. The nozzle bar has openings along each side through which the fuel enters the air stream. When the Venturi throttle is suddenly opened the vacuum is released from the diaphragm and fuel is forced out through the pump spray nozzle. This type of carburetor is equipped with an idle cutoff.

The injector or injection type of carburetor is quite different from the float type and the diaphragm type of carburetors. This carburetor does

AIRCRAFT ENGINES

not have a fuel chamber connected with the outside air, but operates on a closed fuel system. This system remains full of fuel when the engine is stopped by cutting off the fuel flow with the idle cutoff. The air supply in this carburetor is obtained by passing the air from the air scoop through a Venturi opening which contains a large Venturi tube and a small Venturi tube. Opening into the space between the two Venturi tubes are impact tubes, the open ends of which receive the impact of the

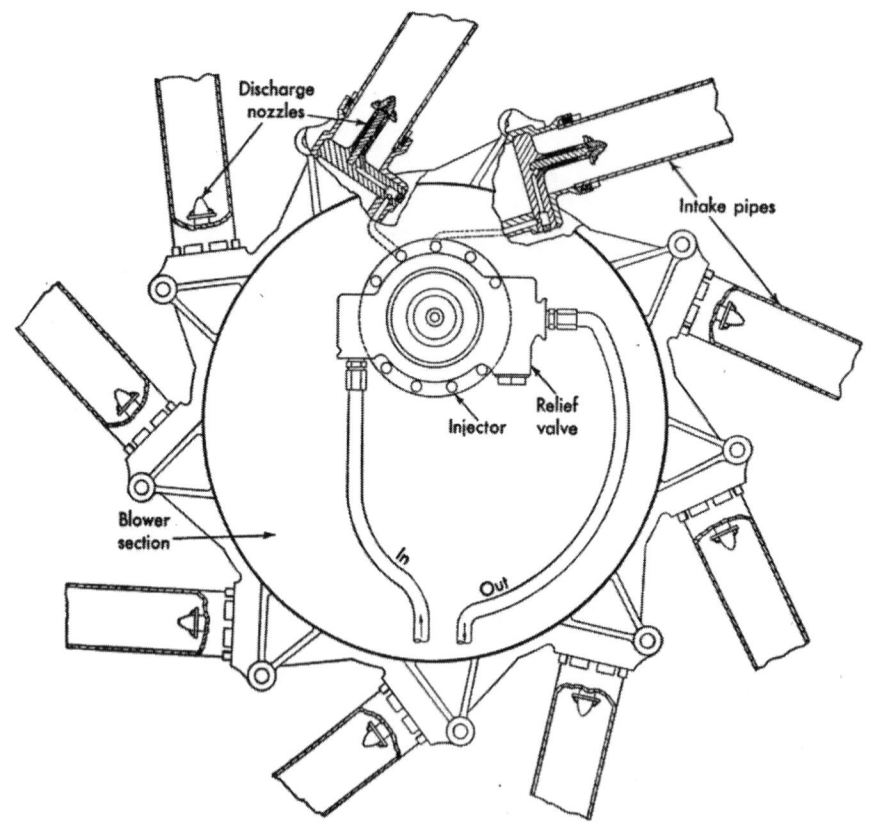

Fig. 165. A schematic drawing showing a fuel-injector system, rear view.

air entering the carburetor. The impact of air on these tubes regulates the pressure on a series of diaphragms which, in turn, regulates the amount of fuel supplied to the discharge nozzles which are located in the intake pipe. The air from the air scoop passes through the Venturi openings, the throttle opening, and past the discharge nozzles. This type of carburetor is not equipped with a carburetor heating arrangement. It is used in conjunction with a supercharger, and the air-fuel mixture which has passed the discharge nozzle is led directly into the super-

CARBURETORS AND SUPERCHARGERS

charger. The supercharger action assists in vaporizing the fuel before it enters the intake manifold.

On many engines a fuel-injection system is used. The fuel is injected into the intake manifold just outside the intake valve port. A measured amount of fuel is injected by a plunger pump. The amount of fuel is regulated by the length of the stroke of the pump plunger. As the throttle valve is opened, increasing the amount of air available to the cylinders, it

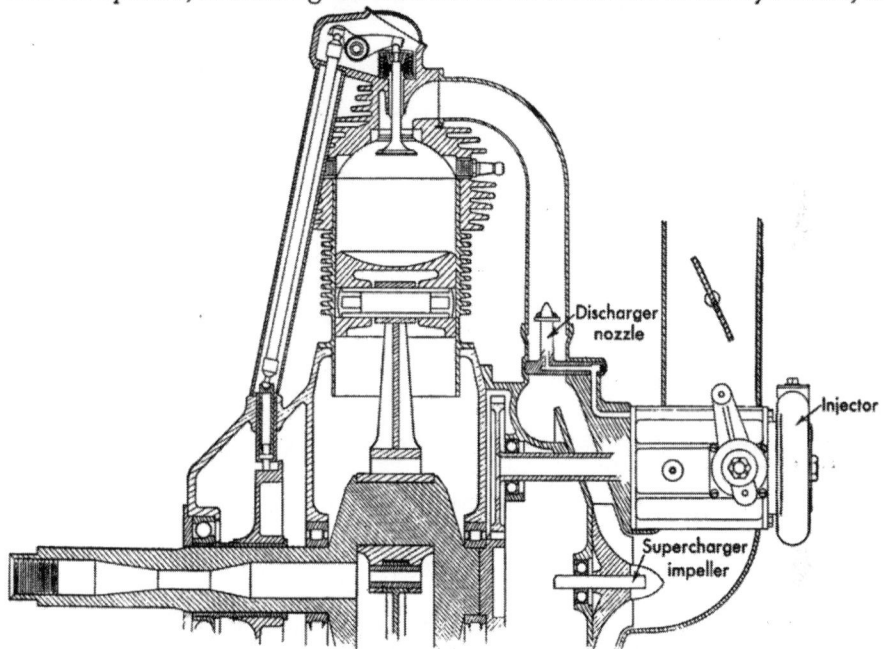

Fig. 166. A schematic drawing showing a fuel-injector system, side view.

also operates the injection system which allows an increasing amount of fuel to become available to each pump. By this method, the proper air-fuel mixture is assured at all throttle settings. The fuel is forced out through a spray nozzle in a finely divided state. This finely divided fuel is vaporized as it is drawn into the cylinder and subjected to the heat therein. Each charge of fuel is supplied to the cylinder at the beginning of the intake stroke. This system eliminates the danger of carburetor icing and, as there is no air-fuel mixture in the intake manifold, decreases the fire hazard.

The power of an aircraft engine depends directly upon the amount of fuel that can be burned with oxygen in the cylinder. The amount of fuel that can be burned is limited by the amount of oxygen contained in any cylinder charge of air-fuel mixture. The charge in the cylinder is

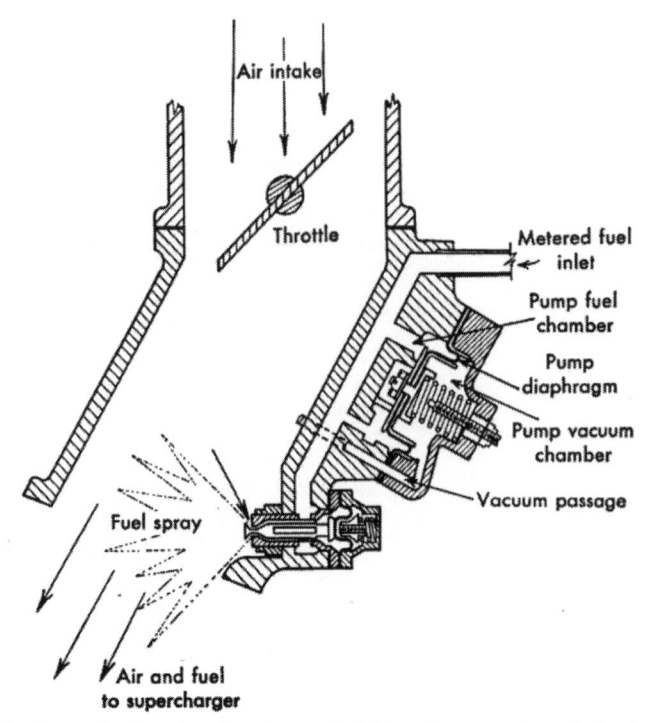

Fig. 167. A schematic drawing showing a fuel-injection nozzle and single-diaphragm accelerating pump.

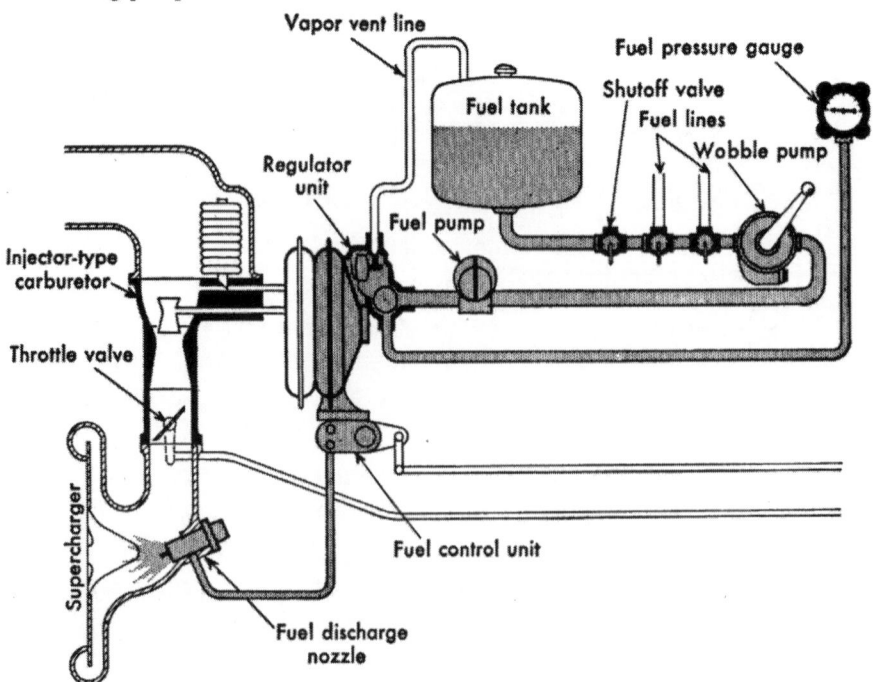

Fig. 168. A schematic drawing of an injection carburetor system.

164

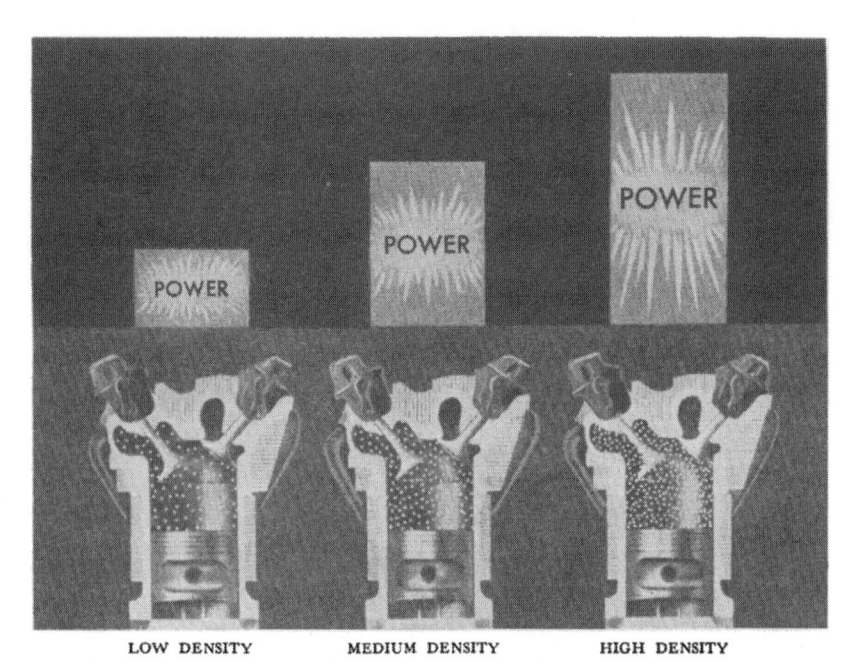

Fig. 169. The density of the atmosphere affects engine power. (Courtesy Wright Aeronautical Corporation)

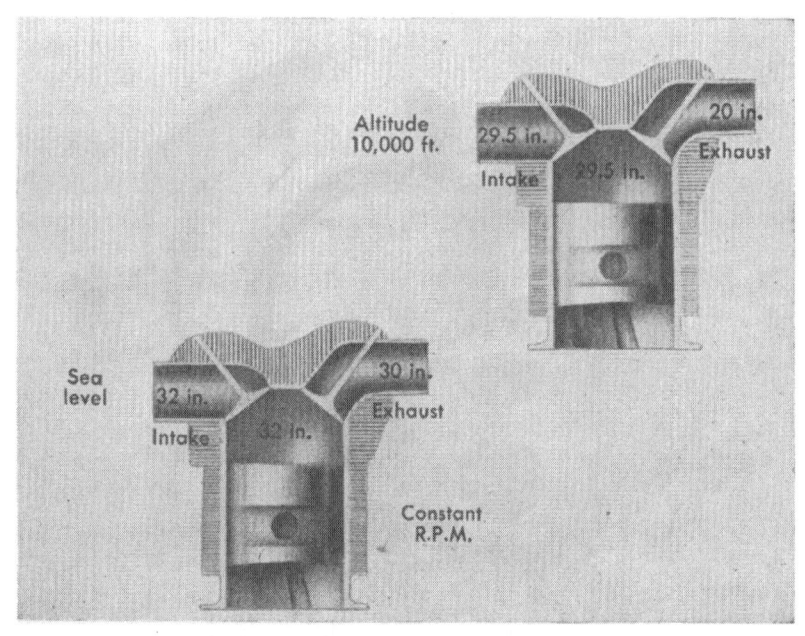

Fig. 170. The effect of altitude on the density of air-fuel mixture and exhaust gases. (Courtesy Wright Aeronautical Corporation)

165

Fig. 171. Showing the effect of supercharging on the density of air-fuel mixture. (Courtesy Wright Aeronautical Corporation)

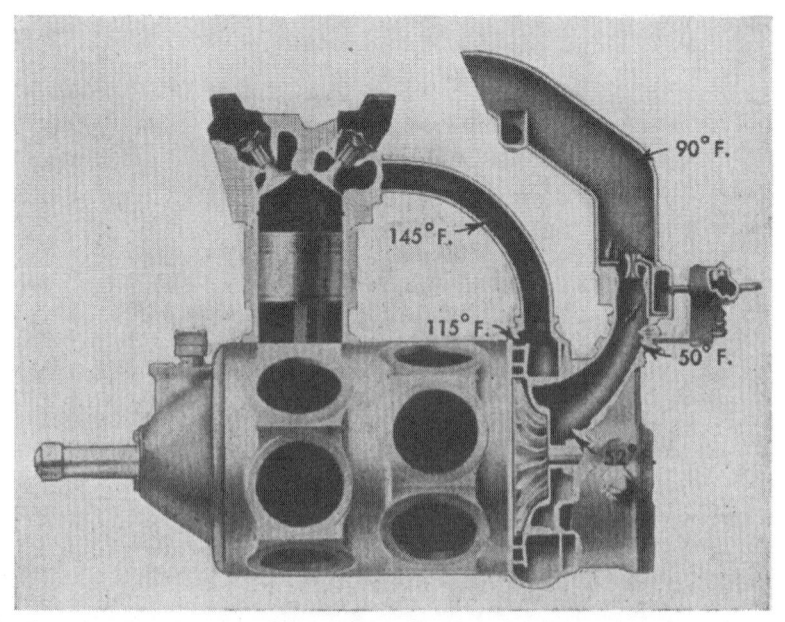

Fig. 172. Showing the effect of supercharging on the temperature of air and the air-fuel mixture. (Courtesy Wright Aeronautical Corporation)

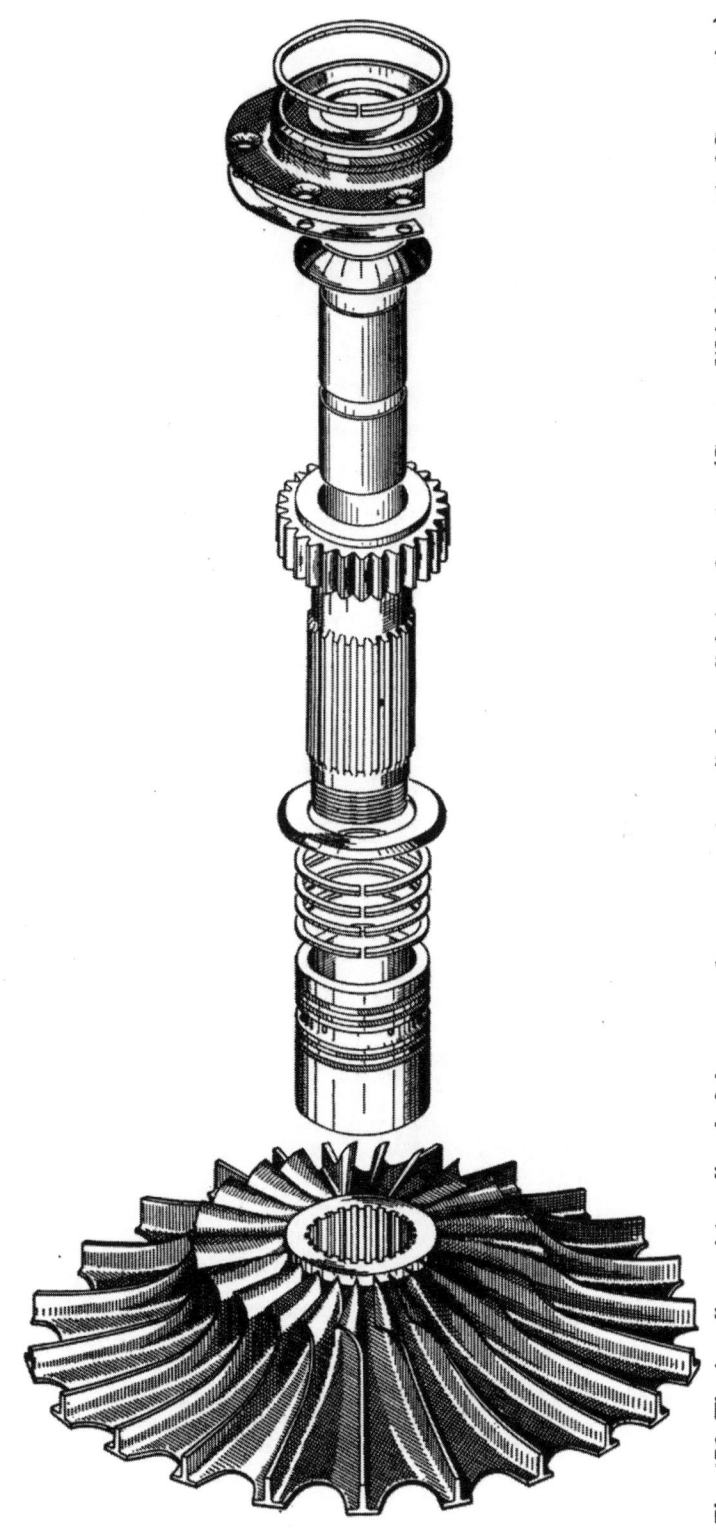

Fig. 173. The impeller and impeller shaft for a supercharger on a nine-cylinder, radial aircraft engine. (Courtesy Wright Aeronautical Corporation)

167

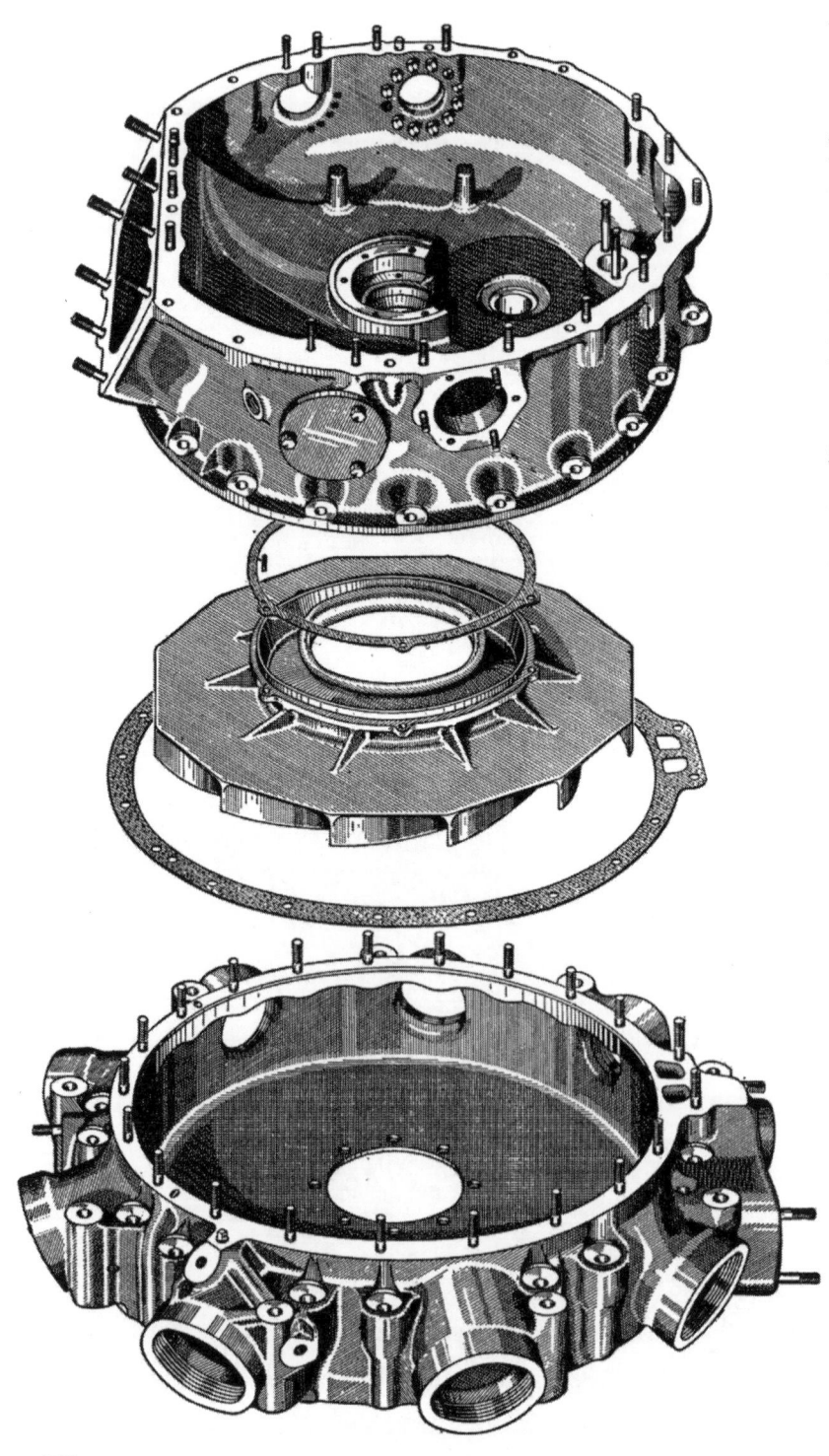

Fig. 174. The supercharger housing and diffuser plate of a nine-cylinder, radial aircraft engine. (Courtesy Wright Aeronautical Corporation)

CARBURETORS AND SUPERCHARGERS

measured in cubic inches. At sea level with the air pressure at 14.7 lb. per sq. in., a cylinder full of the air-fuel mixture contains more oxygen by weight than it does at a 10,000-ft. altitude where the atmospheric pressure is 10 lb. per sq. in. At a 10,000-ft. altitude there is only about 71 per cent as much oxygen by weight in a cubic inch of air as there is

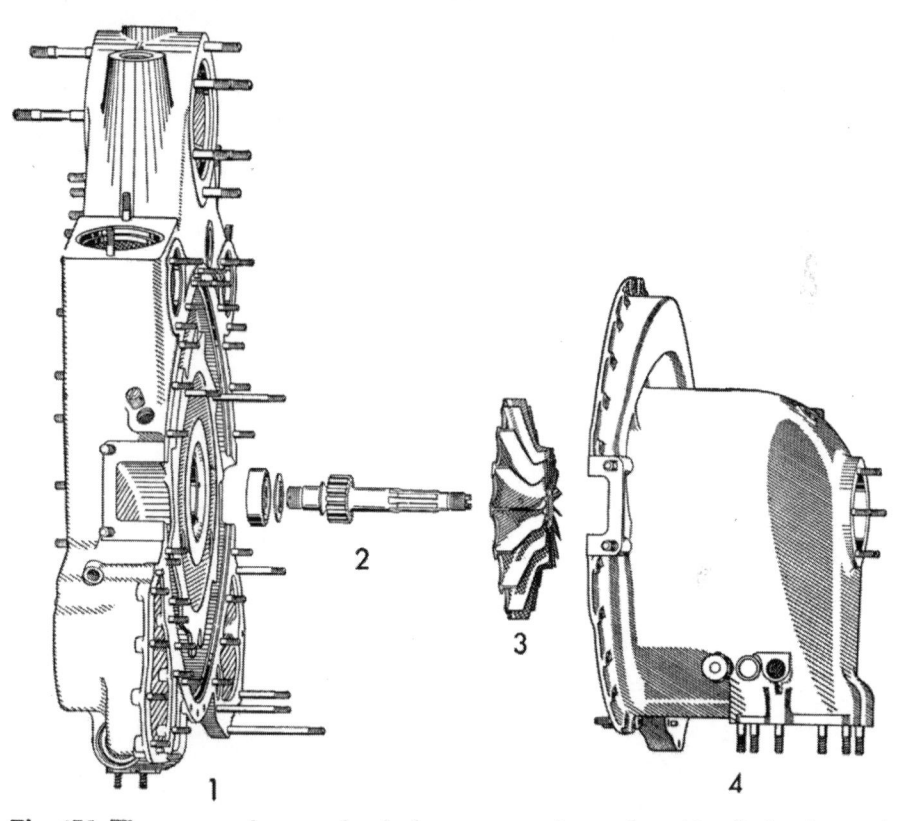

Fig. 175. The rear crankcase and a single-stage supercharger for a 12-cylinder, inverted V-type aircraft engine. (Courtesy Ranger Aircraft Engines) (1) Rear crank case; (2) supercharger drive shaft; (3) impeller; (4) supercharger housing.

at sea level. An unsupercharged engine developing 100 hp. at sea level will deliver approximately 85 per cent of its rated horsepower at an altitude of 5000 ft., if the revolutions per minute are the same at both elevations. Under the same conditions of crankshaft speed, this engine would deliver approximately 71 hp. at 10,000 ft.; 58 hp. at 15,000 ft.; 47 hp. at 20,000 ft.; and 38 hp. at 25,000 ft. If, however, the manifold pressure can be maintained at 14.7 lb. per sq. in., which corresponds to approximately 30 in. of mercury pressure, the engine would deliver its maximum horsepower at high altitudes.

AIRCRAFT ENGINES

A supercharger is an arrangement whereby the manifold pressure can be maintained at sea-level pressure at various elevations. A supercharger is a centrifugal air pump. It consists of a rotating part which has attached to it plates or vanes that force the air-fuel mixture into the manifold at increased pressures. In addition to the rotating part which is the impeller, a number of vanes are fixed to a nonrotating part forming the diffuser section.

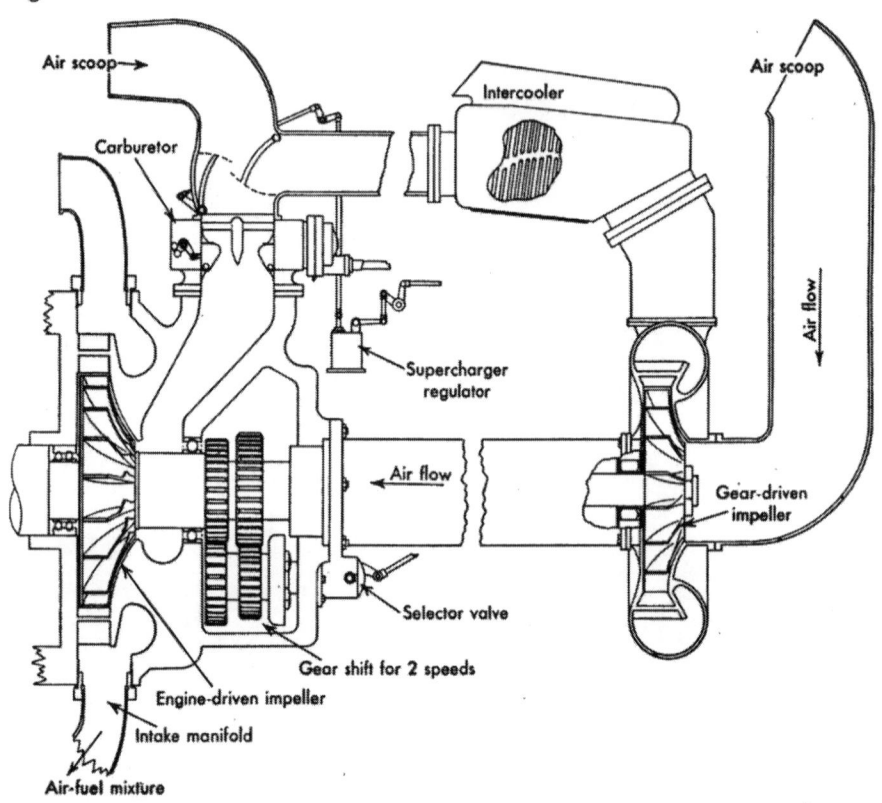

Fig. 176. A schematic drawing of a two-stage engine-driven supercharger.

The air-fuel mixture flows from the diffuser vanes into the distributing chamber which opens into the intake manifold or a manifold ring. The impeller rotates at extremely high speeds.

A simple supercharger is usually built into a rear portion of the crankcase on a radial engine, and in a separate case at the rear of an in-line engine. The simple supercharger has only one speed, and the gear ratios may vary from 6:1 to 14:1. This means that the impeller rotates at 6 to 14 times the crankshaft speed. A standard ratio of approximately 10:1 is commonly used.

Low ratio (7.06:1)

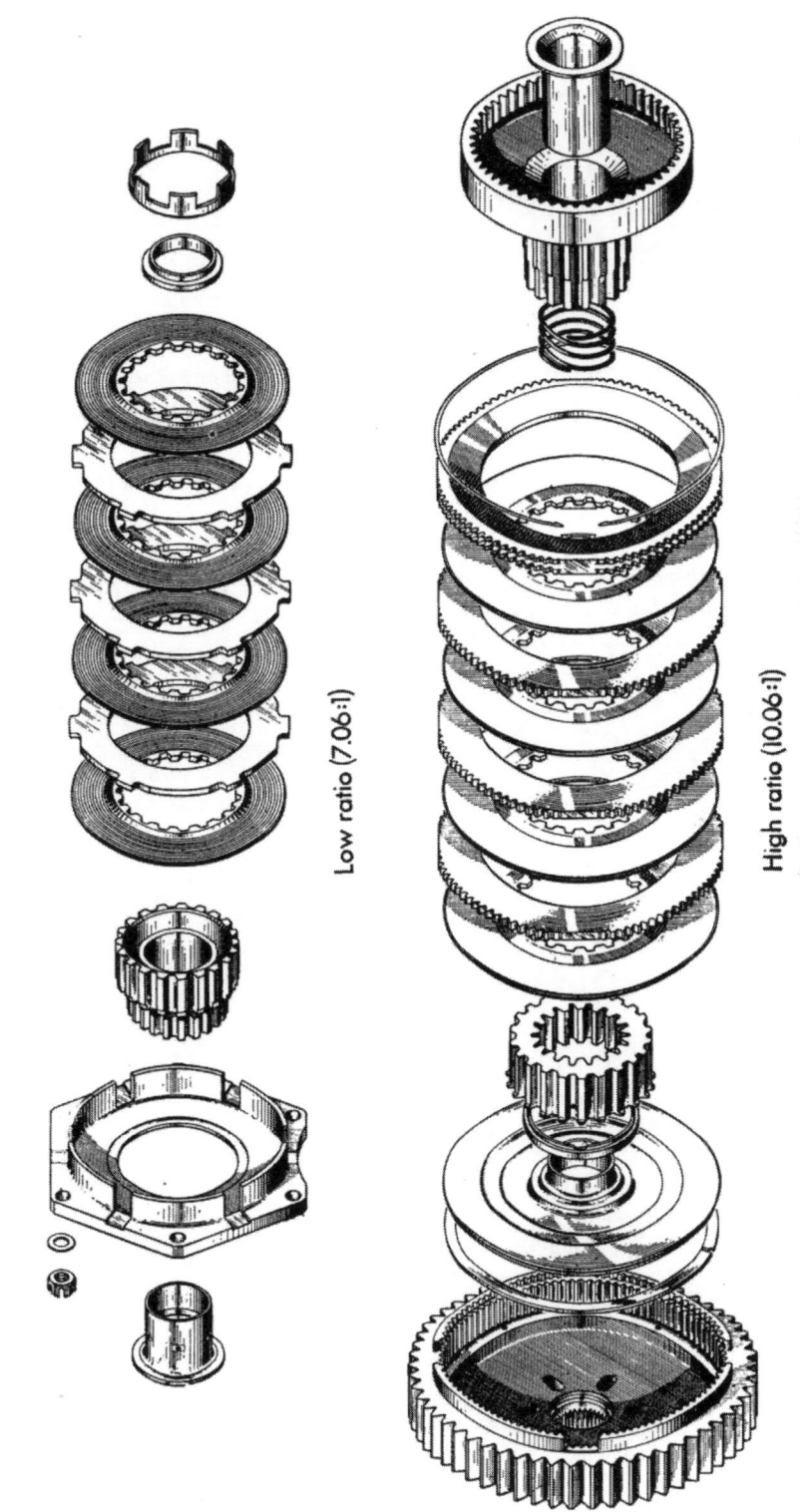

High ratio (10.06:1)

Fig. 177. The impeller drive of a supercharger used on a 14-cylinder, twin-row, radial aircraft engine showing the two-speed clutch parts. (Courtesy Wright Aeronautical Corporation)

AIRCRAFT ENGINES

Some superchargers are designed to operate at two speeds by means of a clutch and gear arrangement. At low speed the supercharger has a ratio of approximately 7:1, which is used on take-off and at altitudes below 8000 ft. As the engine begins to lose power, it means that it has reached its critical altitude. The critical altitude of the engine is the altitude at which it no longer produces its full-rated horsepower at full throttle. At

Fig. 178. Dr. Sanford A. Moss, the inventor of the supercharger, examining the turbine of a record-breaking turbo-supercharger presented to him as a Christmas gift after it had functioned flawlessly for 1004 hours on 102 Flying Fortress bombing missions. Note size of supercharger. (Courtesy General Electric Company)

this point the throttle is partially closed and the supercharger put into high gear. The impeller now rotates at approximately 10 times the crankshaft speed and will develop its rated horsepower to approximately 14,500 ft. This altitude is approximately the critical altitude for a two-stage, gear-driven, centrifugal supercharger. This arrangement simply consists of two superchargers. The first discharges the compressed air into the second. These superchargers are usually operated from the engine, and one acts as a booster for the other.

CARBURETORS AND SUPERCHARGERS

When it became necessary to operate aircraft at extreme altitudes, an additional means of building up the manifold pressure became necessary. The turbo-supercharger enables an aircraft to operate well in excess of 30,000 ft. The engine at this altitude develops approximately full horsepower. The turbo-supercharger consists of a supercharger which is connected to a gas turbine operated by exhaust gases from the engine.

Fig. 179. A production line of airplane turbo-superchargers. (Courtesy General Electric Company)

These gases at high temperatures and at high velocities are led through a gas turbine which, in turn, rotates a supercharger. The amount of air furnished by this supercharger to the regular supercharger in the engine may be regulated automatically or by an escape gate under the control of the pilot. At low altitudes, this gate is opened and the exhaust gases escape into the air without turning the supercharger. When in operation, the air compressed by the turbo-supercharger is led into the built-in engine supercharger, then into the intake manifolds. Oversupercharging at any given altitude would force too large a charge into the cylinder, thus causing detonation. To prevent this, manifold pressure gauges are installed which inform the pilot at all times of the condition of pressure in the manifold.

XV LUBRICANTS AND LUBRICATING SYSTEMS

Lubricants are used to reduce friction. Many kinds of lubricants have been used to reduce friction between the moving parts of machinery.

Lubricants may be classified as vegetable, animal, and mineral. Animal and vegetable lubricants may be used satisfactorily to lubricate slowly moving parts that are not subjected to excessive temperatures. Mineral products are used almost exclusively for the lubrication of internal-combustion engines. Lubricants are almost always used in bearings of various types.

Whenever two moving surfaces of a machine are so designed that they slide over each over, a bearing of some kind becomes necessary. A bearing is an arrangement for the reduction of friction. A plain bearing may be a circular or a flat, smooth surface.

In an in-line engine the bottom of the tappet that comes in contact with the cam is a flat bearing. This bearing is subjected to friction developed by sliding over the cam. A circular, sliding bearing is illustrated by the valve guide. The valve stem slides through this circular bearing. Plain bearings may be of the circular rotating type, such as most connecting-rod bearings. In this bearing a highly polished crankpin rotates inside a bearing that is cylindrical and closely fitting. Almost all sliding bearings have one side of softer material than the other. When both sides of the bearing are of the same hardness, wear is increased. When one side of the bearing is softer than the other, the softer side takes the wear. For example, the valve guides are usually bronze, while the valve stem is hardened steel.

Another type of bearing is the roller bearing which is designed so that the contact surfaces of one part of the bearing roll over the other. This bearing is made up of a number of hard, small, steel cylinders that roll between the two parts of the bearing. This bearing is often used in large aircraft engines for the main crankshaft bearing.

LUBRICANTS AND LUBRICATING SYSTEMS

The ball bearing also makes use of a rolling surface. This bearing has a number of highly polished, hard steel balls that roll over the bearing surface. Ball bearings offer the least resistance of any of the various types of bearings used in aircraft engines.

Fig. 180. A cutaway view of a radial aircraft engine showing oil and air-fuel-mixture passages. (Courtesy Jacobs Aircraft Engine Company)

In some large machines, such as hydro-electric turbines and generators, oil bearings may be used. In this type of bearing the weight of the moving part is supported upon a column of oil.

A good lubricant is a substance that has very little internal friction.

The particles of oil slide freely over each other. When used in a bearing, the lubricant should be such that it forms a thin, clinging film over the parts to be lubricated. The lubricant should have enough body so that it cannot be squeezed out of the bearing, thereby allowing the moving metal parts to come into contact.

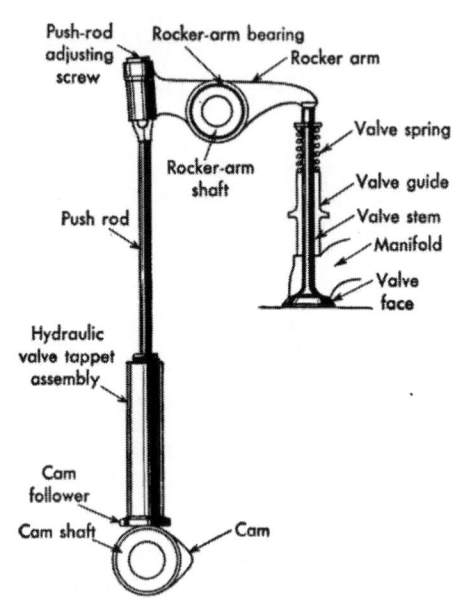

Fig. 181. A drawing to show the various parts of a valve train. (Courtesy Eaton Manufacturing Company)

In successful lubrication, the lubricant forms a layer between the moving surfaces. The particles of lubricant in this layer, sliding freely over each other, reduce friction.

Lubricants have a number of typical physical and chemical properties. Viscosity is a measurement of the friction within the oil itself. Viscosity determines the body of the oil. A heavy-bodied oil has a high viscosity and pours slowly. A light-bodied oil has a low viscosity and pours readily. The higher the viscosity of the oil, the greater pressure it will stand in a bearing without being squeezed from between the moving parts. The lower the viscosity, the more freely the parts can move over each other. In lubricating a bearing, oil of the lowest viscosity practicable should be used. The oil should be just heavy enough to prevent the parts' coming in contact.

The viscosity of an oil is given in terms of S.A.E. (Society of Automotive Engineers) or Saybolt numbers. The S.A.E. viscosity numbers

LUBRICANTS AND LUBRICATING SYSTEMS

are approximately one half the Saybolt numbers for oils having the same viscosity. For example, 30 S.A.E. is equal to approximately 60 Saybolt; 50 S.A.E. is approximately 100 Saybolt. The viscosity of an oil is measured by allowing a specified amount of oil at a given temperature to flow through a definite size of opening. The number of seconds it takes

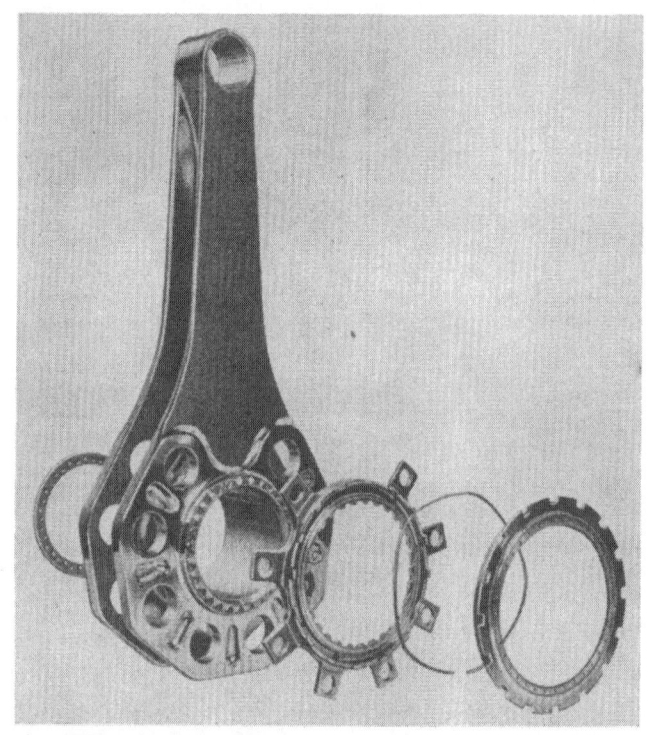

Fig. 182. A master rod having a solid circular plane bearing. (Courtesy Wright Aeronautical Corporation)

for this amount of oil to flow through the opening determines its viscosity number. For example, if 60 cc. of oil flows through the opening in 30 sec., the oil may have a viscosity number of 30. Temperature always has an effect upon viscosity. Oils become thinner when heated. Since oils are subjected to high temperatures in aircraft engines, the most desirable oil is one that shows the least change in viscosity with changes in temperature. This, of course, is but one of the desirable qualities.

The gravity of an oil is the actual weight per unit volume. A hydrometer similar to that used for testing batteries is used to test the gravity of an oil. The gravity of an oil is not of great value in determining its lubricating qualities.

The flash point is the temperature to which an oil must be heated to

177

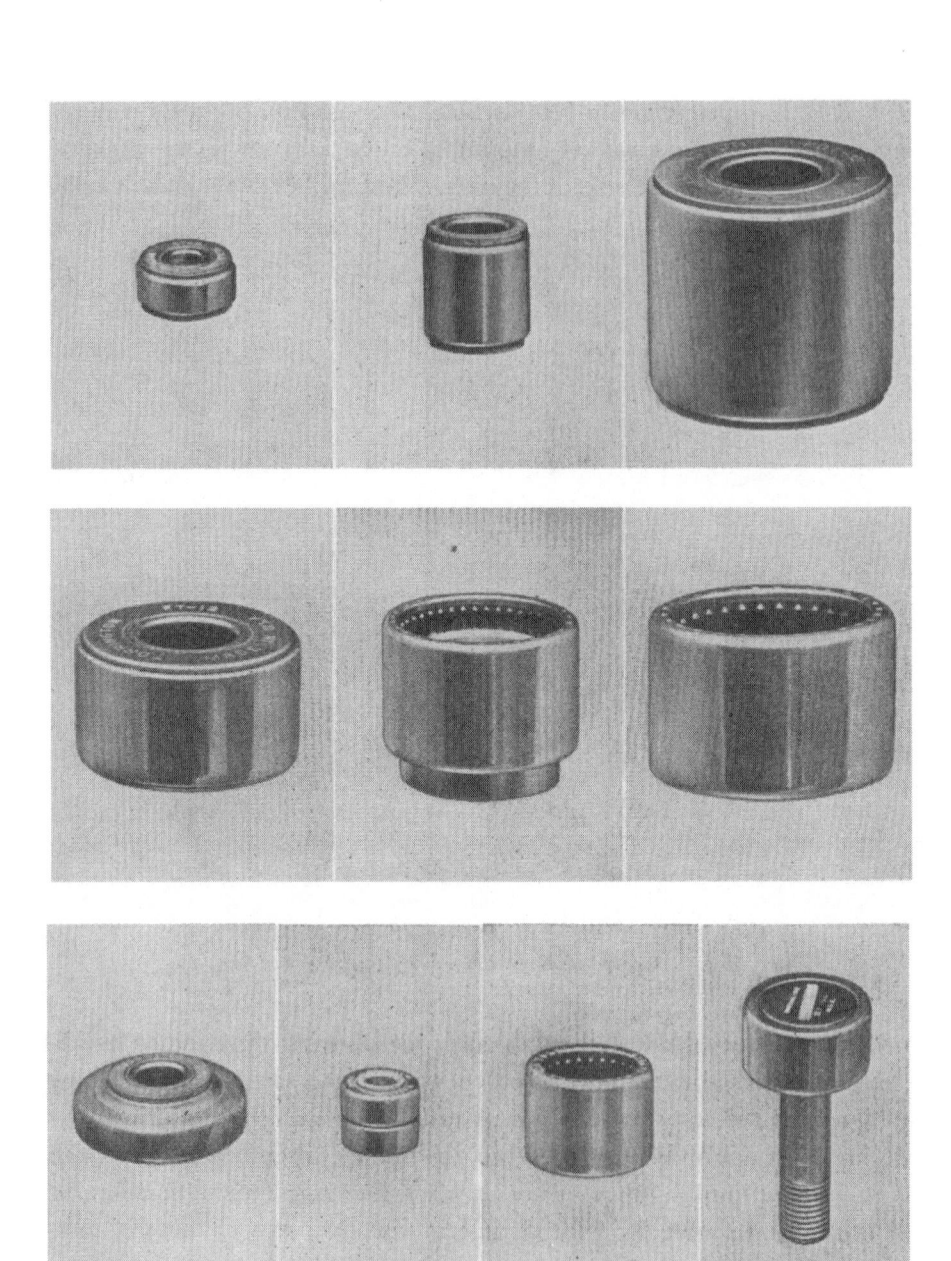

Fig. 183. A typical group of needle roller bearings used in aircraft and aircraft engines. (Courtesy The Torrington Company)

LUBRICANTS AND LUBRICATING SYSTEMS

give off vapors that will ignite when an open flame is passed over the surface of the liquid. This test is of value in determining the volatility of an oil and indicates the temperature at which parts of the oil are turned to vapor.

The fire point is the temperature to which the oil must be heated to continue to burn. At this temperature the oil burns continuously. This test has no particular value in determining the lubricating qualities of the oil.

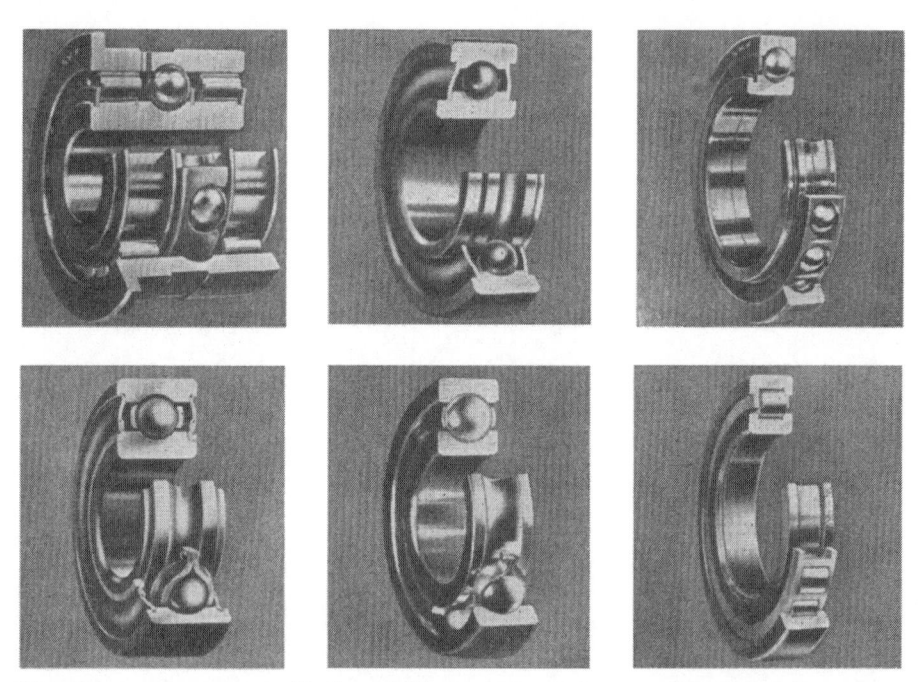

Fig. 184. A group of typical ball and roller bearings used in aircraft and aircraft engines. (Courtesy S K F Industries, Inc.)

The pour point is the lowest temperature at which a lubricant will pour. This test does not determine the lubricating qualities, but indicates the temperature at which the oil must be in order to lubricate a cold engine.

The carbon residue is determined by the amount of solid or semi-solid material left when the oil is completely evaporated at high temperatures.

Color has no particular value in determining lubricating qualities. Emulsification, or the readiness with which the lubricant combines with water, is another test to which the oil may be subjected. Practically

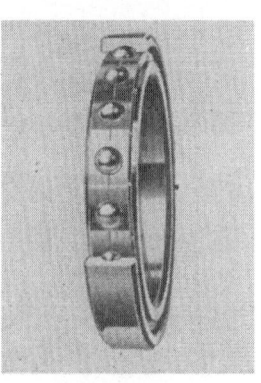

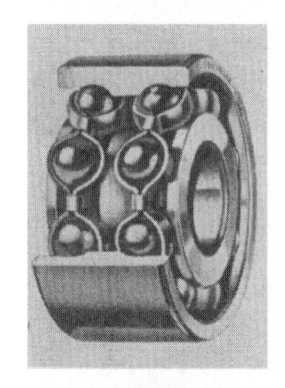

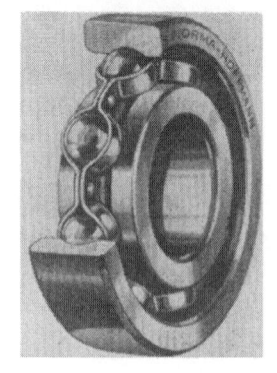

Fig. 185. Another group of ball and roller bearings used in aircraft and aircraft engines. (Courtesy Norma-Hoffmann Bearings Corporation)

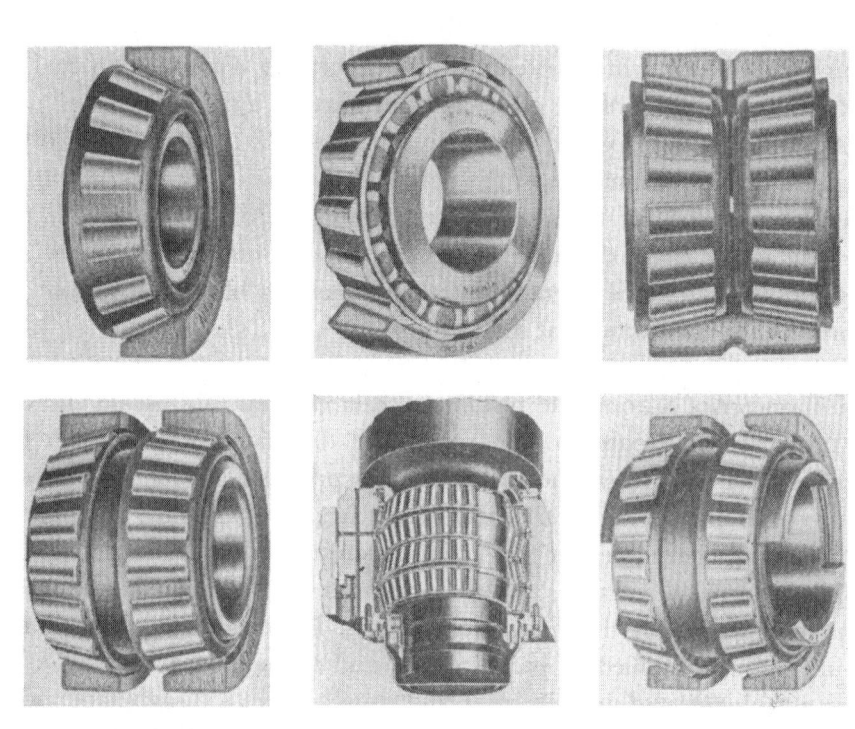

Fig. 186. A group of tapered roller bearings used in aircraft and aircraft engines. (Courtesy The Timken Roller Bearing Company)

AIRCRAFT ENGINES

all mineral oils mix freely with gasoline and may become diluted. This property need not cause serious concern, as the temperatures at which the engine is operated cause the gasoline to evaporate from the oil as soon as the engine reaches operating temperatures.

Some fuel and lubricating systems are equipped with a valve by which the oil entering the engine is deliberately diluted with fuel to reduce the viscosity of the oil to assist in starting a cold engine.

Aircraft engines operate at higher temperatures than do automobile engines, being operated at a high percentage of their horsepower rating over long periods of time. An aircraft may cruise for hours with the throttle set at approximately 75 per cent of the maximum power. An automobile is seldom operated at three quarters of its full power for more than a few minutes at a time.

For this reason and because aircraft pistons are usually at least twice the diameter of automobile pistons, greater clearances must be allowed to take care of expansion. Clearances are the spaces left between the moving parts of the engine to allow for expansion, proper lubrication, and movement between the parts when they are expanded due to the temperatures developed in the engine when operating at full power.

The mineral lubricants, as well as the mineral fuels, are made up of a large number of hydrocarbon compounds. The crude petroleums have been roughly classified as paraffin base and asphalt base. The asphalt-base petroleums are dark in color and usually have a disagreeable odor and a high viscosity. The paraffin-base petroleums are rather light in color, ranging from amber to dark red, having a somewhat pleasant odor and a low viscosity. The final products from either type, both in lubricants and fuels, vary but little in their characteristics as they are furnished to the consumer. The various parts or fractions of the crude petroleum are distilled off, the lighter parts such as naphtha and gasoline coming off first. These are followed by the heavier fuels, such as Diesel fuel and distillate. The light lubricating oils then begin to come off, followed by the heavier lubricating oils and greases. The heavy residues, such as asphalt and paraffin, are left in the still. To produce the finished product, the various fractions are carefully blended to give the desired characteristic to the lubricant or fuel.

The greases consist of the heavier portions of the petroleum and they do not flow at air temperature. Greases may be blended to give the desired characteristics, such as resistance to heat and water, and may have sticky characteristics so that they will cling to the parts being

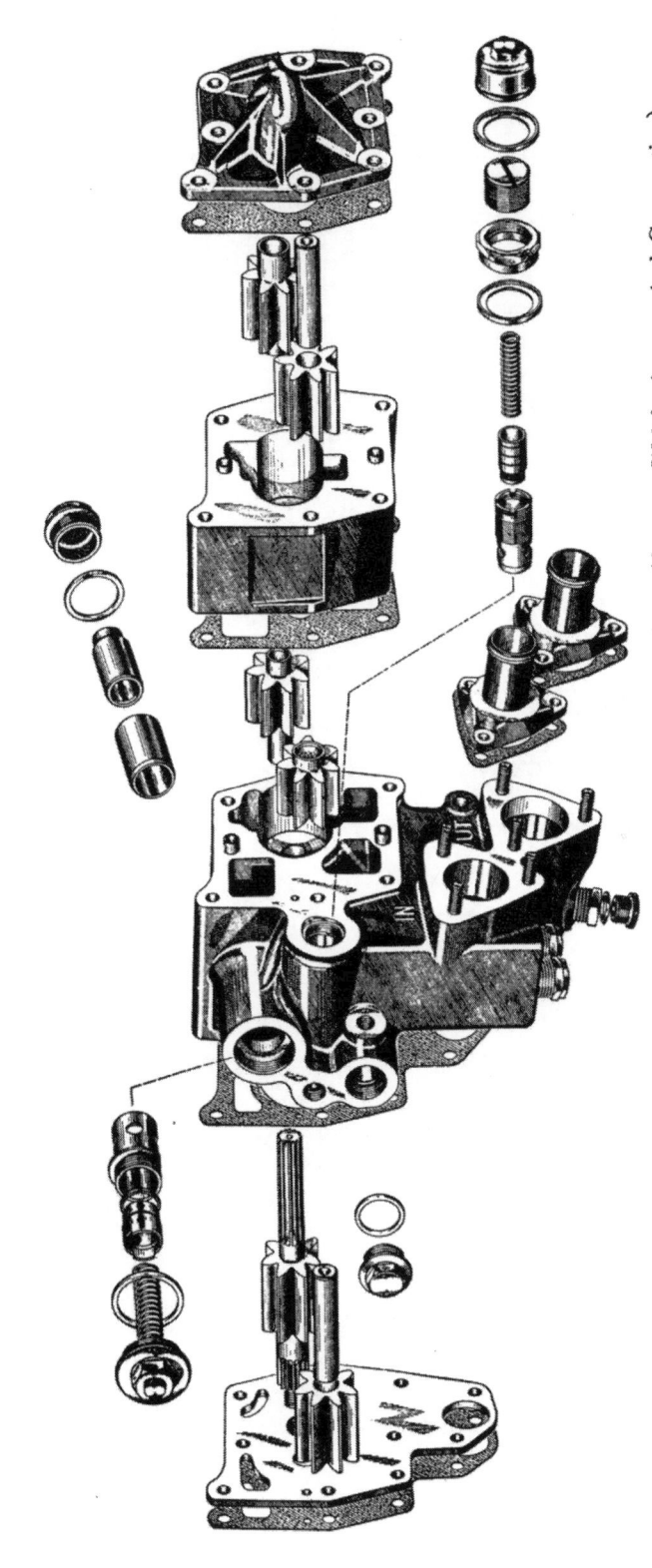

Fig. 187. An exploded view of an oil pump for a 14-cylinder, two-row, radial aircraft engine. (Courtesy Wright Aeronautical Corporation)

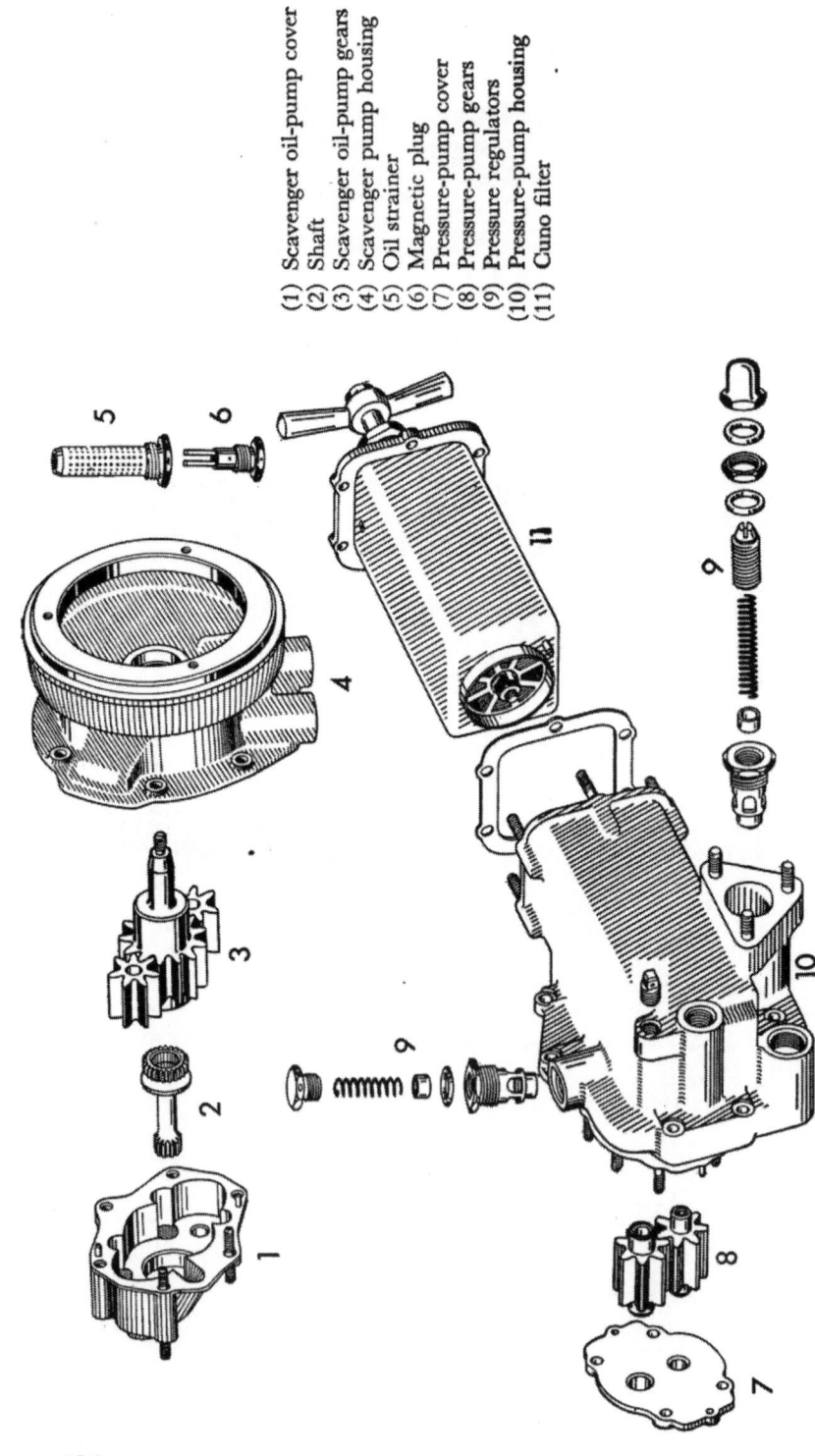

(1) Scavenger oil-pump cover
(2) Shaft
(3) Scavenger oil-pump gears
(4) Scavenger pump housing
(5) Oil strainer
(6) Magnetic plug
(7) Pressure-pump cover
(8) Pressure-pump gears
(9) Pressure regulators
(10) Pressure-pump housing
(11) Cuno filter

Fig. 188. An exploded view of a scavenger oil pump and its accessories for an inverted, in-line, dry-sump aircraft engine. (Courtesy Ranger Aircraft Engines)

LUBRICANTS AND LUBRICATING SYSTEMS

lubricated. The greases vary in consistency from semisolid to very hard. Greases are usually applied by means of an alemite or zerk gun.

There are two main classifications of lubricating systems — the wet sump and the dry sump. In the wet-sump system, the main oil supply is carried in the crankcase or in a special sump attached to the engine.

Fig. 189. A light aircraft engine with the oil supply tank attached to the bottom of the crankcase. (Courtesy Continental Motors Corporation)

This system is used chiefly on aircraft engines of low horsepower. The wet sump may be used successfully when the engine operates at comparatively low temperatures and the aircraft is not used extensively for acrobatic flying. If an engine of this type is inverted, the oil flows from the sump into the cylinders and crankcase, flooding the engine.

The dry-sump system became necessary with the development of the radial and inverted type of engine.

In engines of high horsepower it is necessary that some provision be made to cool the oil. Oil is an important factor in engine cooling.

The heat carried away from the engine by the oil must be eliminated. If the oil were retained in the engine, the temperature would become too high and prevent effective lubrication. In the dry-sump system, the oil tank is separated from the engine and may be located either in the engine nacelle or some part of the aircraft structure. A typical, dry-sump system consists of the oil tank, a pressure line to the engine, a return line from the engine to the tank, oil radiators for cooling the oil, a breather line and proper vents, a pressure pump, a pressure relief valve, scavenger pumps, drain cocks, and oil strainers and cleaners. There should also be instruments to indicate oil pressures and temperatures. Oil tanks should be located above the level of the oil pump and be of such capacity that there is approximately 1 gal. of oil for each 12 gal. of fuel. Oil tanks may be of any suitable shape and are usually made of aluminum alloy or stainless steel. Suitable drain cocks or plugs should be arranged so that the entire oil system may be emptied. The breather line from the crankcase is often connected with the oil tank so that oil vapors may be returned, thereby avoiding waste of oil.

The oil is supplied to the engine under pressure. The oil supply and pressure are maintained by a pump, usually of the gear type. The oil, after passing through the engine, is collected in a suitable sump.

Some engines may have more than one sump. The oil is returned to the oil tank by one or more scavenger pumps which should have a capacity considerably greater than the pressure pump. In inverted flight or acrobatic maneuvers, considerable oil may be collected in the engine which must be removed rapidly by the scavenger pumps when the aircraft is returned to a normal position.

All of the oil lines should be at least equal in diameter to the inlet of the pumps. The oil from the engine, discharged by the scavenger pumps, is usually led to the oil radiator, which is equipped with a by-pass valve or pressure relief valve. This valve operates when the oil is heavy or when the flow through the radiator is restricted. This valve allows the oil to by-pass the radiator and be returned directly to the oil tank. The radiator may be equipped with a thermostatic control, shutters, or other device by which the temperature of the oil may be regulated. A minimum, as well as a maximum, temperature of the oil should be maintained. If the oil is too cold, the engine may be improperly lubricated. An expansion space of approximately 10 per cent of the capacity of the oil tank must be allowed to take care of expansion of the oil and prevent the tank from foaming over. Since the oil being returned to the tank may be

LUBRICANTS AND LUBRICATING SYSTEMS

quite foamy, the return inlet may be equipped with a baffle to spread the oil over the surface of the oil in the tank, preventing the oil supply from becoming full of air bubbles.

Fig. 190. A cutaway view showing the scavenger pump and an oil filter which contains a magnet for collecting metallic particles. Note the fly wheel attached to the crankshaft. (Courtesy Ranger Aircraft Engines)

The filler inlet of the oil tank should be so located that the tank cannot be filled completely. The cap should be plainly marked "oil," and some method should be provided for determining the amount of oil in the tank. Usually a graduated rod similar to that in automobiles is used.

The oil lines are usually copper or aluminum tubing and must be connected to the engine by means of flexible joints to prevent breakage due

AIRCRAFT ENGINES

to vibration. These joints are often made of a special hose that should be removed when damaged or worn. Special fittings or clamps should be used. The oil line should be as free as possible from loads. Whenever it is removed, the line should be annealed. To anneal an aluminum line the proper temperature may be obtained as follows. Using an oxy-acetylene

Fig. 191. A cutaway view showing oil passages and oil holes in a hollow shaft. (Courtesy Ranger Aircraft Engines)

blowpipe, adjust the mixture strongly on the acetylene side. With this smoky flame, coat the entire tube with soot. Adjust the flame to neutral and burn off the soot coating. This method will properly anneal the tube without overheating. A copper tube may be annealed by heating to a cherry red and then quenching in water.

The engine is lubricated by a pressure system. Crankshafts, camshafts, and accessory drive shafts are usually hollow or contain oil passages. On

LUBRICANTS AND LUBRICATING SYSTEMS

some engines, particularly the radial type, the pressure oil is fed into the hollow crankshaft by means of a plain bearing. Both the bearing and the crankshaft are drilled with suitable holes. The oil is supplied to this bearing under pressure and led by the crankshaft through suitable oil passages to the connecting-rod bearing. Holes in the crankpin allow the

Fig. 192. A cutaway view showing an oil drain tube in the rear of the crankcase of an inverted in-line aircraft engine. Note the tachometer drive gears and the starter dogs on the rear of the crankshaft. (Courtesy Ranger Aircraft Engines)

oil to escape into the connecting-rod bearings. This oil is sprayed from around the edge of the connecting-rod bearing, lubricating the cylinder walls and the bearings in the piston. Other parts, such as accessory gearing and rocker arms, are lubricated by special oil passages.

On some of the horizontally opposed engines, the oil is supplied to the rocker arm and valve stems by being forced into the rocker-arm housing through one push-rod housing, flowing back to the sump through the other push-rod housing.

Fig. 193. A cutaway view showing the main oil strainer, pressure relief valve and the pressure oil pump. (Courtesy Ranger Aircraft Engines)

In-line engines usually have the engine supplied by pressure oil through a hollow camshaft or accessory drive shaft. Oil from various engine parts drains into a sump from which, in the case of dry-sump engines, it is returned to the oil supply tank by means of the scavenger pump. If the oil lacks sufficient viscosity, too much oil will escape past the piston rings and be burned in the cylinder. If the oil is too heavy, enough oil may not reach the bearings or the cylinders.

LUBRICANTS AND LUBRICATING SYSTEMS

The oil pressure gauge indicates the pressure of the oil furnished to the engine and is not an indication of how much oil is being supplied. The oil pump is so arranged that it maintains a constant pressure as determined by the by-pass valve. A thin oil would indicate the same pressure as a heavy oil. It is necessary, therefore, that the oil temperature

Strainer

Oil filter

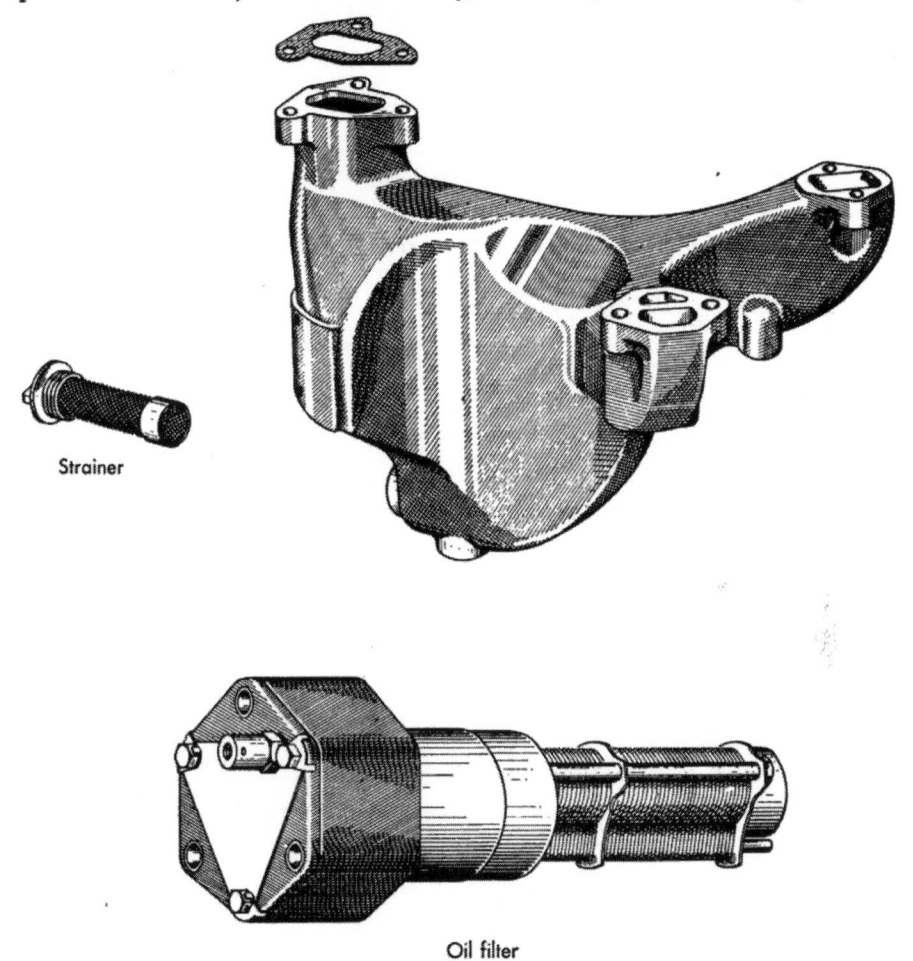

Fig. 194. An exploded view showing the oil sump and Cuno oil filter used on a nine-cylinder radial aircraft engine. (Courtesy Wright Aeronautical Corporation)

be maintained within the range recommended by the manufacturer. The oil temperature gauge usually indicates the temperature of the oil as it reaches the scavenger pump. Excessive oil temperatures indicate either a high operating temperature or a low oil supply.

All oil systems should be equipped with the proper strainers or oil cleaners. Oil strainers may be of fine wire screen or other straining devices. The mechanism may consist of a centrifugal arrangement or a disk cleaner such as the Cuno. The Cuno cleaner consists of a number of disks so arranged that the space between them is only about 0.003 in. Oil passes between these disks, and large particles of foreign matter collect on the outer edge and are removed by wipers. The cleaner may be arranged so that the disks are continually rotated by means of the oil pressure, or the disks may be rotated by means of a suitable handle or wrench. All strainers and cleaners should be removed and cleaned at regular intervals.

The consumption of oil depends upon the normal amount required to lubricate properly the engine and cylinders. To lubricate the cylinders properly, a certain amount of oil is expected to work by the piston rings and be lost. Excessive consumption of oil may be caused by leaks either in the oil lines or the engine itself, by improper functioning of the rings, or the use of improper oil.

Oil pressures to the engine vary from about 60 lb. per sq. in. to 90 lb. per sq. in. The normal consumption of oil varies from about 0.01 lb. per brake horsepower per hour to about 0.025 lb. per brake horsepower per hour, depending upon the horsepower of the engine and the throttle position.

The greatest single factor influencing oil consumption, besides leaks, is faulty operation of the piston rings. At high speeds, if the viscosity is too high, the rings have a tendency to slide over the film of oil, much as a high-speed boat skims over the surface of the water. This allows too much oil to work to the top of the cylinders and be consumed. Properly designed oil rings assist in scraping the oil downward on the cylinder walls, returning it to the crankcase. Some engines are equipped with warning devices that signal when the oil pressure or oil level has dropped to the danger point.

XVI PROPELLER FUNDAMENTALS

The propeller, while not a part of the engine itself, is considered a part of the power plant. It is by means of the propeller that the power developed by the engine is changed from rotary motion to thrust. Thrust is the forward force that pulls the aircraft through the air. The thrust developed by the propeller is due to displacement of air. The propeller blades are set at an angle so that the air is forced backward as they rotate.

Early propellers consisted of flat paddle-like blades that acted much the same as electric fans or blowers. These blades bit off quantities of the air, throwing it backward. The force necessary to move the air was transferred to the propeller shaft and thus pulled the aircraft forward.

Modern propeller blades have a cambered section similar to an airplane wing. The aircraft propeller is composed of two or more narrow airfoil-like blades which rotate in a vertical plane. The face of the blade is the side opposite to the direction of flight. The back of the blade is the side toward the direction of flight. The face of the blade is almost flat, and the back of the blade is cambered.

The lift of an aircraft wing is largely on the top because of the cambered surface. Also most of the thrust of an aircraft propeller is developed by the cambered surface, which literally pulls the propeller forward through the air. The thrust developed by a propeller corresponds to the lift developed by an aircraft wing. More than 75 per cent of the thrust may be on the back of the blade.

The blade is set at an angle to its plane of rotation. This angle is known as the "pitch angle" of the blade. The hub of the propeller is attached to the crankshaft or propeller shaft. The thick part of the wooden blade surrounding the hub is called the "boss." The streamlined cap covering the hub and boss is the spinner. The outer end of the blade is the tip, and the inner end, the root. The part of the blade that first meets the air when rotating is the leading edge, and the opposite edge is

AIRCRAFT ENGINES

the trailing edge. The distance from the leading edge to the trailing edge is the chord of the blade. The distance that the blade would screw forward in a semisolid material is known as the "theoretical pitch" of the propeller. When rotating, the propeller does not usually move forward a distance equal to the theoretical pitch. The distance that the propeller actually moves forward with each rotation is the "effective pitch." The difference between the theoretical pitch and the effective pitch is

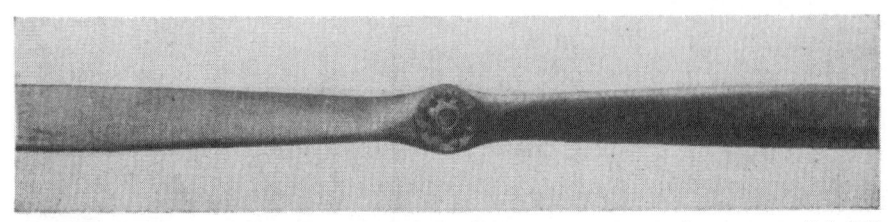

Fig. 195. A 1918 propeller. This type of propeller was commonly known as a "club." (Courtesy Hamilton Standard Propellers)

Fig. 196. The first propeller used by the Northwest Airways, Inc., October 1, 1926. (Courtesy Hamilton Standard Propellers)

the "propeller slip." The slip usually amounts to about 20 per cent of the theoretical pitch while in level flight. The pitch of the blade depends upon the designed speed of the aircraft. The propeller on an airplane designed to fly at 100 m.p.h. would have a theoretical pitch of approximately one half the theoretical pitch of a propeller on an airplane designed to fly at 200 m.p.h.

When the propeller is rotating and the aircraft is stationary, the slip equals 100 per cent, and the effective pitch is zero; the theoretical pitch never changes unless the angle of blade setting is changed. The effective pitch may vary from zero to over 80 per cent, depending upon the distance the aircraft moves through the air for each revolution of the propeller. Pulling itself through the air, the blade follows a path similar to the screw threads on a bolt.

If a propeller rotates at the rate of 1800 r.p.m. and the speed of the aircraft is 120 m.p.h., or 2 miles per minute, the effective pitch may be

PROPELLER FUNDAMENTALS

found by dividing 1800 into twice 5280 (5280 ft. equals 1 mile). The effective pitch would, therefore, be 5.8 ft.

If the theoretical pitch is 7 ft., the slip is found by subtracting the effective pitch from the theoretical pitch, which would be 7 minus 5.8, making the slip 1.2. The efficiency of this propeller is found by dividing the theoretical pitch into the effective pitch and, in this instance, would be 82.8 per cent.

Fig. 197. An adjustable-pitch propeller having aluminum alloy blades, built in 1923. (Courtesy Hamilton Standard Propellers)

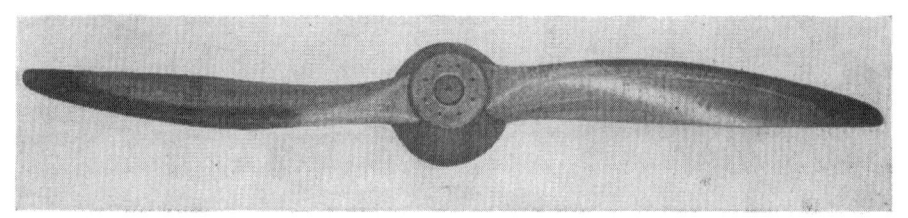

Fig. 198. A wooden propeller used on the Douglas World Cruisers which were flown around the world by the United States Army in 1924. (Courtesy Hamilton Standard Propellers)

On a fixed-pitch propeller, the theoretical pitch is usually stamped on either the hub or the boss. The faster the airplane is designed to travel, the greater the pitch of the propeller. If the pitch is too low for the speed of the airplane, the propeller speeds up above the maximum r.p.m. allowed. If the pitch is too great for the designed speed of the aircraft, the propeller churns and loses its effectiveness.

The tip speed limits the length of the propeller blade. A propeller having long blades and turning at a relatively low speed is more effective than a propeller having short blades turning at a high speed. As the tip of the propeller approaches the speed of sound, it loses its effectiveness.

The blade-pitch angle varies from the root to the tip, the pitch being greater near the root and less near the tip. This is necessary because the tip travels at a much greater speed than the part of the blade near the hub.

Propellers with more than two blades are used to obtain more propeller area without having blades of excessive length. Propellers having

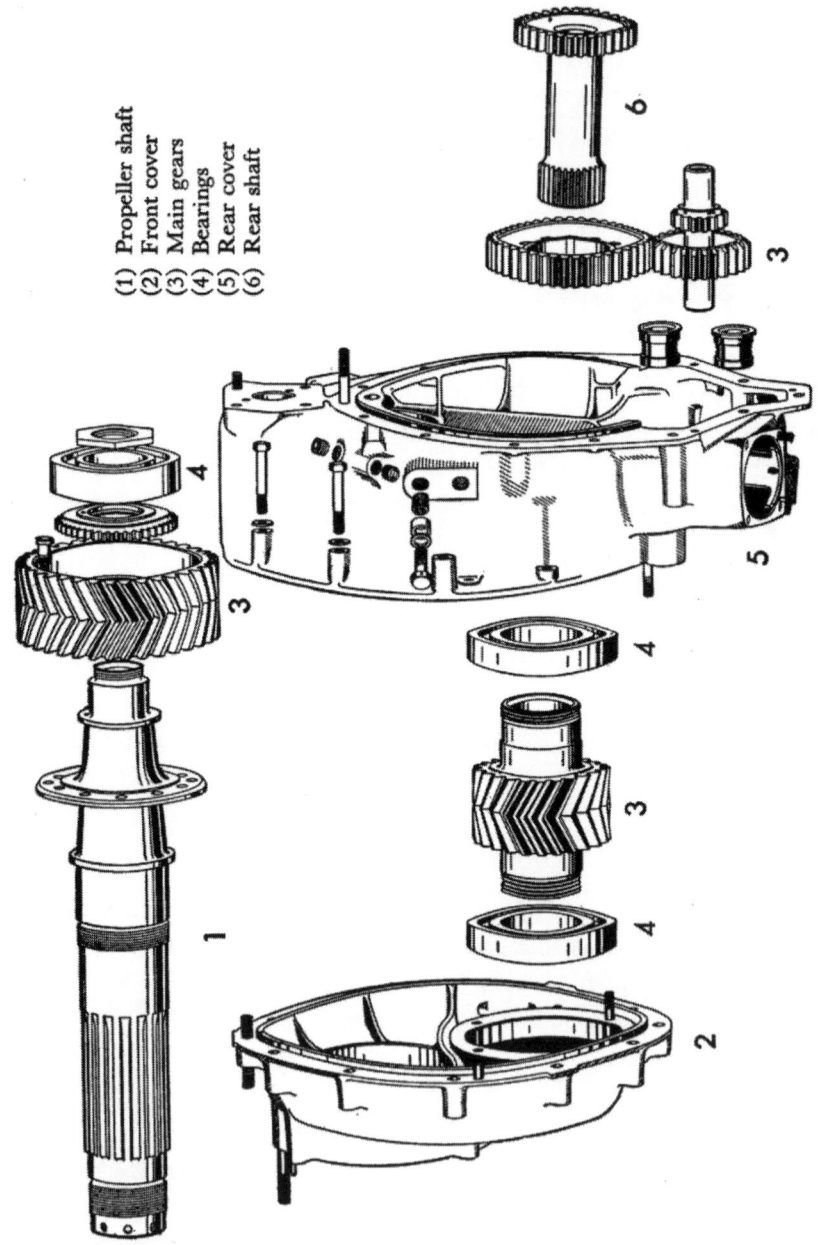

(1) Propeller shaft
(2) Front cover
(3) Main gears
(4) Bearings
(5) Rear cover
(6) Rear shaft

Fig. 199. A propeller reduction gear for an inverted aircraft engine. (Courtesy Ranger Aircraft Engines)

PROPELLER FUNDAMENTALS

only one blade have been used on some low-powered aircraft. Propellers having four or more blades are used on some of the high-powered engines.

The power of the engine depends largely upon its revolutions per minute. To develop the full power of the engine, many propellers are

Fig. 200. A phantom view of the reduction gear used on large radial aircraft engines. (Courtesy Wright Aeronautical Corporation)

equipped with reduction gears. These gears allow the propeller to turn 1800 or 1900 r.p.m. while the engine turns several hundred r.p.m. faster.

Propellers are classified as fixed-pitch, two-pitch, adjustable-pitch, constant-speed, controllable-pitch, and feathering. The adjustment of the controllable-pitch and feathering types may be mechanical, hydraulic, or electric. On some propellers, the control is automatic, but on most propellers the control is exercised by the pilot. On fixed-pitch propellers, the angle of blade setting cannot be changed. In this classification are included the one-piece wooden propeller and the one-piece metal propeller. These types of propellers are used on most low-powered aircraft. The disadvantage of the fixed-pitch propeller is that the engine output cannot be readily changed to meet the various conditions under which

197

AIRCRAFT ENGINES

the aircraft may be flown. During the take-off and climb extra power is needed, and a propeller of low pitch is desirable. While cruising, a propeller of high pitch and lower engine output are most effective. Many times, in the early days of aviation, when a pilot was forced to land at high elevations or in small fields, a propeller of low pitch was substituted

Fig. 201. A cutaway view showing propeller reduction gears used on a large, high-powered, radial aircraft engine. Note the cam plates and cam followers. (Courtesy Wright Aeronautical Corporation)

for the standard propeller in order to take off. After the aircraft was in the air, the engine had to be throttled back to prevent overspeeding with a corresponding loss of power and speed.

An adjustable-pitch propeller is constructed so that the angle at which the blades are set can be changed while the airplane is on the ground. This type of propeller has the blades separate from the hub. The root of the blade is fastened to the hub by means of clamps or other suitable

PROPELLER FUNDAMENTALS

fastenings. These fastenings may be loosened and the blade set at the desired angle.

Controllable-pitch propellers are equipped with blades that may be rotated while the airplane is in flight.

When feathering a propeller, the blades are rotated until the leading

Fig. 202. An exploded view of the propeller reduction gears used on a high-powered radial aircraft engine. This type of reduction gear has a ratio of about 16 to 9; the crankshaft rotating 16 times while the propeller rotates 9 times. (Courtesy Wright Aeronautical Corporation)

edge of the blade is in such a position that the chord of the blade is approximately parallel to the air flow. In case of damage to an engine equipped with a non-feathering propeller, the propeller tends to windmill, due to the air flow which causes it to continue rotating. By quickly feathering the blades, the engine may be stopped to prevent further damage. Some of the early propellers were designed with a brake arrangement by which the pilot could stop the rotation of the propeller in case of engine failure.

Wooden propellers (Fig. 198) are usually built up of a number of laminations of hardwood. Birch and mahogany seem to be the most satisfactory, although oak and other woods have been used. The laminations are carefully cut to the desired thickness and dried to a low moisture content. These laminations are glued together under heavy pressure, using either casein glue or one of the synthetic glues. The glued propeller is allowed to dry thoroughly before being roughed out on a propeller lathe. A further period of several days is then allowed for drying before the propeller is carefully finished to exact size.

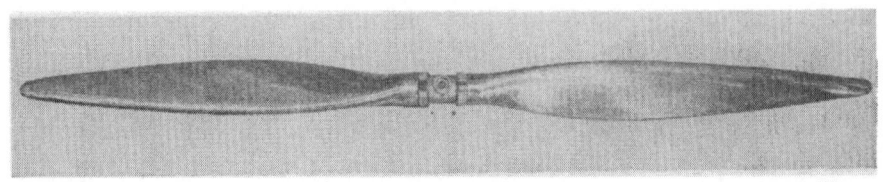

Fig. 203. An adjustable propeller having aluminum alloy blades built in 1924. These blades have threaded ends and are clamped to the hub. (Courtesy Hamilton Standard Propellers)

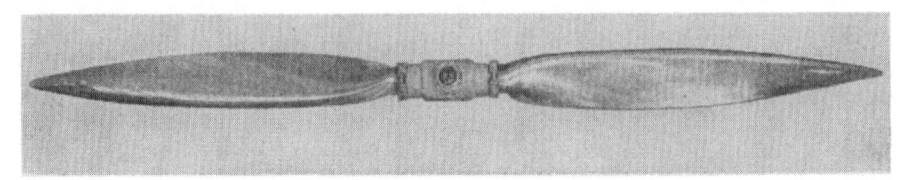

Fig. 204. An adjustable propeller having aluminum alloy blades built in 1925. These blades have retaining shoulders which are clamped into a two-piece hub. Note the pointed blades. (Courtesy Hamilton Standard Propellers)

One type of blade is built up of thin sheets of plywood fastened together with a hot-setting synthetic glue. These blades are then compressed and allowed to set at comparatively high temperatures. A chemical reaction takes place in the glue, forming a permanently hard mass. Some of these blades are fitted with a metal root which fits into a metal hub, thus allowing the pitch to be changed.

The tip and part of the leading edge are usually protected from pebbles, hail, and other hard objects by means of a metal covering. Small holes are drilled in the extreme end of the metal tip to allow the escape of moisture which may collect under the covering and which is thrown to the outer end of the blade by centrifugal force.

Fixed-pitch, metal propellers may be built of a single piece of metal. Metal blades, due to their higher strength, are thinner than wooden blades and have the advantage of allowing the pitch to be changed by careful twisting of the blade. Propeller blades may be made of alloys of aluminum, magnesium, or steel. Metal blades may be either hollow or solid.

The pilot and aircraft-engine mechanic should understand the basic

PROPELLER FUNDAMENTALS

theory upon which the variable-pitch propellers operate. The loss in performance of aircraft engines having fixed-pitch propellers led to the development of the controllable-pitch, constant-speed, and quick-feathering propellers. With fixed-pitch propellers, the propeller blades must be set at such an angle that the propeller will not exceed its rated r.p.m. with a wide-open throttle in level flight. This is necessary to

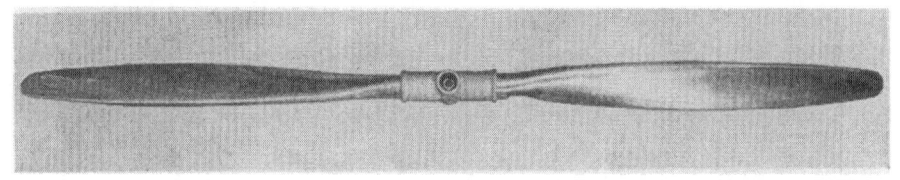

Fig. 205. An adjustable propeller having aluminum alloy blades built in 1928. These blades are screwed into the hub and locked by threaded taper pin. Note the rounded ends of the blades. (Courtesy Hamilton Standard Propellers)

Fig. 206. An adjustable propeller having magnesium alloy blades built in 1930. (Courtesy Hamilton Standard Propellers)

prevent overworking the engine and is a requirement of the Civil Aeronautics Administration and engine manufacturers. This means that the engine is held down to about 80 per cent of its normal r.p.m. during take-off. With a fixed-pitch propeller, an engine rated at 300 hp. would deliver only about 240 hp. on take-off. During climb, this same engine would develop about 85 or 90 per cent of its rated r.p.m.

This loss in engine performance was largely overcome by the development of the two-position, controllable-pitch propeller. With this propeller it is possible for the pilot, by means of cockpit controls, to change the blade setting from high pitch to low pitch or from low pitch to high pitch while the engine is in operation. This operation corresponds to that of shifting gears in an automobile. During take-off and climb, the propeller is placed in low pitch. Low pitch allows the engine to develop full power during take-off and climb. When the desired altitude has been reached, the pilot, by operating the proper control, places the propeller in high pitch for cruising performance. High pitch is usually set for the desired air speed and engine r.p.m. at cruising altitudes.

AIRCRAFT ENGINES

The constant-speed propeller and the full-feathering propeller are adaptations of the two-position, controllable-pitch propellers. The constant-speed propeller is automatically operated to maintain predetermined engine r.p.m. The angle of blade setting varies with the

Fig. 207. A cutaway view of the hub of a counterweight-type propeller. (Courtesy Hamilton Standard Propellers)

engine power. Most quick-feathering propellers are constant-speed propellers with a special attachment which allows the blade to be rotated into the feathered position.

It is important to remember that the two-position, controllable propeller gives the pilot a choice of two blade-angle settings — low pitch and high pitch. The constant-speed propeller gives the pilot a choice of engine r.p.m. and power. The constant-speed control automatically changes the pitch of the propeller to maintain any r.p.m. selected by the

PROPELLER FUNDAMENTALS

pilot. Due to the rarefied atmosphere at high altitudes, a fixed-pitch propeller tends to speed up as altitude increases. The speed of a fixed-pitch propeller properly adjusted for sea level will, with the same throttle setting, increase its speed approximately 40 per cent at 20,000 ft. altitude. A constant-speed propeller can be set for 1800 r.p.m. at sea level with a 75 per cent throttle opening and will continue to turn at the same r.p.m. at 20,000 ft. The angle of blade setting will, however, be increased. The greater the angle of blade setting, the more power is required to turn the propeller at any given r.p.m.

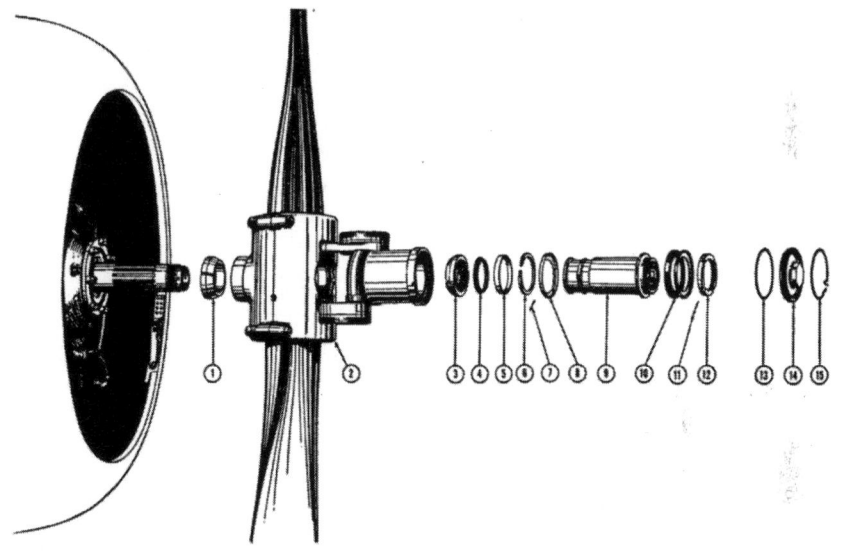

Fig. 208. An exploded view of a counterweight-type propeller extended off the propeller's shaft. (1) Rear cone; (2) hub and blades assembly; (3) front cone; (4) front cone packing washer; (5) front cone spacer; (6) hub snap ring; (7) cotter pin; (8) piston lock ring; (9) piston; (10) piston gaskets; (11) cotter pin; (12) piston gasket nut; (13) cylinder head gasket; (14) cylinder head; (15) cylinder head lock ring. (Courtesy Hamilton Standard Propellers)

The counterweight type of propeller may be used to operate either as a controllable or a constant-speed propeller. When used as a controllable propeller, the pilot selects either the low-blade angle or the high-blade angle by a two-way valve. This valve permits engine oil to flow into or drain from the propeller. If an engine-driven governor is used, the propeller will operate as a constant-speed type. The engine speed will remain constant at any r.p.m. setting within the operating range of the propeller. The governor supplies and controls the flow of oil to and from the propeller, thereby changing the blade angle to meet changes in flight and power conditions.

AIRCRAFT ENGINES

Blade-angle changes are accomplished by the use of two forces. One force is hydraulic and the other is centrifugal. Centrifugal force tends to rotate the blade to high pitch. The oil pressure is obtained from the engine. The flow of this oil is controlled by a manually operated valve.

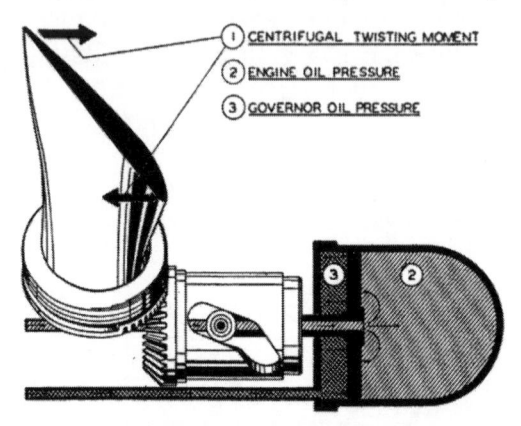

Fig. 209. A diagram to show the three fundamental forces acting upon a hydraulic propeller blade to bring about control. (Courtesy Hamilton Standard Propellers)

When the valve is set for low pitch, oil under pressure from the engine is forced into a cylinder in the propeller hub, thus forcing the blades into low pitch. The pressure of this oil holds the blades in the low-pitch position. When the valve is set for high pitch, the oil is allowed to drain from the cylinder in the propeller hub back into the engine. The action of centrifugal force on the blades, combined with the action of the counterweight and springs, then rotates the blades into the high-pitch position.

XVII CONTROLLABLE-PITCH PROPELLERS

Controllable-pitch propellers fall into three different classes, depending upon the type of operating mechanism. These classes are as follows:

1. The mechanically operated type by which the blades are rotated about their axis by mechanical means;
2. The hydraulically operated type in which the blades are rotated by means of liquid under pressure;
3. The electrically operated type in which the blades are rotated by means of an electric motor.

The mechanically operated type falls into two general classes: one in which the blade is rotated by means of flyweights and springs; the other in which the blade is operated mechanically from the rotation of the crankshaft by means of gears mounted on the engine itself. In the latter type, the propeller gear operates only during the time when the pitch is being changed. The propeller blades are rotated on this type by gears which are engaged by means of an electric solenoid. The gear train has an extremely high gear ratio, approximately 17,000 to 1, which develops a movement slow enough that the blades may be adjusted to any desired angle during flight.

This type of propeller has a hub of chromium vanadium steel forging in one piece. The blade assembly is fastened to the hub by blade nuts screwed into the ends of the blade barrels of the hub and locked to the hub.

The operating mechanism for rotating the blades is enclosed in a gear housing which is assembled in the rear cavity of the hub. Blades are chromium vanadium steel of hollow construction. A counterbalance is used on each blade partially to overcome the centrifugal moment. This twisting moment is present in all propeller blades.

The blades are mounted in ball bearings to increase the ease with which they may be rotated. A stationary worm gear is mounted on the engine. When the control worm gear rotates the blade stop gear, this gear rotates the tube on which the gear is mounted through a ratchet clutch. A stationary threaded shaft is mounted inside the tube carrying two adjusting nuts keyed to the tube. When the tube rotates, the nuts rotate with it and travel along the stationary shaft until one nut pushes the driving clutch out of engagement with the blade stop gear. The gear may continue to turn, but the clutch tube and worms mounted on the tube stop rotating. Reversal of the direction of rotation of the blade stop gear drives the tube through a second ratchet clutch having opposite threads on the other end of the gear. The nuts on the stationary shaft travel together toward the second clutch and push it out of engagement, stopping the rotation of the tube and worms.

Adjustments may be made which will limit the operation of the propeller gear train within definite blade-angle settings. The distance between the adjusting nuts determines how far the blades may be rotated in either direction.

A stationary control worm concentric with the propeller shaft is mounted on the engine front crankcase. This stationary control worm has a rear righthand threaded worm and a front lefthand threaded worm. When this control worm is in the neutral position, the worm threads are out of mesh with the propeller hub gears. By shifting this control worm, which slides forward and backward about the propeller shaft, one or the other of the worm threads may be engaged with the teeth of the blade stop gear in the propeller hub. The blade stop gear is rotated the space between two teeth by the stationary control worm with each revolution of the propeller. The rotation of the blade stop gear is transmitted through the blade stop worms to the three-blade stop worm shafts which mesh with the blade gears attached to the ends of the blades. The movement of the blade gear nut turns the blades in their sockets on the hub. The nut on the end of the blade shank has gear teeth cut around its circumference to engage the blade worm shaft in the final stage of the propeller gear train. The blade gear is screwed directly on the buttress threads on the blade shank and is locked securely by a lock blade doweled and screwed to the blade end.

Two-Position Propellers. The two-position propeller is a controllable propeller which may be adjusted to low pitch or to high pitch. An oil pressure line from the main engine oil supply is conducted to a three-way valve. This valve is connected through a collector ring into the interior of the front end of the crankshaft, then into the pitch-operating cylinder.

CONTROLLABLE-PITCH PROPELLERS

The pitch-changing mechanism consists of a piston and cylinder arrangement, the cylinder being moved by oil pressure.

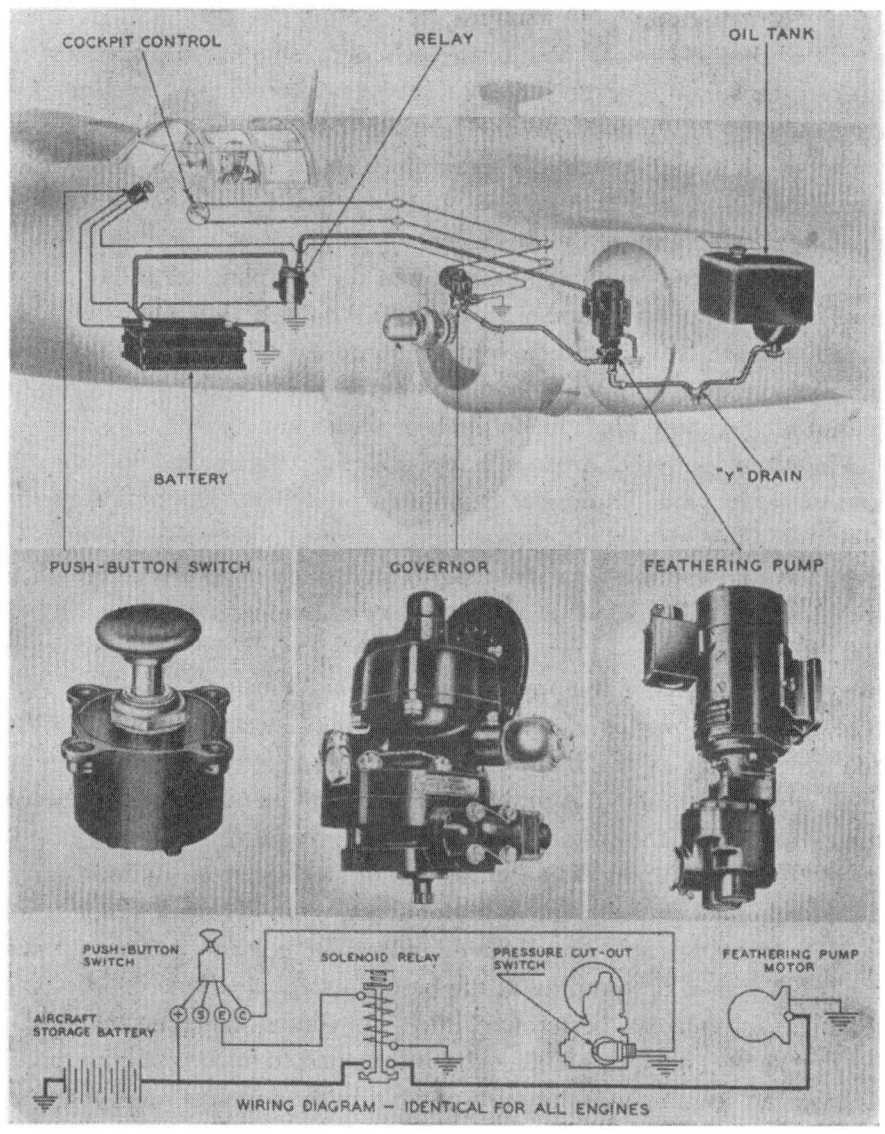

Fig. 210. A typical propeller-governor and accessory installation of a hydraulically operated propeller. (Courtesy Hamilton Standard Propellers)

The three-way valve is so arranged that in one position the oil will flow under pressure into the cylinder within the propeller hub, and in the other position the oil drains directly back into the crankcase.

When the valve is set to the pressure supply position, the oil under pressure flows into the propeller cylinder causing the cylinder to move forward on the piston. The piston remains stationary.

As the cylinder moves forward, the counterweight bearing shaft, which is attached to the base of the cylinder, moves in the cam slot of the counterweight bracket. This bracket is attached to the blade bushing by means of index pins. As the bearing shaft moves up the cam, it causes the bracket to turn. This movement is transmitted to the blade, rotating it into high-r.p.m. or low-pitch position. The low-pitch position produces high r.p.m. of the propeller.

When the three-way valve is placed in the OFF position, the oil pressure in the propeller cylinder is released. Attached to the arm of each blade bracket is a counterweight or flyweight. The action of these weights, due to centrifugal force, rotates the blade to the low-r.p.m. or high-pitch position. High blade pitch produces low r.p.m.

Most high-powered engines have built in attachments for the oil lines necessary for hydraulically operated propellers. The action of the hydraulic pressure and the counterweight control is such that the needed extra force is available for movement to high pitch or to low pitch when the revolutions are below or are above normal value.

The spider of these propellers is made from a heat-treated steel forging. The arms of the spider transmit the principal force from the blades to the propeller shaft. The barrel is of steel and takes up the centrifugal force of the blades. The piston is of steel and threaded to fit the engine crankshaft. A gasket is used to form an oiltight seal between the cylinder and the piston and serves as a guide to the cylinder as it moves back and forth. The cylinder is of steel-lined aluminum alloy. The steel liner protects the cylinder from wear. Threaded holes are located in the flanged base of the cylinder. There is also a hole opposite each spider arm. A stop plug at the base of each hole limits the distance the bearing shaft can be screwed into the cylinder. The counterweight bearing shaft is of steel, and one end is threaded to fit the cylinder holes. The other end is a cone which fits the seat in the cap race and an extension of the shaft which contacts adjusting nuts and limits the travel of the counterweight bearing shaft.

The counterweight bracket is of steel, and the outer end contains a cam slot in which the counterweight moves. The other end fits around the blade bushing. The portion of the bracket next to the blade bushing is scalloped with 40 semicircular holes. Four of these semicircles can be matched with 4 of the 36 semicircles which compose the base circum-

CONTROLLABLE-PITCH PROPELLERS

ference of the blade bushing. Index pins tapped into the 4 matched sets of semicircles make sure that any sideways turning movement of the bracket is changed into a rotating motion of the blade. A steel counterweight is fastened to the outer face of the bracket by screws. A slot in the counterweight corresponds to the cam in the bracket. A cylindrical extension of the bearing shaft moves up and down this slot. Along one side of this slot is a scale upon which degree graduations are stamped. These graduations correspond to protractor measurements of the blade at the 42-in. station. Near one end of the slot is a lead insert upon which is stamped the base setting of the blade.

The propeller is changed to low pitch by the oil pressure moving the cylinder. The blades are returned to high pitch by the action of centrifugal twisting forces acting on the blade, assisted by the action of the counterweights. As the blades move to the high-pitch position, the oil drains from the propeller cylinder back into the crankcase of the engine.

In testing the action of the pitch-changing mechanism, the propeller control should be operated during the running-up time. After starting the engine, the action of the pitch-changing mechanism may be seen by watching the travel of the cylinder on the front of the propeller or by watching the change in tachometer reading. The speed with which the blade angles are changed varies with the speed of the engine, the action being more rapid at high engine speeds. Both the twisting force, due to the centrifugal action on the blade, and the force exerted on the counterweights are greater at high engine speeds.

Before stopping an engine equipped with a two-position controllable propeller, the blades should be set in high-pitch position, allowing the oil to drain from the cylinder back into the crankcase of the engine. In cold weather, if the blades are left in low pitch, the oil might congeal in the cylinder, making the change from low pitch to high pitch impossible.

The engine is always started with the propeller in high pitch to avoid possible robbing of the master-rod bearing of oil. After the engine has run for a minute or two in the high-pitch position, the propeller may be shifted to the low-pitch position and warmed up with this blade setting.

The engine r.p.m. should be checked with the blade in the low-pitch position at full throttle. The take-off and climb are made with the propeller in the low-pitch position. Unless level flight is assumed at a lower altitude, the blade should be changed to the high-pitch setting at about 1000 ft. of elevation. During glides and landings, the propeller should be placed in low-pitch position.

AIRCRAFT ENGINES

Constant-Speed Propellers. The pitch-changing mechanism of the constant-speed propeller is similar to that of the two-position (or controllable) propeller, except that a spring return assembly is added. The spring return assembly is installed inside the propeller piston and is compressed when the propeller shifts to low pitch. This spring arrangement aids the

Fig. 211. A hydromatic propeller assembly showing the engine-driven governor. (Courtesy Hamilton Standard Propellers)

counterweights in returning the propeller to high pitch. The greatest compression is on the spring when the blades are in full low pitch. The compression on the spring decreases as the pitch shifts toward the high-pitch position. The spring operates through approximately two thirds of the distance from full low pitch to full high pitch. During the last third of the distance from low pitch to high pitch, the operating force is

CONTROLLABLE-PITCH PROPELLERS

furnished by the counterweights and by the twisting force on the blades due to centrifugal force.

The constant-speed propeller is usually equipped with a booster pump which takes engine oil from the lubricating system and increases its pressure to 180 to 200 lb. per sq. in. A built-in relief valve regulates the pressure and returns all excess oil to the pump. Just enough oil is used to shift the propeller pitch and take care of any leakage. Only a small amount of oil is needed when changing to low pitch.

The constant-speed control takes the place of the three-way valve used with the two-position propeller. The constant-speed control is a governor arrangement attached to the front crankcase of the engine. This governor is gear-driven directly from the engine. It will automatically maintain any engine r.p.m. selected by the pilot. This regulation of engine speed is accomplished by changing the angle of the propeller blade setting. This type of propeller permits the independent setting of engine speed and engine power. An engine may be set to develop any selected power at any selected speed, and the automatic propeller will maintain that power and speed through all flight conditions.

The pilot may, at will, adjust the controls for different engine r.p.m. There are limits, of course, to both the engine r.p.m. and to the angle of blade setting.

The angle at which the propeller blades may be set is regulated by mechanical stops on the counterweights and the limit and range through which the governor will operate.

When operating at the limit of the blade range, either top or bottom, the propeller will continue to operate similarly to a fixed-pitch propeller, and the r.p.m. will be regulated by the throttle setting and by the manifold pressure. The governor is an independent unit, the sole purpose of which is to regulate the flow of oil to the propeller. The governor contains a pair of flyweights balanced by a spring, the tension of which is determined by the r.p.m. setting as adjusted by the pilot. By the operation of the propeller controls, a tension is placed upon the flyweight speeder spring to predetermine the r.p.m. of the engine. When the engine is rotating at the desired r.p.m., the flyweights are in the neutral position and no oil flows to or from the propeller cylinder.

Engine oil enters the governor unit at the low-pressure side of the booster pump. From the high-pressure side of the pump, the oil is led past the relief valve and into the hollow, drive-gear shaft through ports located in the upper portion of the shaft. The position of the pilot valve within the drive shaft determines whether the oil will pass into the propeller-operating mechanism or be by-passed back to the inlet side of the pump. The action of the flyweights regulates the position of the pilot valve. When the flyweights are in neutral position, which means that the engine is rotating at the predetermined r.p.m., the oil from the booster pump is by-passed back to the inlet side of the pump. With the pilot valve in this position, oil can neither enter the propeller cylinder nor drain from it.

If, for any reason, the engine speed falls below the predetermined r.p.m., the flyweights acted upon by the speeder spring move inward against centrifugal force. As the weights move inward, the pilot valve is pressed downward by the spring. As the valve is lowered, high-pressure oil, instead of being by-passed to the inlet side of the pump, is allowed to flow into the propeller cylinder, moving the blades to a lower pitch position.

The lower pitch of the blades allows the r.p.m. of the engine to increase enough to cause the flyweights to return to the neutral position. As the flyweights move to the neutral position, they raise the pilot valve against the spring pressure, stopping the flow of oil to the propeller cylinder. If the r.p.m. of the engine increase above the predetermined speed, centrifugal force swings the flyweights outward. As the flyweights swing outward, they lift the pilot valve against the spring pressure. This lifting of the valve allows oil to drain from the propeller cylinder, moving the blades to a higher pitch position. This high pitch decreases the r.p.m. of the engine, allowing the flyweights to return to the neutral position.

In actual operation, the flyweights do not remain for extended periods of time in the ON speed position. Variations in throttle setting, bumpy air, and other conditions cause the flyweights to move irregularly from overspeed to neutral to underspeed and back again. The advantage of this type of automatic control of engine r.p.m. allows the pilot to use the desired engine power without increasing its r.p.m. Each movement of the throttle, however, causes a movement of the blades of the propeller to higher or lower pitch. Changes in altitude also bring about changes in propeller-blade pitch setting.

Hydromatic Quick-Feathering Propellers. Many of the principles of construction and operation of the hydromatic quick-feathering propeller are similar to those of the counterweight propeller of the constant-speed type. In addition, however, the hydromatic propeller incorporates the full-feathering feature which permits the rotation of the blades to a high

CONTROLLABLE-PITCH PROPELLERS

pitch. This pitch is great enough to stop the engine rotation and reduce the propeller drag to a minimum.

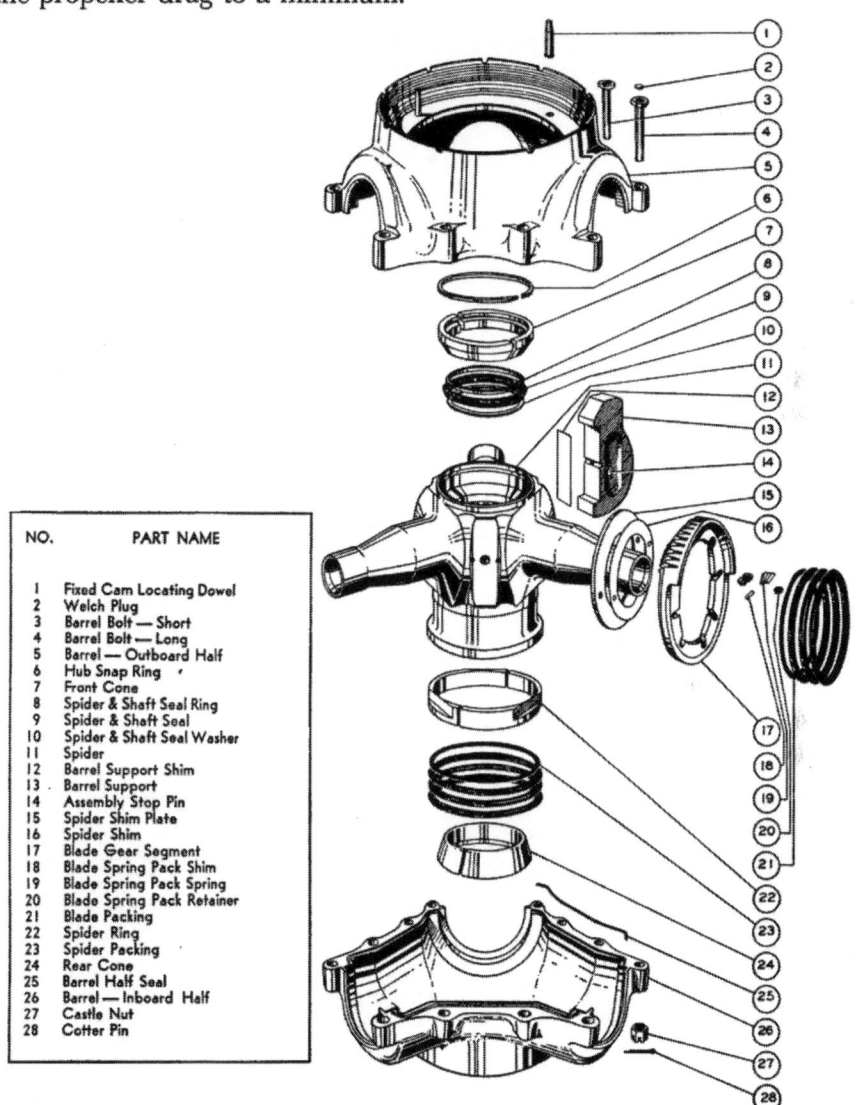

NO.	PART NAME
1	Fixed Cam Locating Dowel
2	Welch Plug
3	Barrel Bolt — Short
4	Barrel Bolt — Long
5	Barrel — Outboard Half
6	Hub Snap Ring
7	Front Cone
8	Spider & Shaft Seal Ring
9	Spider & Shaft Seal
10	Spider & Shaft Seal Washer
11	Spider
12	Barrel Support Shim
13	Barrel Support
14	Assembly Stop Pin
15	Spider Shim Plate
16	Spider Shim
17	Blade Gear Segment
18	Blade Spring Pack Shim
19	Blade Spring Pack Spring
20	Blade Spring Pack Retainer
21	Blade Packing
22	Spider Ring
23	Spider Packing
24	Rear Cone
25	Barrel Half Seal
26	Barrel — Inboard Half
27	Castle Nut
28	Cotter Pin

Fig. 212. The barrel assembly of a hydromatic, controllable-pitch propeller. (Courtesy Hamilton Standard Propellers)

Feathering is the term applied when the blades of a propeller are turned to such a high pitch that the chord of the blade lies in or near the direction of flight. When in this position, the propeller not only offers the least resistance to the air stream, but also acts as a powerful brake in stopping the rotation of the engine. To return from the feathering position to the normal operating position, the blades are turned from the feathered position to a lower pitch. This operation is commonly called "unfeathering." When the blades are turned to this low-pitch position, the pressure of the air flow causes the propeller to "windmill" and crank the engine.

The operation of this propeller makes use of the same fundamental forces acting to control the blade angle as are used in the other constant-speed propellers. These forces are: (1) the centrifugal twisting moment of the blades toward high pitch, which is used when increasing the blade angle; (2) oil under pressure which may be used to change the blade angle from high pitch toward low pitch; and (3) engine oil under booster pressure from the governor which moves the blades toward high pitch. The balance between these forces is maintained by the propeller governor. The propeller governor measures the amount of oil to and from the propeller to maintain the constant-speed operation and boosts the engine oil pressure. When it is desired to feather the blades, an auxiliary pressure supply system is necessary. This consists essentially of an independent oil supply which is under the manual control of the pilot to provide pressure up to 400 lb. per sq. in. for feathering and 600 lb. per sq. in. for unfeathering. When the propeller is being unfeathered and the blades are rotated into the flight operating range, the normal constant speed automatically becomes available.

The hydromatic propeller is furnished with an operating pitch range of 35° for constant-speed control. The feathering feature makes it possible to stop the rotation of an engine in a few seconds.

The operating mechanism of this propeller is lubricated by engine oil pressure. Constant-speed control is provided for all models. The propeller consists of three major assemblies: the hub and blade assembly, the dome assembly, and the distributor-valve assembly.

The hub and blade assembly consists of three major parts, which are the spider, the barrel, and the blades. The spider is the foundation of the entire propeller. Its central bore is splined to fit the engine shaft. It is through the spider that all power is transmitted from the engine to the propeller. The spider is equipped at either end with an accurately ground cone seat. At the outer end, provision is made for the propeller retainer nut and the front cone by means of which the spider is attached rigidly to the engine shaft. The spider arms support the blades and take the greater part of the thrust and torque loads from the blades. The barrel is supported on a spider by means of phenolic blocks located

CONTROLLABLE-PITCH PROPELLERS

"E" SHANK BLADE ASSEMBLY

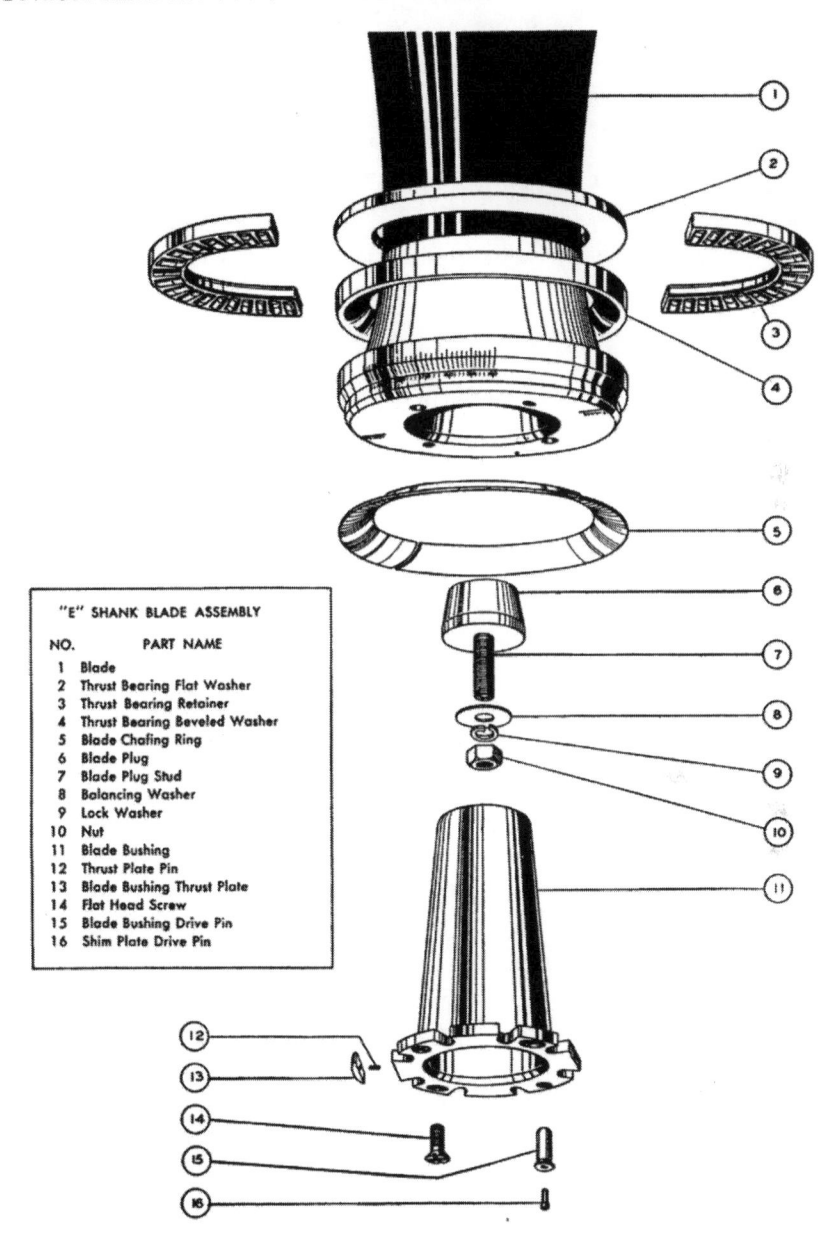

NO.	PART NAME
1	Blade
2	Thrust Bearing Flat Washer
3	Thrust Bearing Retainer
4	Thrust Bearing Beveled Washer
5	Blade Chafing Ring
6	Blade Plug
7	Blade Plug Stud
8	Balancing Washer
9	Lock Washer
10	Nut
11	Blade Bushing
12	Thrust Plate Pin
13	Blade Bushing Thrust Plate
14	Flat Head Screw
15	Blade Bushing Drive Pin
16	Shim Plate Drive Pin

Fig. 213. The blade shank assembly of a hydromatic controllable-pitch propeller. (Courtesy Hamilton Standard Propellers)

between the spider arms. Shoulders machined at the barrel take the centrifugal loads from the blades. These loads are transmitted to the barrel through heavy roller bearings. Oil seals are used between the blades and the barrel and between the spider and the barrel. The barrel

AIRCRAFT ENGINES

forms an oiltight casing which houses the entire hub mechanism and provides support for the attachment of the dome unit.

The dome assembly comprises the pitch-changing mechanism by means of which oil under pressure acting on a double-acting piston controls the blade-twisting moments. The dome assembly consists of four major parts: two cylindrical coaxial cams, a double-walled piston, and a dome cylinder which serves also as a housing for the entire unit. The piston and the cylinder are machined from aluminum alloy forgings.

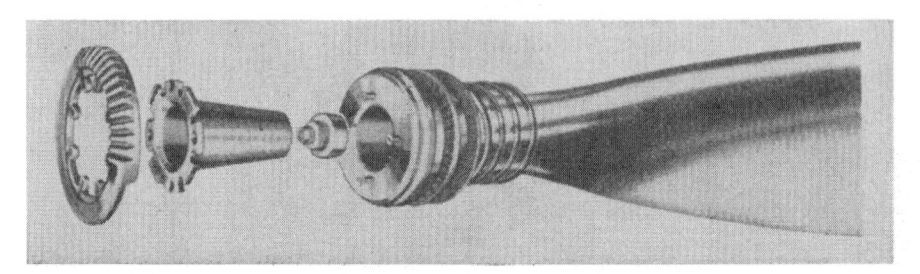

Fig. 214. A blade shank with the thrust bearing and oil seal of a controllable-pitch propeller. (Courtesy Hamilton Standard Propellers)

When the dome unit is installed in the hub assembly, the outer, or stationary, cam becomes fixed in the barrel and provides support for the remaining parts of the dome unit. The inner, or rotating, cam with which the main drive is integral, is supported within the stationary cam by means of ball bearings. These ball bearings take the gear reaction and piston oil forces. The piston motion is transmitted to the rotating cam by means of four cam rollers carried on shafts supported by the inner and the outer walls of the piston.

The distributor-valve assembly serves two purposes. (1) During the constant-speed operation of the propeller, the distributor-valve assembly provides a passage through which engine oil, boosted in pressure and measured by the governor, is led to or from the inboard side of the propeller piston. It also provides a passage through which oil under engine pressure is led to or from the outboard end of the cylinder. During feathering, the same two passages provide means for delivering high-pressure oil (from the auxiliary pressure system) to the inboard side of the piston and a means of conducting oil from the outboard end of the cylinder to the engine lubricating system. The difference in pressure which exists across the piston moves it toward the outboard end of the cylinder and feathers the blades. During constant-speed operation, or feathering, there is no movement of the distributor valve, and

CONTROLLABLE-PITCH PROPELLERS

the assembly merely provides a passage through which oil may flow to and from the cylinder. (2) In unfeathering, the function of the distributor valve is to reverse the above-mentioned passages. The high-pressure oil from the auxiliary system is then led to the outboard side of the piston,

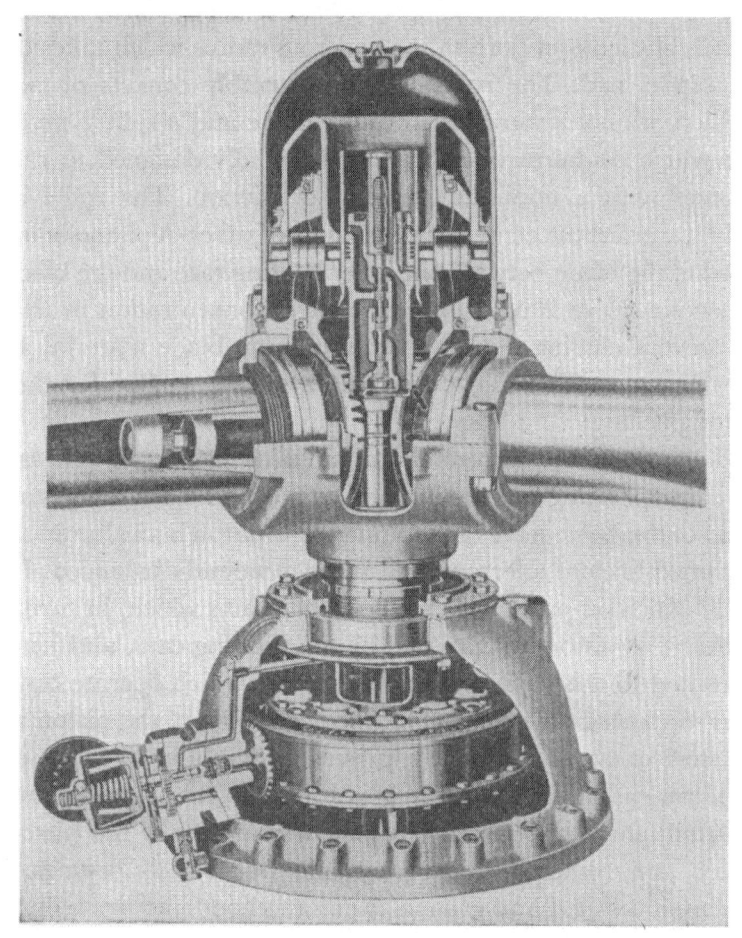

Fig. 215. A cutaway view showing the construction of a hydromatic, quick-feathering propeller and the engine-driven governor. (Courtesy Hamilton Standard Propellers)

and the inboard end is connected to the engine lubricating system. This reverses the pressure difference and moves the piston toward the inboard end of the cylinder, thus unfeathering the blades.

The blades used with hydromatic propellers are similar in basic design to the blade used on the controllable type of propeller. They differ, however, in slight detail at the inner end and are not interchangeable between the two propeller types. The blades are of semihollow construction and are manufactured from high-strength, aluminum alloy forgings. The blades have a thin solid tip and a hollow upset shank for attaching the blade to the hub. The upset shank has an internal aluminum bronze bushing which supports the blade on the spider arm and transmits the blade thrust and torque load to the spider. Machined on the outside of the upset portion is a generous radius which takes the centrifugal blade load. The roller-bearing assembly consists of two steel races which are not removable from the blade and a split type bearing retainer which contains roller bearings especially designed to carry the large forces with a minimum amount of friction. The roller-bearing assembly transmits the centrifugal load to the barrel. A phenolic material is molded to the blade between the inner bearing race and the blade butt. This phenolic sleeve is used to reduce stress concentration in the blade and to prevent chafing between the aluminum blade material and the steel bearing race. This sleeve furnishes a sealing surface for the blade and blade packing.

The dome is machined from aluminum alloy forging. It acts as a case for the cam-operating mechanism and as a cylinder for the piston. The outer end of the dome serves as a spinner. The piston is an aluminum forging machined to close tolerance and is independently balanced. It is the means by which oil-pressure forces move the cams which, in turn, rotate the blades. The piston is supported by the rotating cam, and its motion is transmitted to this cam by four steel rollers which operate on bronze bushings supported on steel shafts. The inner bore of the piston is fitted with a steel sleeve permanently pressed into place. A large double-acting piston gasket is held firmly in place by means of a ring nut which forms an oiltight seal between the pressure surfaces of the piston. The stationary and rotating cams are made from steel forgings. These forgings are machined and ground to ensure smooth, accurate operation. Attached to the inboard end of the rotating cam is a bevel gear which meshes with the gear segments attached to each blade butt. This bevel gear in the rotating cam is made of a single piece of material. The slots of the rotating cam slope in the opposite direction from the slots of the stationary cam. The four steel rollers, located on shafts at the inboard end of the piston, operate in the cam slots of the rotating cams and of the stationary cams. At the inboard end of the stationary cam, the stop locating plate is secured by three unequally spaced dowels and six screws. The inner diameter of the plate has a series of fine teeth which mesh with similar teeth on the outside diameter of the angle-range, limit-

CONTROLLABLE-PITCH PROPELLERS

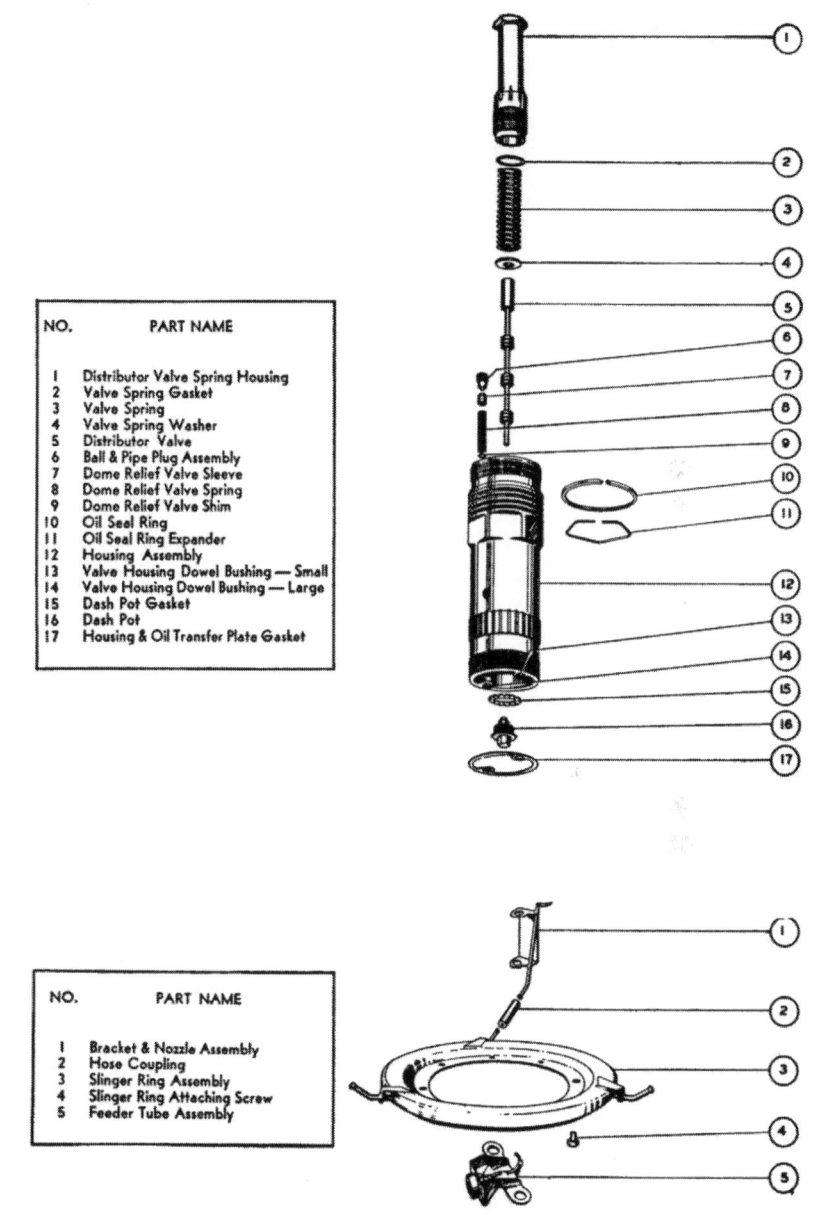

NO.	PART NAME
1	Distributor Valve Spring Housing
2	Valve Spring Gasket
3	Valve Spring
4	Valve Spring Washer
5	Distributor Valve
6	Ball & Pipe Plug Assembly
7	Dome Relief Valve Sleeve
8	Dome Relief Valve Spring
9	Dome Relief Valve Shim
10	Oil Seal Ring
11	Oil Seal Ring Expander
12	Housing Assembly
13	Valve Housing Dowel Bushing — Small
14	Valve Housing Dowel Bushing — Large
15	Dash Pot Gasket
16	Dash Pot
17	Housing & Oil Transfer Plate Gasket

NO.	PART NAME
1	Bracket & Nozzle Assembly
2	Hose Coupling
3	Slinger Ring Assembly
4	Slinger Ring Attaching Screw
5	Feeder Tube Assembly

Fig. 216. A hydromatic distributor valve assembly and hydraulic de-icing device assembly. (Courtesy Hamilton Standard Propellers)

stop rings. Three dowel holes equally spaced in the inboard flange of the stationary cam and in the stop locating plate are used to align the dome assembly in the barrel. The stop rings are machined from steel forgings. The dome assembly includes two of these rings which are identical except for markings which indicate the blade angle-range setting. Due to a 5:4

AIRCRAFT ENGINES

gear reduction between the rotating cam and the blades, the stop plate and the stop rings have 288 teeth each arranged so that indexing one tooth of the stop changes the blade angle 1°. Two lugs, located 180°

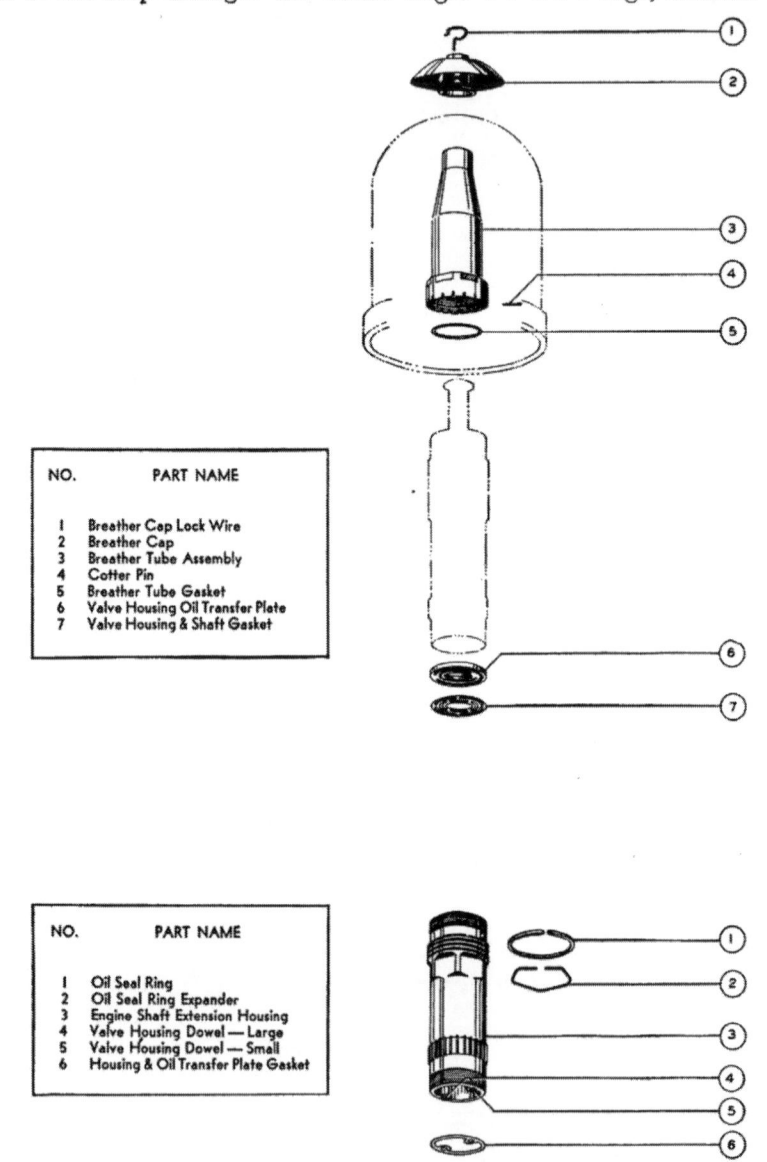

NO.	PART NAME
1	Breather Cap Lock Wire
2	Breather Cap
3	Breather Tube Assembly
4	Cotter Pin
5	Breather Tube Gasket
6	Valve Housing Oil Transfer Plate
7	Valve Housing & Shaft Gasket

NO.	PART NAME
1	Oil Seal Ring
2	Oil Seal Ring Expander
3	Engine Shaft Extension Housing
4	Valve Housing Dowel — Large
5	Valve Housing Dowel — Small
6	Housing & Oil Transfer Plate Gasket

Fig. 217. Hydromatic, propeller-shaft, breathing parts and engine-shaft extension assembly. (Courtesy Hamilton Standard Propellers)

apart on each stop ring, engage with the stops on the rotating cam. The rotating cam movement is limited to the desired range by adjusting the position of the stop rings in the stop plate.

CONTROLLABLE-PITCH PROPELLERS

The barrel is made from steel forgings which have been carefully heat-treated. The edges of the barrel are machined in pairs to ensure perfect machining. The shoulders of the barrel blade bores carry the centrifugal blade loads, and provision is made for chevron-type packing for oil tightness.

The barrel is carefully balanced as a unit. The spider is machined from a steel forging, heat-treated. The splines and cone seats are finished to standards which ensure satisfactory installation on the proper engine shaft. Phenolic blocks are provided for alignment for support of the barrel on the spider. Solid brass shims are used for adjustment of the spider and barrel concentrically during assembly.

Shim plates and solid brass shims are installed between the face of the blade bushings and the spider arm shoulder. The shim plates are installed on shim-plate drive pins mounted on the blade bushing. Solid brass shims are placed between the shim plate and the face of the blade bushing for adjustment of the blade torque during assembly.

Oil packings are provided to ensure oil tightness between the barrel and the blades and between the barrel and the spider. The sealing assembly consists of three V-shaped rings constructed of synthetic rubber material which is resistant to oil. These rings are supported at each end by split header and follower rings. The oil packings are held in place by packing retainers. The barrel halves are provided with grooves in the parting surfaces which carry synthetic-rubber oil seals. The distributor valve housing is an aluminum alloy casting provided with cored passages for the oil operating pressures. A steel sleeve shrunk into the center bore of the housing contains ports which align with oil passages in the housings. In order to unfeather, the normal flow of oil through these ports is redirected by a change in the position of the distributor valve which operates within the sleeve. Change in position of the distributor valve is accomplished by applying to it a higher pressure than that required for normal constant-speed operation or feathering. The valve is equipped with oil-seal rings. Cored breather passages are incorporated in some housings for those engines which still breathe through the propeller shaft.

The blades are manufactured from aluminum alloy forgings, heat-treated for high tensile strength. The butt end is provided with a tapered hole for the bushing and a shoulder to carry the thrust-bearing loads. The thrust bearing consists of two thrust races, which are not removable from the blade, and a slit bearing retainer. The centrifugal blade loads

AIRCRAFT ENGINES

are transmitted from the shoulders on the blade butt to the inner or beveled thrust washer through the phenolic sleeve molded on the blade.

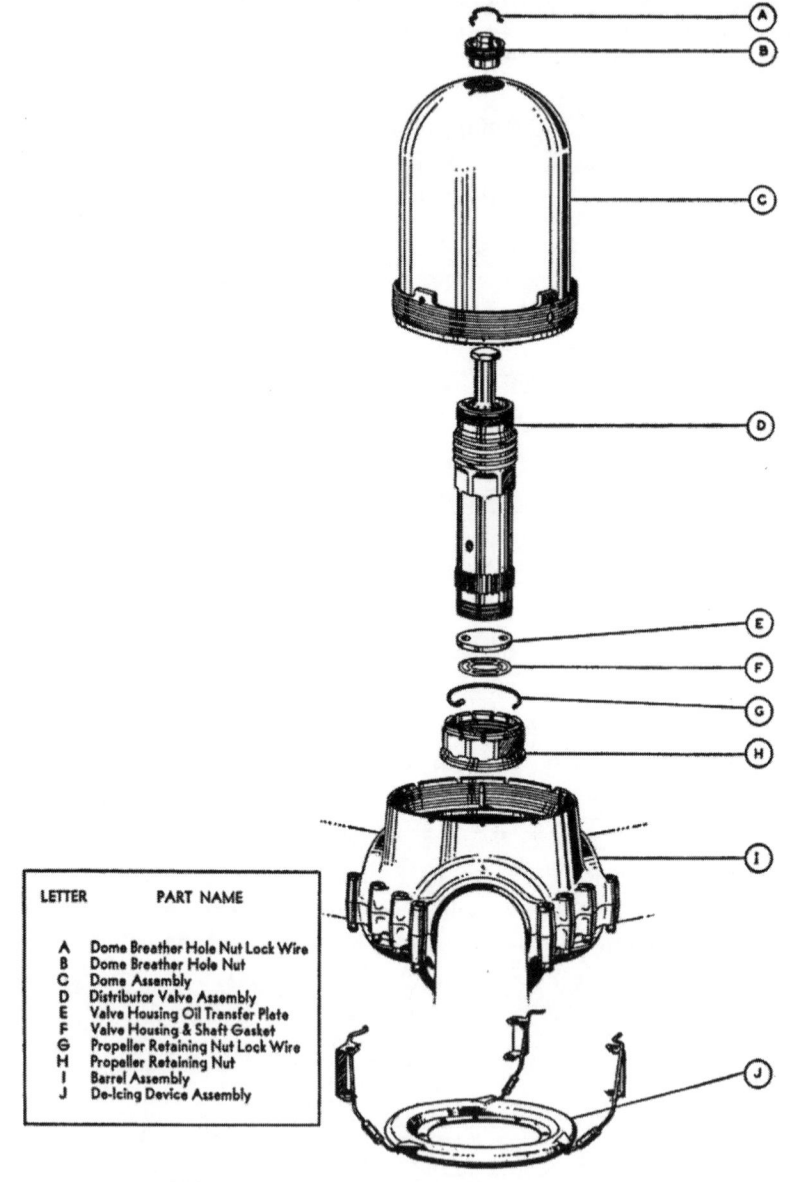

LETTER	PART NAME
A	Dome Breather Hole Nut Lock Wire
B	Dome Breather Hole Nut
C	Dome Assembly
D	Distributor Valve Assembly
E	Valve Housing Oil Transfer Plate
F	Valve Housing & Shaft Gasket
G	Propeller Retaining Nut Lock Wire
H	Propeller Retaining Nut
I	Barrel Assembly
J	De-Icing Device Assembly

Fig. 218. A complete hydromatic propeller assembly. (Courtesy Hamilton Standard Propellers)

The blade bushings are made of aluminum bronze alloy. The bushings are shrunk into the tapered holes in the blade ends and secured by two drive pins and two lock screws. Each blade bushing is provided with

CONTROLLABLE-PITCH PROPELLERS

eight slots which carry the spring packs for attachment of the gear segment. Two of the spring pack slots are offset to give an initial preload to

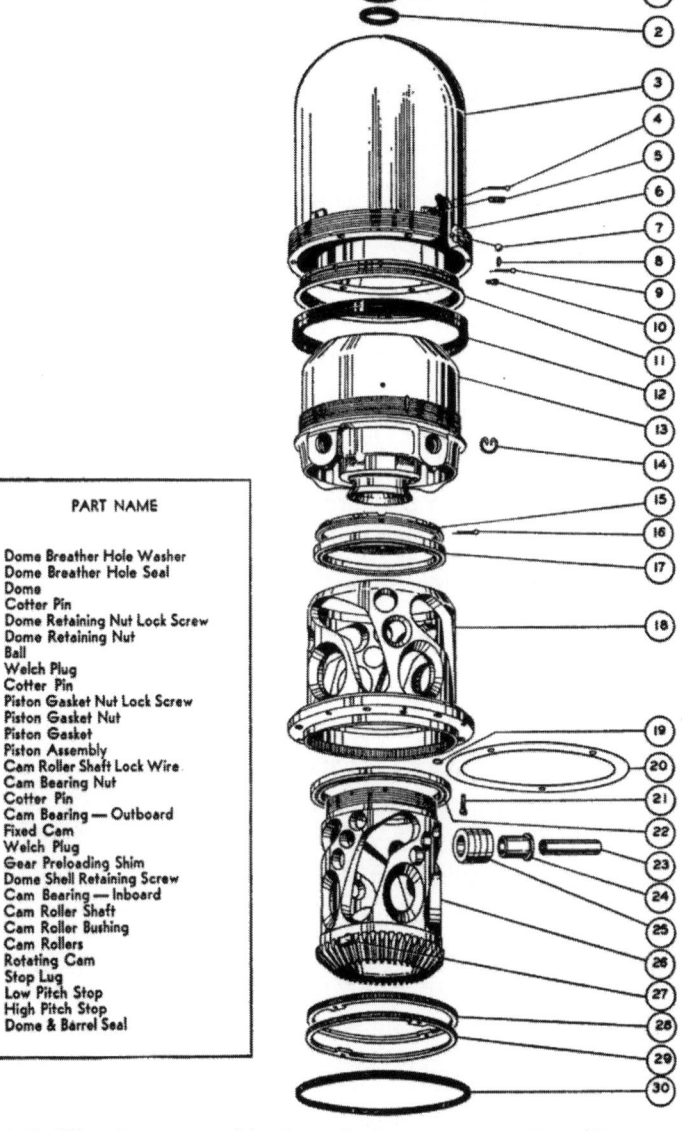

NO.	PART NAME
1	Dome Breather Hole Washer
2	Dome Breather Hole Seal
3	Dome
4	Cotter Pin
5	Dome Retaining Nut Lock Screw
6	Dome Retaining Nut
7	Ball
8	Welch Plug
9	Cotter Pin
10	Piston Gasket Nut Lock Screw
11	Piston Gasket Nut
12	Piston Gasket
13	Piston Assembly
14	Cam Roller Shaft Lock Wire
15	Cam Bearing Nut
16	Cotter Pin
17	Cam Bearing — Outboard
18	Fixed Cam
19	Welch Plug
20	Gear Preloading Shim
21	Dome Shell Retaining Screw
22	Cam Bearing — Inboard
23	Cam Roller Shaft
24	Cam Roller Bushing
25	Cam Rollers
26	Rotating Cam
27	Stop Lug
28	Low Pitch Stop
29	High Pitch Stop
30	Dome & Barrel Seal

Fig. 219. The dome assembly for a hydromatic propeller. (Courtesy Hamilton Standard Propellers)

the gear teeth. There is but one angular position for assembly of the gear segment on the blade bushing for righthand tractor propellers.

Each spring pack assembly consists of 2 spring retainers and approximately 34 spring leaves. The purpose of the spring packs is to provide a flexible coupling between the gear segment and the blade bushing. The spring packs also permit preloading of the gears to prevent back lash.

The blade gear segments are made from steel forgings. The gear segment is attached to the blade bushing by 8 spring pack slots equally spaced. Each segment contains 15 teeth which engage with the teeth of the rotating cam. In this manner, the actuating forces are utilized for the movement of the blades within the operating range. Each blade is fitted with an aluminum plug which is wedged into the hollow shank at a point just beyond the inner end of the blade bushing. This plug prevents oil from escaping into the extreme end of the taper. The blade plug is provided with a stud on which washers may be installed for balancing purposes.

The de-icer slinger ring is attached to the rear barrel by means of 8 screws. Bracket and nozzle assemblies are provided for distribution of de-icer fluid from the slinger ring to each blade. A shield is included for attachment to the thrust-bearing cover plate of the engine which carries an adjustable discharge tube. The discharge tube is connected to the de-icer fluid supply by suitable piping.

Electric Adjustable Propellers. This type of propeller has all of the advantages of the hydromatic propeller. The blades of this propeller are adjustable over a wide range of angles to meet all operating conditions.

The blades on this propeller may be adjusted to feather the propeller or may be turned through low pitch to a negative angle, which assists in handling large flying boats on the water and greatly reduces the distance required for landing by acting as an aerodynamic brake. This reverse pitch operation also makes for safer landings on icy or wet runways.

The blade angles of this propeller are regulated by a small reversible electric motor. Energy for the operation of this motor is derived from the electrical power supply of the aircraft. The current passes from brushes mounted in a housing attached to the engine nose to slip rings on the rear of the propeller hub. These slip rings have connector leads passing through the propeller hub to the motor.

The electric motor controls the angle of the blade setting through a two-stage planetary type of speed reducer which drives a master bevel gear. The master bevel gear meshes with a bevel gear attached to the shank of each blade.

Depending upon the direction of the rotation of the motor, the angle of the blades of the propeller is increased or decreased. The planetary type speed reducer has the capacity to convert efficiently the high speed

CONTROLLABLE-PITCH PROPELLERS

of the electric motor into the relatively slow but powerful force required to adjust the blades.

Attached to the front end of the electric motor is a brake which stops the rotation of the motor when the pitch-changing current is interrupted. This brake locks the blades at any desired position by locking the motor.

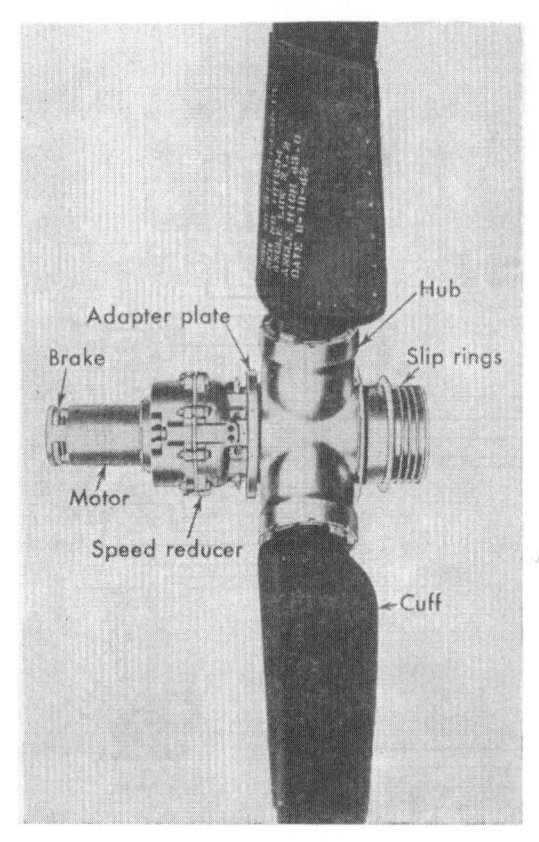

Fig. 220. The propeller assembly of an electric controllable-pitch propeller. (Courtesy Curtiss-Wright Corporation)

To keep the blade angle within an effective flying range, switches are provided in the hub end of the power unit. One limit switch determines the maximum blade angle within the range, while a second switch determines the minimum angle. A third switch stops the pitch change at the feathered position.

In the reverse-thrust propeller, an additional switch limits the reverse angle to that required for such an operation. These limit switches are cam operated and are connected to the proper leads of the electric motor.

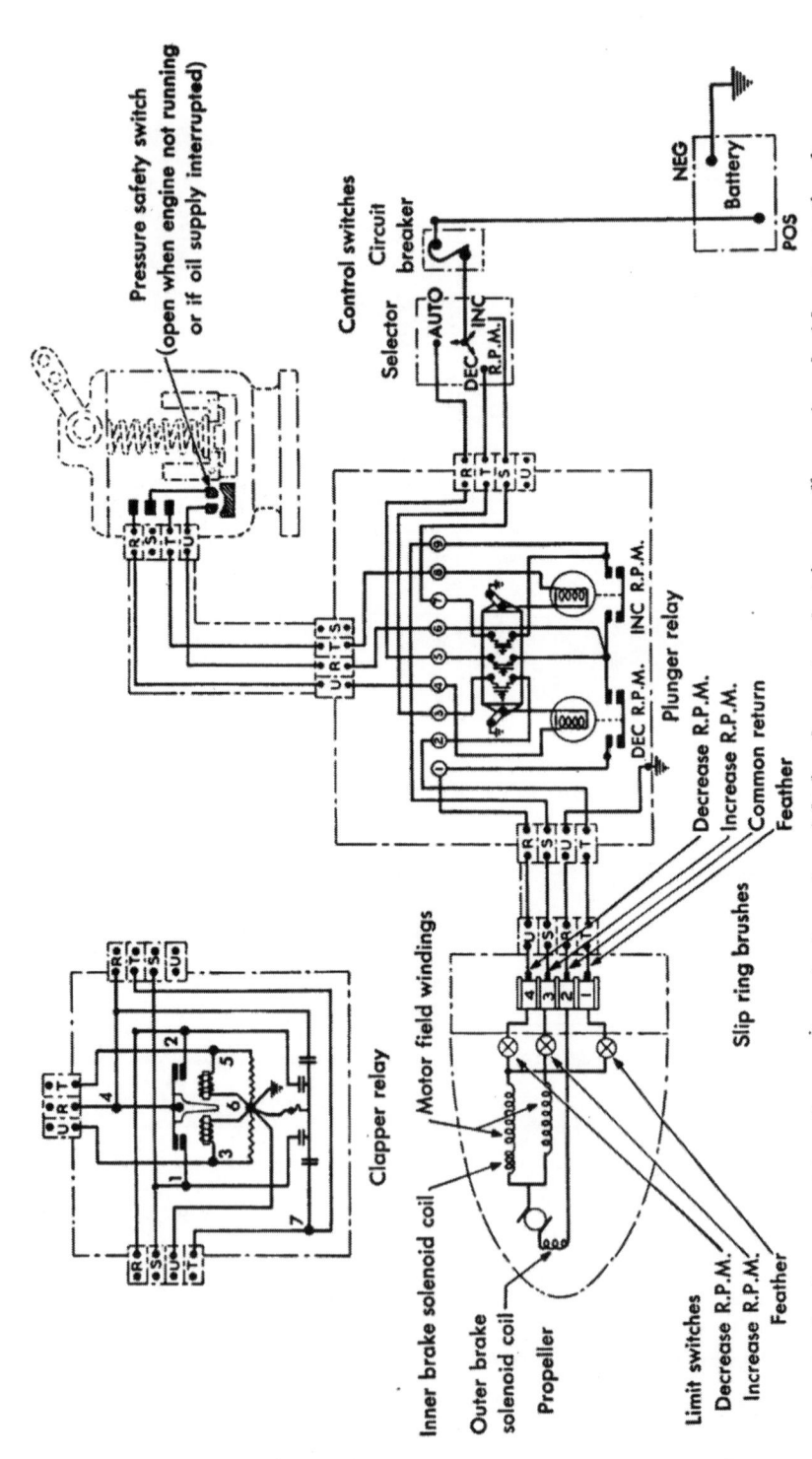

Fig. 221. A schematic wiring diagram for a single-engine, right-hand, rotation electric propeller equipped with a proportional governor. (Courtesy Curtiss-Wright Corporation)

CONTROLLABLE-PITCH PROPELLERS

The control system operates the proper switching of the electric current to the propeller motor. A constant-speed governor, a relay, a cockpit governor control, suitable switches, and wiring make up the governor type of control system.

An automatic synchronizer control system is now used on many types of multi-engine airplanes. This system provides automatically synchronized, constant-speed governing and the desirable feature of simultaneous selection of identical r.p.m. of all engines by the manipulation of the single control knob or lever.

The control selector switch is a single-pole, triple-throw switch having four positions, namely, (1) off, (2) automatic, (3) increase r.p.m., and (4) decrease r.p.m. A new design of selector switch incorporates feathering by a fifth position.

The OFF position on the control selector switch is the neutral position. When in this position, the electric circuits to the propeller are open. In this position, the setting of the propeller blades remains fixed. By holding this switch lever normally in either the increase-r.-p.-m. position or in the decrease-r.-p.-m. position, the desired change in blade-angle setting is brought about. Upon being released, the switch lever will return to the OFF position causing the blades to remain fixed at the setting brought about by the operation of the switch.

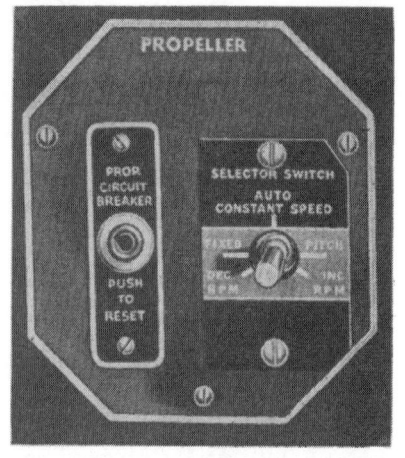

Fig. 222. A single-engine, electric-propeller switch panel. (Courtesy Curtiss-Wright Corporation)

If the selector switch is placed in the automatic position, the automatic control units take over and the propeller acts as a constant-speed propeller. The synchronizer or governor, through the pitch changing mechanism, automatically maintains the engine r.p.m. at the preselected speed.

A feather switch may be used to bring about feathering of the propeller. In one position, this switch completes the normal propeller circuits. In the other position, it completes only the feathering circuit.

An additional switch is used when the reverse-pitch propeller is installed. This switch is similar in action to the feather switch and is used to break the normal circuit and complete the reverse-pitch circuit at

AIRCRAFT ENGINES

the same time. A push button is pressed to actuate the reverse circuits. Another type of reversing installation incorporates a switch on the engine-throttle quadrant levers. This switch actuates the reversing and return-from-reverse circuits when the throttle quadrant lever is operated in a specified manner.

In order to provide a more rapid pitch change during feathering and reverse operations, most installations use a voltage booster. This booster increases the voltage of the current delivered to the propeller mecha-

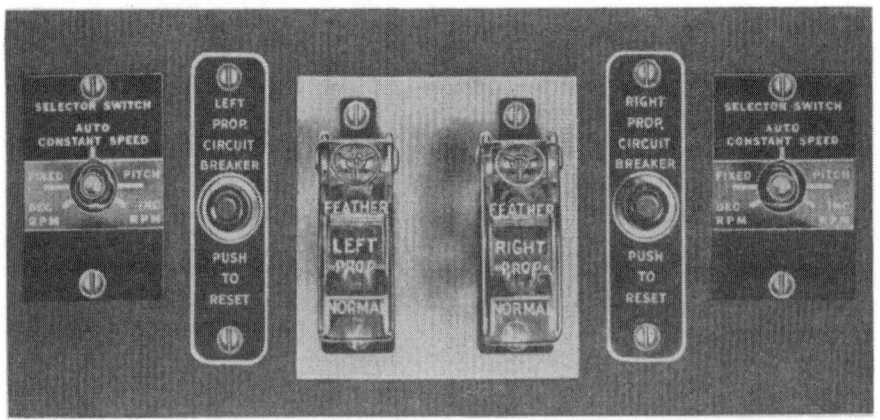

Fig. 223. A twin-engine, electric-propeller switch panel. (Courtesy Curtiss-Wright Corporation)

nism. The increased voltage speeds up the blade-angle change. This faster rate of pitch change prevents engine overspeeding when the blades are turned through zero angle for reversing and return-from-reverse operations.

A solenoid switch in the booster is connected in series with the feathering or reverse circuits. This switch introduces the current required to operate the booster and cuts out when the feathered or reverse setting position is reached.

If for any reason there is an electrical failure while in operation, the blades will remain set at the position at which they were operating when the failure occurred.

When the current fails, the break on the motor is automatically set. The circuit breaker must be in the closed position whenever the propeller is operated. The circuit breaker has the function of protecting the propeller circuit against overloading. When an excessive load opens the circuit breaker, it may be reset by positioning it again.

The propeller may be made to operate as a fixed-pitch propeller by

CONTROLLABLE-PITCH PROPELLERS

manually operating the selector switch. Whenever the selector switch is in the OFF position, the propeller operates as a fixed-pitch propeller.

When the feathering switch is placed in the feather position, the blades will assume the full-feathered position regardless of the setting of any other switch. The blades will assume this position without further attention on the part of the pilot.

Automatic constant-speed control should be used during all take-offs, landings, and maneuvers. It may be used for all flying, if desired.

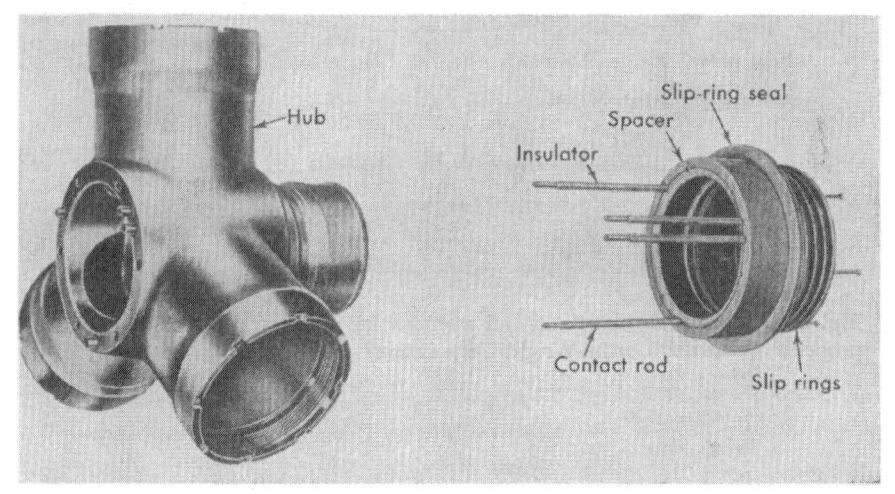

Fig. 224. The forged steel hub and insulated contact rods for an electric controllable-pitch propeller. (Courtesy Curtiss-Wright Corporation)

Selective fixed-pitch control may be used during steady cruising flight.

Feathering may be used whenever desired, as there is no limit to the number of times this operation may be performed.

The hub of the electric propeller is machined from a solid forging of alloy steel. The blade assemblies are inserted into the hub barrel and held in place by retaining nuts. On the rear of the hub are mounted the bronze slip rings, insulated from the hub and from each other. Insulated brass connector rods carry the electric circuits through the passages from the slip rings to contacts at the face of the hub. The hub and blade assembly parts are lubricated by filling the hub cavity with a semifluid lubricant. A pressure grease fitting is located on the rear of the speed-reducer housing.

The blades are of hollow-steel construction and are made by welding together two formed sheets of alloy steel. The shank and cambered face are formed in one sheet and the thrust face in another. The root of the

AIRCRAFT ENGINES

blade is flanged to transmit the centrifugal blade loads directly to the retaining bearings. A splined blade gear is located in the internally splined blade root by means of a splined adapter. The gear is held in place by a single bolt. A retaining nut and stack of antifriction bearings secure the blade in the hub socket.

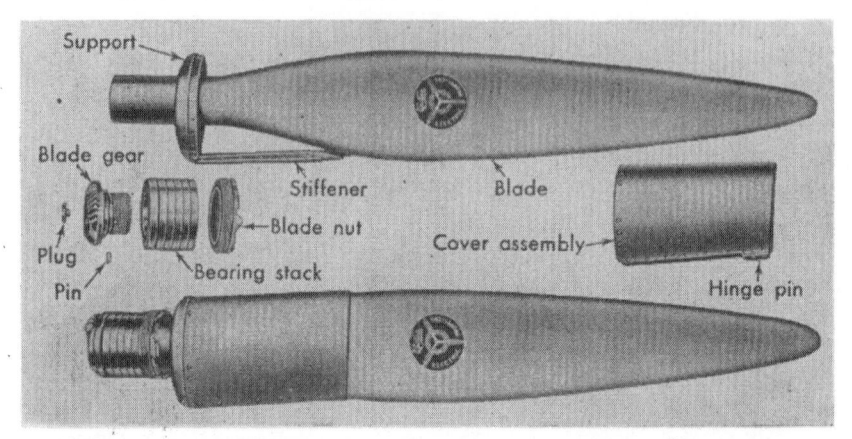

Fig. 225. A propeller blade and cuff assembly for an electric controllable-pitch propeller. (Courtesy Curtiss-Wright Corporation)

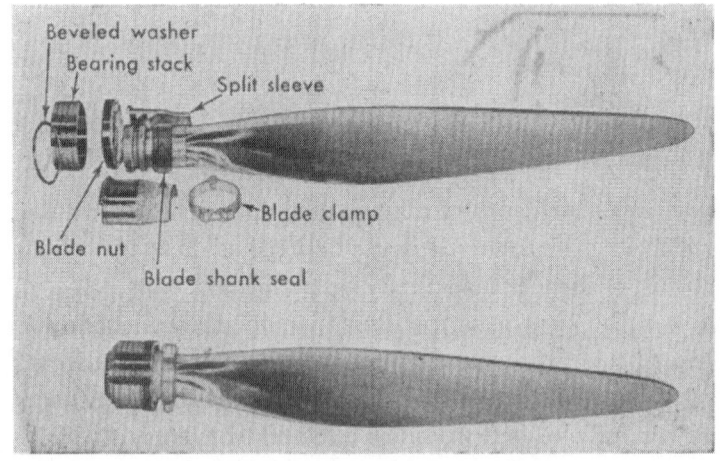

Fig. 226. Aluminum-alloy blade assembly for an electric controllable-pitch propeller. (Courtesy Curtiss-Wright Corporation)

The bearings are designed to permit free rotation of the blade in the hub under high operating loads. The cuff consists of a cast magnesium former and a trailing-edge support over which is fitted a properly shaped aluminum alloy sheet. The outer end of the cuff is equipped with a molded synthetic-rubber chafing strip which makes a close fit between the blade and the cuff.

CONTROLLABLE-PITCH PROPELLERS

When aluminum alloy blades are used, the blade is machined from a forging of high-grade aluminum alloy. The root is shaped to receive a split steel sleeve which is clamped to the blade. Bevel gear teeth, by which the blade is rotated, are machined on the end of one of the halves of the split sleeve. A blade nut and bearing stack arrangement is used to hold the blade assembly to the hub. This arrangement is similar to that used for the hollow steel blade.

The power unit is the mechanism which controls the rotation of the propeller blades to change the pitch. It is made up of several subunits

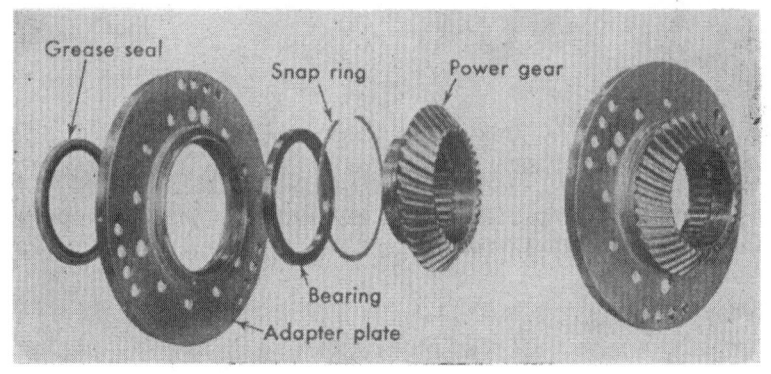

Fig. 227. The power gear assembly for an electric controllable-pitch propeller. (Courtesy Curtiss-Wright Corporation)

which are enclosed within the power unit cover. The power gear assembly includes a bevel gear which meshes with the blade gears. It also contains an angular contact type of ball bearing which absorbs the power gear thrust. Included in this assembly is a steel adapter plate in which the power gear and bearing are mounted. The adapter plate serves as a mounting support for the power unit cover. The speed reducer is of two stages of planetary type reduction gearing.

The speed reducer is contained in a housing made from aluminum alloy. All moving parts are fitted with ball bearings to reduce friction to a minimum. All gear teeth are surface-hardened.

Oiltight seals are used to eliminate the necessity of frequent lubrication. The unit is partially filled with an oil having an extremely low pour point. This oil provides a continuous oil bath for the speed-reducer parts.

Inserted at the hub end of the speed reducer are blade-angle limit switches. These switches of the snap-action type are operated by pivot arms which ride on a cam attached to the main drive gear of the speed reducer.

AIRCRAFT ENGINES

The limit switches, which are connected in the electric-motor leads, are spring-loaded electrical contacts. When installed in the power unit, these contacts mate with the fixed contacts on the front face of the hub.

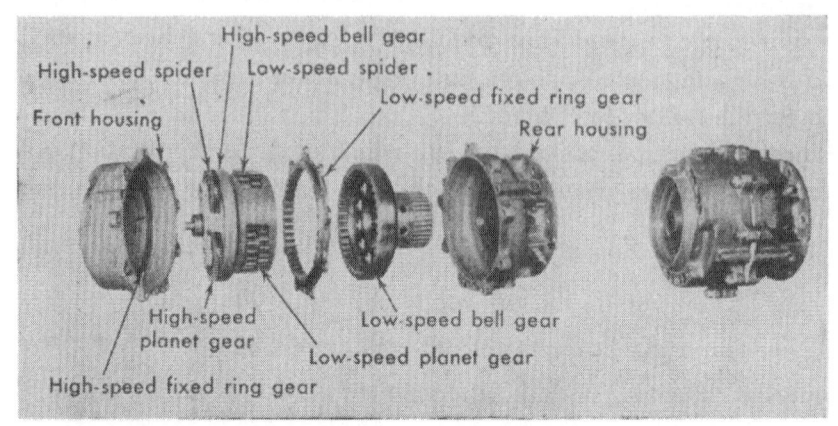

Fig. 228. The speed reducer assembly for an electric controllable-pitch propeller. (Courtesy Curtiss-Wright Corporation)

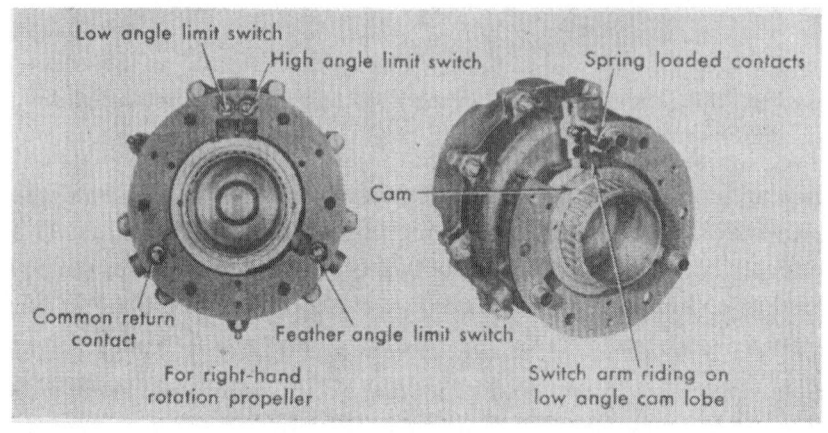

Fig. 229. A cutaway view of a limit switch for an electric controllable-pitch propeller. (Courtesy Curtiss-Wright Corporation)

As the pivot arms ride on their respective cam lobes, the contacts are snapped open and the circuits are opened. The cam lobes accurately control the low, high, and feather blade angular settings.

Reverse-thrust propellers have an additional limit switch to stop the pitch change when the negative angle of blade setting has been reached.

The motor used to change the pitch of the blades is attached to the front housing of the speed reducer. The motor shaft is splined to the driving pinion of the high-speed stage of the reduction unit. This motor

CONTROLLABLE-PITCH PROPELLERS

is of the series type and has two field windings to enable it to rotate in either direction.

The motor brake assembly is contained in the front end of the pitch-change motor. This assembly consists of a front brake plate, a splined hub keyed to the motor shaft, a floating, splined, disc brake facing, and a rear brake-plate assembly. The rear brake plate is held against the splined disc facing by means of three brake springs.

The end shield assembly also contains a brake solenoid coil which is located behind the rear brake plate. This coil is connected in series with the motor. Whenever the motor is operated, the solenoid is energized by the current flowing to the motor and releases the brake. When no current is flowing through the motor, the solenoid becomes inoperative and the brake is automatically applied by the brake springs.

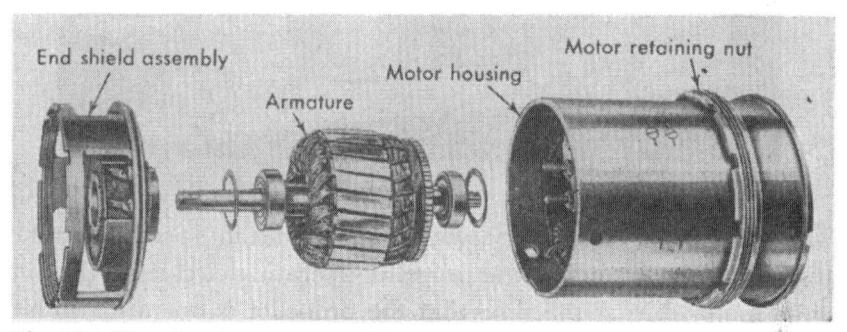

Fig. 230. The electric motor assembly for an electric controllable-pitch propeller. (Courtesy Curtiss-Wright Corporation)

An aluminum alloy housing bolted to the nose section of the engine forms a mounting for the slip-ring brush holder.

The brush assembly is equipped with a connector plug for connecting the electric wires and conduit.

The governor is usually mounted on the nose section of the engine and acts to maintain the engine at a selected speed by changing the propeller blade angle in a similar manner to the constant-speed, hydraulically operated propeller. The governor acts as a single-pole, double-throw switch by means of which the current to the pitch-change motor is automatically directed.

The center switch contacts in the governor are moved by means of an oil servo mechanism which is controlled by the action of flyweights. A servo unit consists of a cylinder containing a piston which is moved back and forth by oil pressure. The flyweights are driven directly by the engine, and any change in engine speed causes the flyweights automatically to operate the switch contacts directing the flow of current in the proper direction to the motor.

The electric current may be made to flow in the proper direction to cause the motor to rotate in either direction, depending upon whether the blades are to be moved to a higher or lower pitch.

Fig. 231. An electric relay. (Courtesy Curtiss-Wright Corporation)

The relay is the heavy-duty switch in the constant-speed, control circuit. Its purpose is to close the propeller-operating circuits to the pitch-change motor during the time that the propeller is operating in automatic control.

Fig. 232. A voltage booster. (Courtesy Curtiss-Wright Corporation)

A current of low amperage is controlled by the governor contacts and energizes the relay solenoid coils. The energized solenoid acts upon a soft-iron-core armature of the relay and causes the heavy-duty, relay contact point to close the propeller circuits.

CONTROLLABLE-PITCH PROPELLERS

A resistor is connected across each coil of the relay to damp out small fluctuating currents. This damping action prevents chattering and causes the point to make solid contact and break evenly.

A condenser is connected between the movable contacts and the ground return to prevent arcing. This condenser also smooths out electrical pulsations which may be set up when the contact points break the propeller circuit.

Fig. 233. A solenoid switch. (Courtesy Curtiss-Wright Corporation)

The voltage booster is a single dynamotor unit placed in the propeller circuit to increase the normal voltage. Increasing the voltage produces a rapid rate of pitch change for feathering and reversing operations.

A solenoid switch is used in the booster in conjunction with the feather or reverse switch in the cockpit to control the operating current. Operation of the feather or reverse circuits causes the solenoid switch to put the booster in operation automatically. The solenoid switch is connected in series with these circuits.

When the blade reaches the feather or reverse-pitch position, the

AIRCRAFT ENGINES

limit switch in the propeller releases the booster solenoid switch and all propeller circuits are opened.

The thermal circuit breaker in the propeller circuit is provided to protect the circuit against electrical overloads. This circuit breaker, if opened, may be closed by pressing in the button on the circuit breaker.

The feather switch is used only on multiengine aircraft and is a double-pole type for two positions. The purpose of this switch is to break the normal propeller circuit and, at the same time, to complete the feather circuit.

Fig. 234. A selector switch for an electric controllable-pitch propeller. (Courtesy Curtiss-Wright Corporation)

A switch guard is provided to prevent accidental operating of this switch which, during normal propeller operations, remains in the normal position.

The synchronizer control system provides automatic, constant-speed control by electrically matching the engine speed with the speed of an

Fig. 235. A feather switch for an electric controllable-pitch propeller. (Courtesy Curtiss-Wright Corporation)

Fig. 236. A circuit breaker for an electric controllable-pitch propeller. (Courtesy Curtiss-Wright Corporation)

independent standard or master. Any variation from the matched condition produces a corrective action, thus synchronizing all engines with the master.

CONTROLLABLE-PITCH PROPELLERS

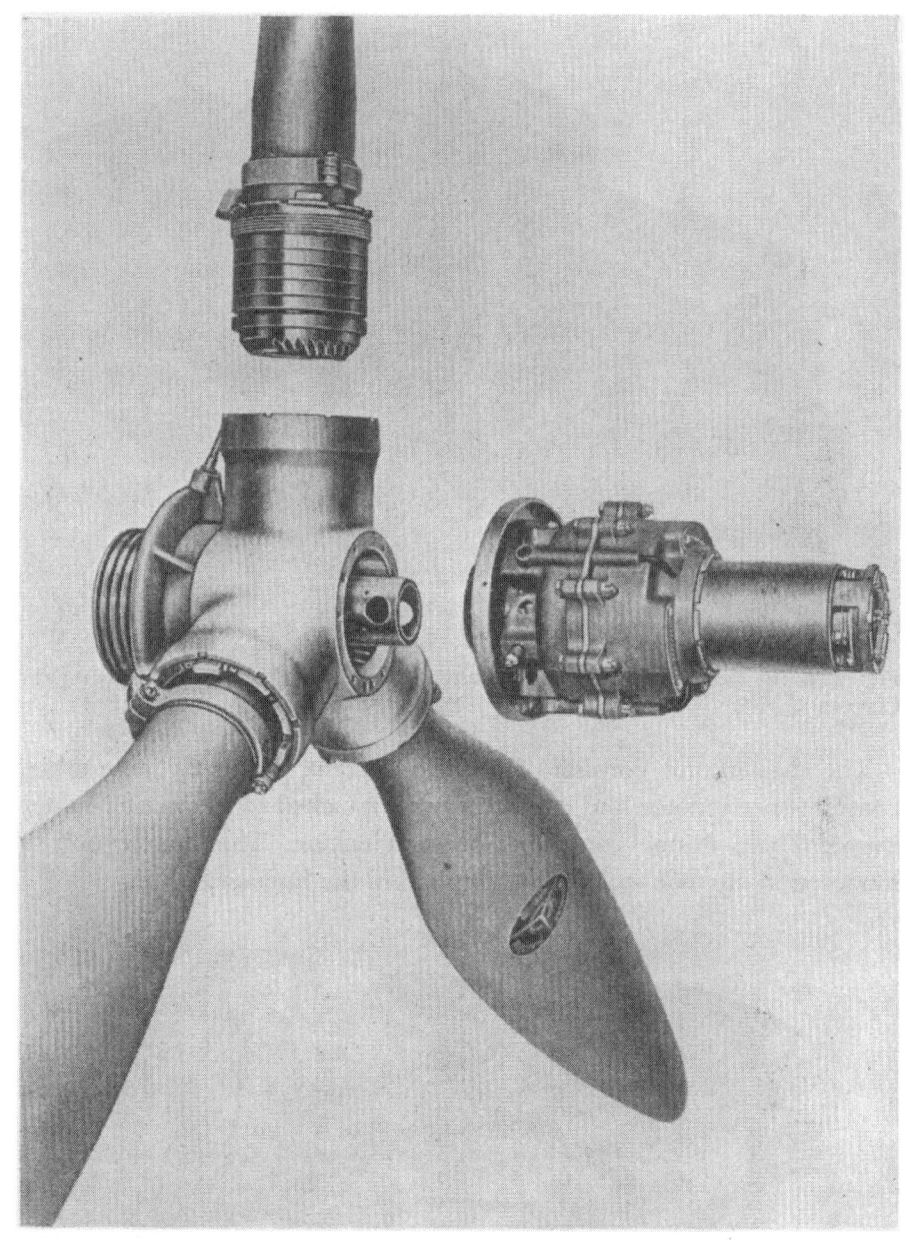

Fig. 237. Unit construction of an electric controllable-pitch propeller. (Courtesy Curtiss-Wright Corporation)

The engine-mounted alternator, which replaces the conventional governor, contains only one moving part: a single-piece permanent-magnet rotor. This simple and rugged unit is not affected by engine vibration.

AIRCRAFT ENGINES

Fig. 238. An electric propeller synchronizer-control unit. (Courtesy Curtiss-Wright Corporation)

The master unit consists of a direct-current, manually adjustable, constant-speed motor and contactor units attached to the master-motor frame. There is one contactor for each engine. The master unit is mounted in any convenient location within the fuselage.

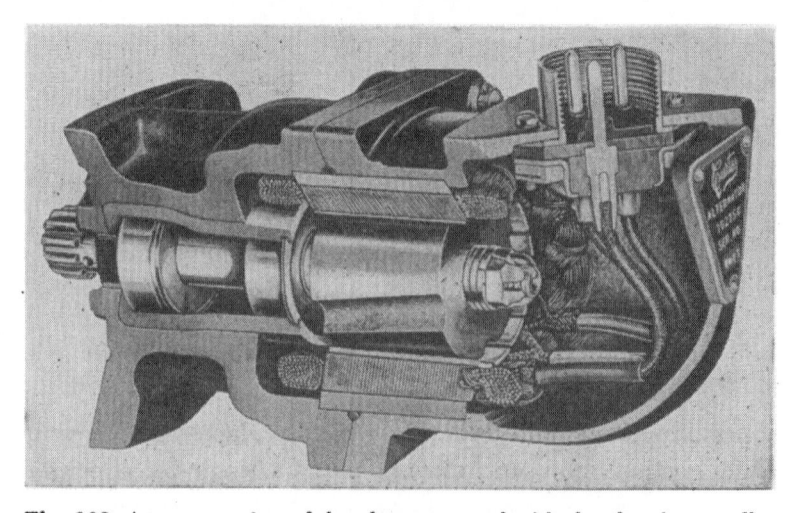

Fig. 239. A cutaway view of the alternator used with the electric propeller synchronizer-control unit. (Courtesy Curtiss-Wright Corporation)

CONTROLLABLE-PITCH PROPELLERS

The alternators, which are permanent-magnet-type, three-phase generators, are mounted on the engine-governor-drive pad and are connected by wiring to the master-unit contactors. The alternator output is delivered to the master-unit contactors at frequencies proportional to crankshaft speeds.

The operational principle of the synchronizer is based on the comparison of the engine speed with the master-motor speed. This comparison takes place in the contactor unit in the following manner.

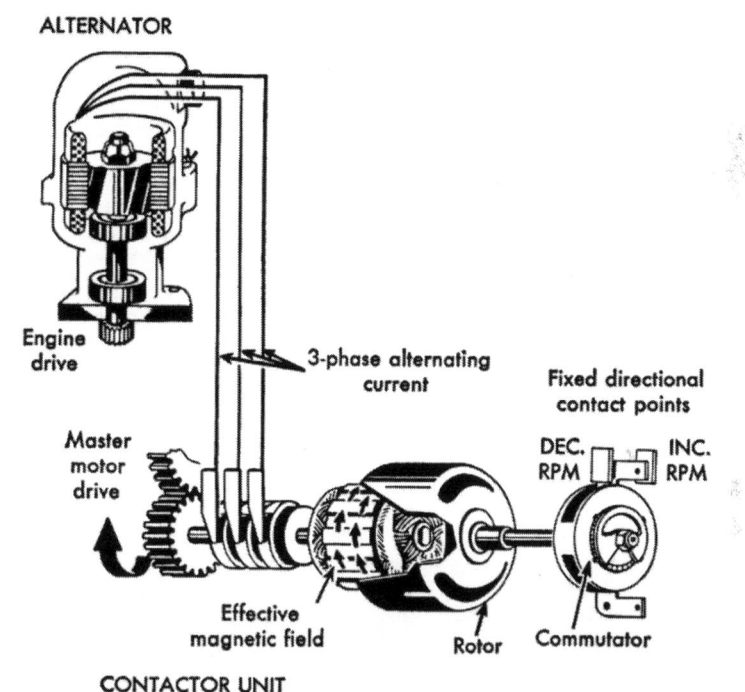

Fig. 240. A diagrammatic view of the electric propeller synchronizer-control system. (Courtesy Curtiss-Wright Corporation)

Each contactor contains an armature which is geared to and driven by the master motor. The alternators corresponding to each contactor unit are connected through slip rings to the armature winding in such a manner that the magnetic field built up in the armature windings will have a phase rotation opposite to that of the direction of rotation of the armature itself. If the speed of the engine is exactly equal to that of the master motor, the magnetic field will rotate at a speed equal to that of the armature. Since the directions of rotation are opposite, the magnetic field in space will be stationary. Thus, the magnetic, bell-shaped rotor, which enshrouds the armature windings and which is magneti-

AIRCRAFT ENGINES

cally locked in the field of the armature, will also stand motionless. Mounted on the opposite end of this rotor shaft is the contact mechanism which energizes the relays which, in turn, control the propeller-pitch-change motor. When the magnetic field is in a stationary condition, as just described, the motionless rotor assumes a position such that the contact mechanism remains opened. This is the "on-speed" condition.

When the speed of the engine deviates from the master-unit setting, it follows that this difference in speed will manifest itself in both the magnetic field and the rotor of the engine's contactor unit, thus forcing the rotor to rotate. Turning of the rotor shaft then closes the contact mechanism which brings about a change in pitch to correct for the speed difference. An important feature incorporated in the unit provides a proportional correction for an off-speed condition; i.e., the frequency and duration of pitch-change impulses vary with the degree of off-speed. Increase or decrease pitch control is determined by the direction of rotation of the magnetic field and rotor, this being dependent upon whether the engine speed is greater or less than the master-motor speed.

TABLE II. ACTIONS BROUGHT ABOUT IN THE SYNCHRONIZER CONTROL SYSTEM BY VARIOUS SPEED CONDITIONS

	STATOR ROTATION	EFFECTIVE MAGNETIC FIELD ROTATION	ROTOR AND COMMUTATOR ROTATION	RESULT
Engine speed exceeds master-motor speed	Clockwise	Counterclockwise (faster than stator rotation)	Counterclockwise	Decrease r.p.m. correction
Master-motor speed exceeds engine speed	Clockwise	Counterclockwise (slower than stator rotation)	Clockwise	Increase r.p.m. correction
Master-motor and engine speeds identical	Clockwise	Counterclockwise (same rotational speed as stator)	Stationary	On-speed

If for any reason the speed setting of the master unit selected by the pilot is not maintained, a protective device places all propellers in fixed pitch. It is then possible to control the blade angles of any of the propellers manually by use of the selector switches.

The synchronizer-control system retains the standard electric-propeller features of individual control of the propellers for feathering, reversing, and manual selective-pitch operations. The synchronizer master unit will continue to govern and synchronize any one or more engines, irrespective of whether the propellers on the remaining engines are being manually controlled, operated in fixed-pitch, feathered, or reversed.

CONTROLLABLE-PITCH PROPELLERS

Self-Contained Controllable Propellers. This propeller is of simple design, being hydraulically operated and entirely contained within a single unit. No special fittings or engine alterations are necessary for the installation. A control lever is connected to the cockpit by a standard airplane fitting.

This propeller has available a wide range of blade angles controlled by a sensitive and fast-acting governor which provides instant and accurate control. The propeller has an exceptionally fast rate of blade-angle change which is adequate to maintain constant engine speed throughout the most extreme maneuvers.

Fig. 241. A four-blade propeller installed on a high-powered aircraft. (Courtesy Aeroproducts Division, General Motors Corporation)

The blade is of hollow-steel construction, incorporating a longitudinal strengthening rib. It is composed of two members, namely, the thrust member and the camber sheet. These members are brazed together along the leading and trailing edges and along the longitudinal rib. This construction eliminates any tendency to "oil can," a type of vibration which is sometimes encountered with thin, unsupported camber sheets.

The thrust member is a machined steel forging which forms the blade shank, thrust face, longitudinal rib, and leading- and trailing-edge reinforcements. The camber sheet forms the camber side of the blade. It is formed from a steel sheet and brazed to the thrust member. Both members are completely ground and polished before brazing.

On some models, a cuff ring is machined on the shank of the blade for the purpose of retaining a blade cuff, if required. The external surfaces

AIRCRAFT ENGINES

of the blade are treated to prevent corrosion and rust. The internal surfaces are rust-proofed and hermetically sealed. This sealing is accomplished by the balance cup which is installed in the shank of the blade. In manufacture, the blade is balanced vertically and horizontally against a master blade by adding lead in the proper quantity and location in the balance cup. All blades are, therefore, in perfect balance

Fig. 242. A four-blade, hydraulic, controllable-pitch propeller for high-powered engines. Note paddle-shaped tip of blade. (Courtesy Aeroproducts Division, General Motors Corporation)

with each other. To allow for the final balancing of the propeller assembly, a stud is provided in the center of the balance cup on which balance washers may be added.

The regulator is a doughnut-shaped unit which serves as a reservoir for the hydraulic oil and contains a pump, a pressure-control valve, a governor, filter screens, and a manual r.p.m. control mechanism. The unit is composed of a cover and a housing. This housing is a machined aluminum casting on which are mounted the oil pump, pressure-control valve, governor, and filter screens. A removable plug is threaded into the cover to permit the addition and removal of oil, and to allow access to the governor adjusting screw.

CONTROLLABLE-PITCH PROPELLERS

The oil pump rotates with the regulator and is driven by a stationary gear which is a part of the adapter assembly. Due to the centrifugal force created by propeller rotation, oil is forced to the outer diameter of the regulator where it enters the oil pump. This ensures a constant supply of oil to the pump which produces the hydraulic pressure necessary to operate the torque unit.

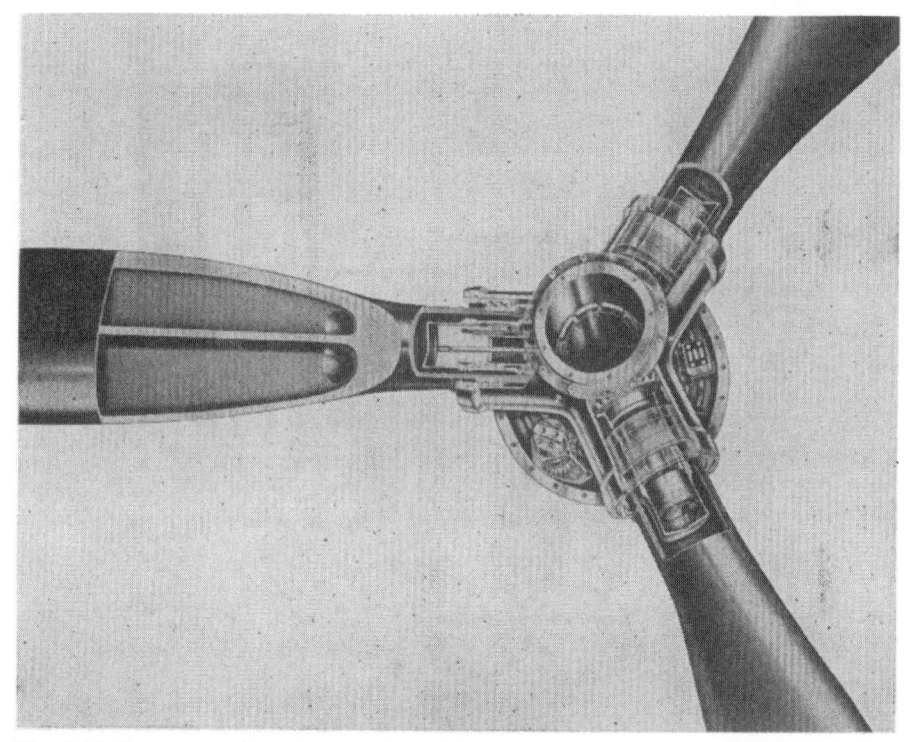

Fig. 243. A cutaway view showing the pitch-changing mechanism and blade construction of a hydraulically operated, constant-speed, controllable-pitch propeller. (Courtesy Aeroproducts Division, General Motors Corporation)

Oil from the pump is forced through a filter to the pressure-control valve and governor. The pressure-control valve is located in the hydraulic system between the oil pump and the governor to control maximum pressure within the system.

In this design, the flyweights of a conventional type of governor are replaced by a piston which also distributes the oil to the blade-angle-changing mechanism. The governor rotates with the regulator, and its piston is mounted radially to the axis of the propeller rotation. The piston is mounted on one end of a lever, a movable roller fulcrum is located at the other end, and a spring is placed between the two.

AIRCRAFT ENGINES

Since the governor rotates with the regulator unit, centrifugal force tends to throw the governor piston outward from the center of rotation. At a selected r.p.m., the piston will be in a central position in the governor cylinder. At this r.p.m., there is no flow of hydraulic fluid to the

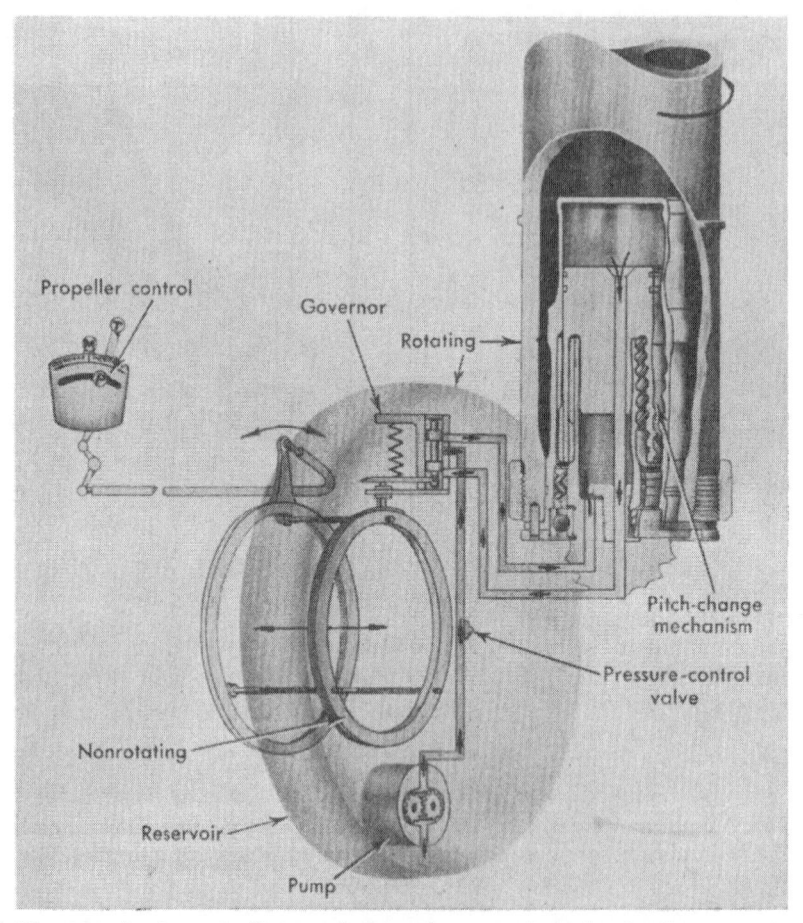

Fig. 244. A schematic diagram showing the parts of a hydraulically operated, constant-speed, controllable-pitch propeller. (Courtesy Aeroproducts Division, General Motors Corporation)

torque units in the hub and, therefore, no change in blade angle. With an increase in r.p.m., the piston will move outward, creating what is called an "overspeed" condition, opening ports which allow hydraulic fluid under pressure to flow to the inboard side of the torque pistons in the hub. This results in a change to a higher blade angle. The r.p.m. will then return to normal. With a decrease in r.p.m. below normal, the governor piston will be forced nearer the axis of rotation by the

CONTROLLABLE-PITCH PROPELLERS

spring to an "underspeed" condition, opening ports which allow hydraulic fluid under pressure to flow to the outboard side of the torque pistons in the hub. This decreases the blade angle. The r.p.m. will again return to normal, and the governor piston, due to the increased centrifugal force, will move out to its central position, on-speed.

The r.p.m. at which the on-speed condition will occur can be selected manually by the pilot. This is accomplished by varying the position of the roller fulcrum.

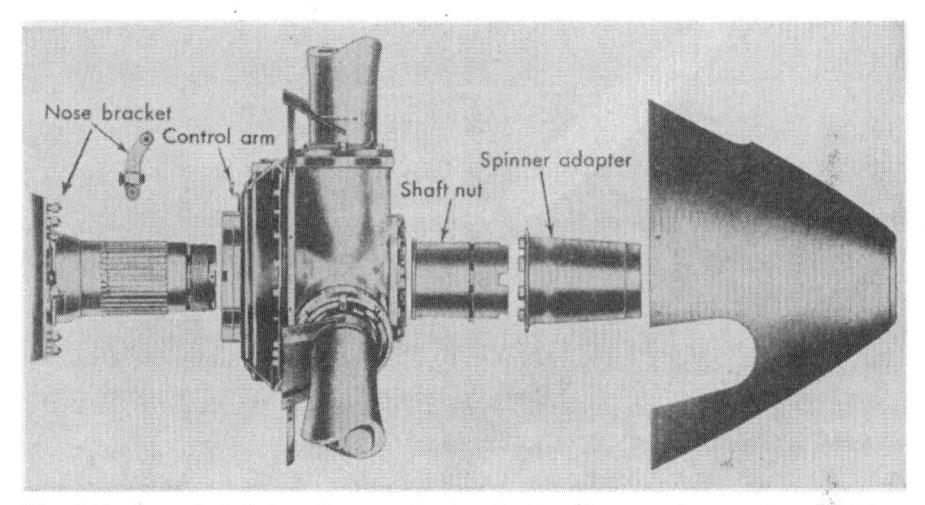

Fig. 245. An exploded view of a propeller installation. (Courtesy Aeroproducts Division, General Motors Corporation)

The adapter assembly is stationary and is carried on bearings, one in the regulator housing and one in the cover. Rotation of this assembly is prevented by an adapter stop which is bolted to the engine reduction-gear case. This stop engages in a notch in the regulator adapter plate.

Steel tubes and pressure pads brazed into one assembly of circular form are cast into the regulator housing. This assembly consists of one tube which conducts the oil from the pump to the pressure-control valve and then to the governor. Two other tubes distribute the oil from the governor to the increase- and decrease-r.p.m. ports of the hub at each socket from which point the fluid is carried to each torque unit through drilled passages in the hub.

The hub is a one-piece, machined, steel forging having drilled passages to provide for the transfer of hydraulic pressure.

A master gear meshes with the blade gears and synchronizes the movement of all blades. This gear is installed in the hub.

AIRCRAFT ENGINES

The spinner adapter is mounted on the front of the hub. It supports the spinner nose, provides a blast tube for cannon on military models, and serves as a shaft nut lock.

Changing the blade angle is accomplished by a hydraulically operated torque unit incorporated in each blade socket of the hub. A torque unit consists of a fixed spline, a piston having a splined skirt, and a splined

Fig. 246. A two-blade, hydraulically operated, constant-speed, controllable-pitch propeller mounted on a test stand. (Courtesy Aeroproducts Division, General Motors Corporation)

cylinder. The fixed spline is mounted solidly on the hub with a hollow bolt. The piston, which is splined inside and outside, works between the fixed spline and the inside of the blade cylinder. To increase the blade angle and reduce the r.p.m., hydraulic pressure is transferred to the fixed spline through an offset hole at the base and forces the piston outward toward the cylinder head. To decrease the blade angle and increase the r.p.m., hydraulic pressure is transferred through the hollow, fixed, spline bolt and tube to the outboard side of the piston head, and forces the piston inward. Since both sets of splines are helical, outward or inward movement of the piston rotates the blade cylinder in relation

CONTROLLABLE-PITCH PROPELLERS

to the fixed spline. The blade cylinder engages the dowel pins in the blade butt and transmits the turning action to the blade.

The design of this propeller is readily adaptable to the production of dual-rotation propellers. A dual-rotation propeller consists of two

Fig. 247. A dual-rotation six-blade propeller mounted on a test stand. This is a hydraulically operated controllable-pitch propeller, three blades of which rotate in one direction with three blades rotating in the opposite direction. A cooling fan is mounted back of the blades to assist in cooling an air-cooled engine. (Courtesy Aeroproducts Division, General Motors Corporation)

separate complete propellers, coaxially mounted, which are rotated in opposite directions at equal r.p.m.

The inner of the two dual-rotation propellers is practically the same as a single-rotation propeller. The blade-angle change of the outboard propeller is controlled by a coordinating mechanism which links the two propellers.

Feathering and unfeathering this propeller is accomplished by a gas-charged accumulator which forces oil through a feathering valve to the torque units. The feathering valve is actuated when the propeller control lever in the cockpit is moved to the feathering position, which is to a position beyond the normal low-r.-p.-m. position. When this is done, the gas pressure in the accumulator forces part of the oil in the accumu-

Fig. 248. A controllable-pitch, constant-speed propeller for light airplanes. Note the compact governor unit. The control rod on the right selects the desired r.p.m. (Courtesy Thomson Industries, Inc.)

lator to the torque units, thus turning the blades to the feathered angle. When the cockpit control is moved from the feathering position, the feathering valve is again moved. The valve in this position directs oil to the opposite side of the torque pistons and unfeathers the blades. The accumulator is recharged with oil within approximately 30 sec. of normal propeller operation. There may be incorporated in this propeller a negative-pitch mechanism. This mechanism will allow the propeller to be placed in a negative-pitch position for braking purposes to decrease the landing run or to help maneuver an airplane on the water or on ice-coated fields.

CONTROLLABLE-PITCH PROPELLERS

Light Airplane Controllable-Pitch Propellers. While controllable-pitch propellers have been used for some time on large aircraft, it is not until recently that a successful controllable-pitch, constant-speed, full-feathering propeller has been developed which is suitable for installation on light aircraft. Propellers of this type for large aircraft are operated by

Fig. 249. Side view of a controllable-pitch, constant-speed propeller mounted on a 185-horsepower engine. (Courtesy Thomson Industries, Inc.)

means of either a hydraulic or electric system. This propeller operates by means of a variable-ratio, V-belt drive.

The variable ratio is brought about by using a split pulley. When the halves of a pulley are moved together, the diameter is increased. When they are separated, a decrease in diameter results.

The propeller uses two drives, one having a fixed ratio and the other a variable ratio made up of two split pulleys. The propeller pitch is changed by the variation between the ratios of the two drives, as shown in Fig. 250.

The governor shaft, which is close to and parallel to the propeller

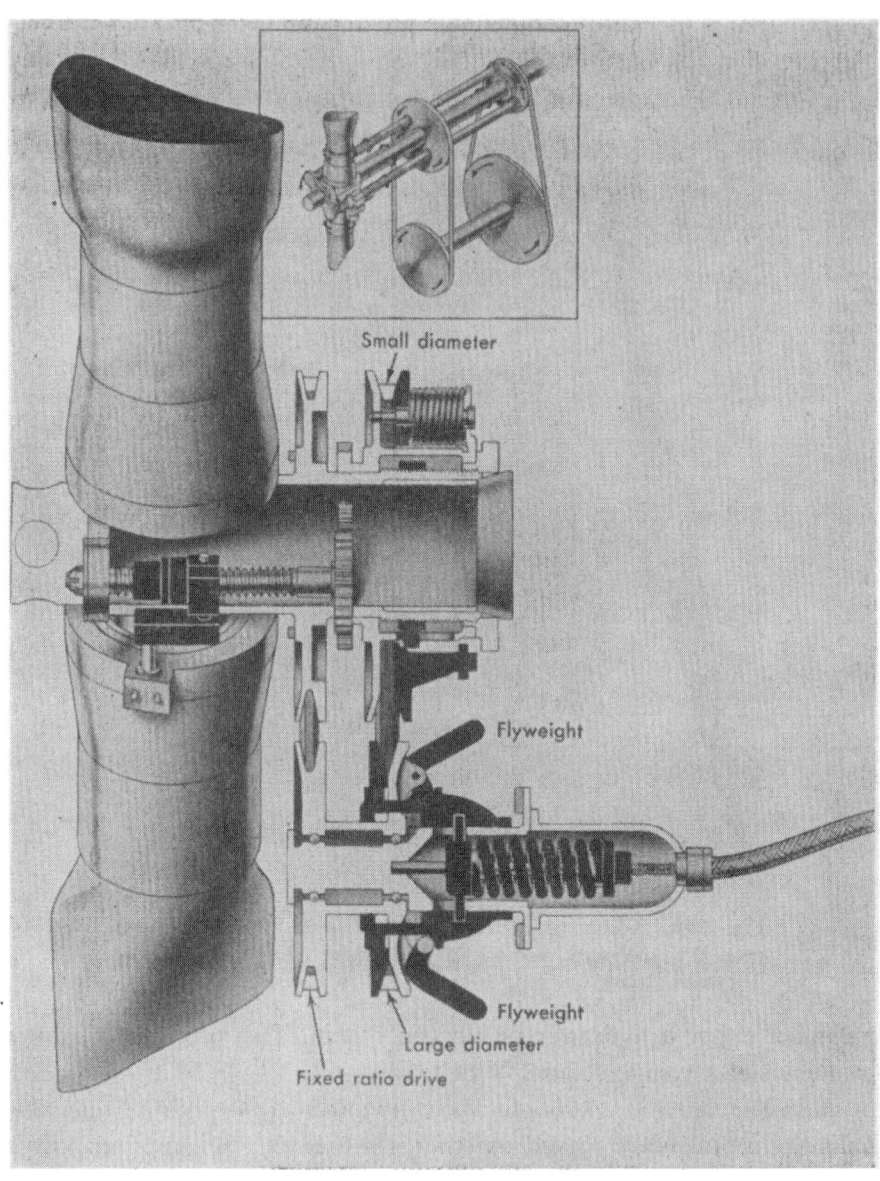

Fig. 250. A schematic drawing showing the operation of the constant-speed propeller for light airplanes. Note flyweights are in the over-speed position. (Courtesy Thomson Industries, Inc.)

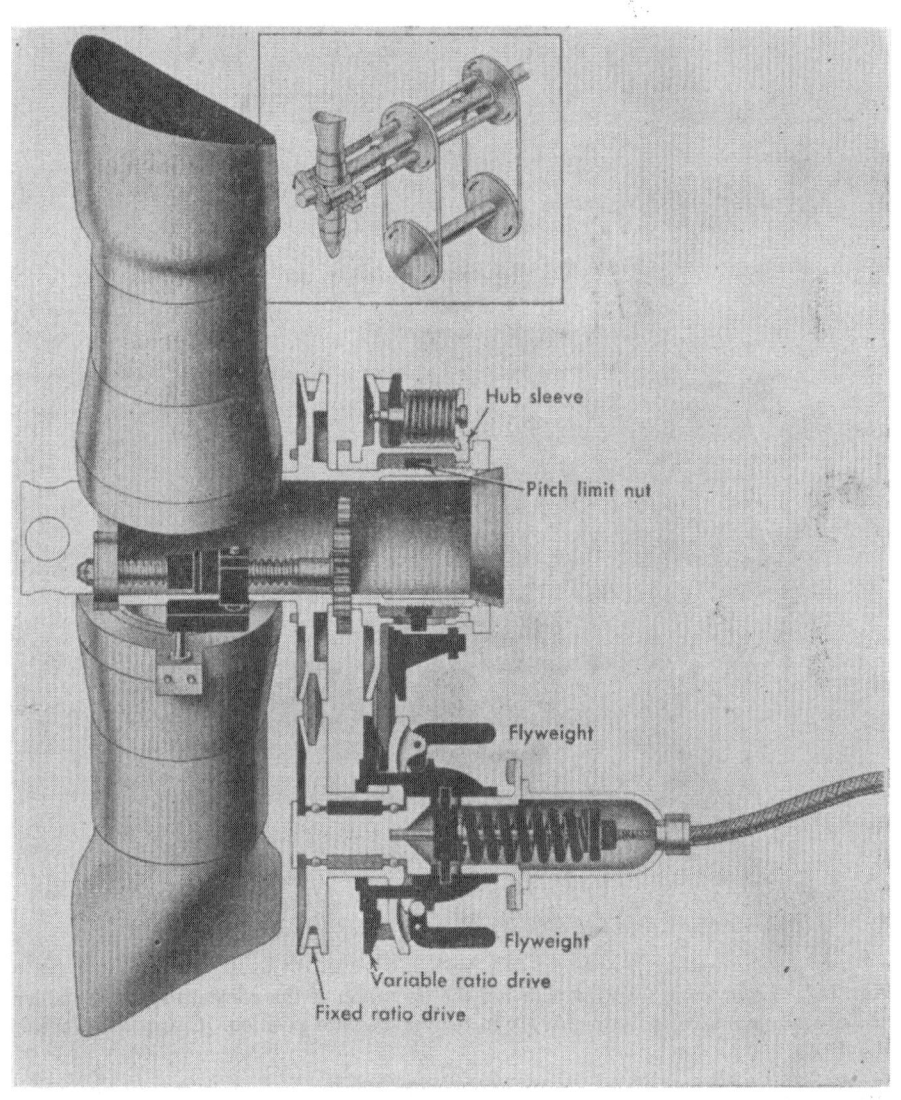

Fig. 251. A schematic drawing showing the operation of the constant-speed propeller for light airplanes. Note flyweights are in the neutral position. (Courtesy Thomson Industries, Inc.)

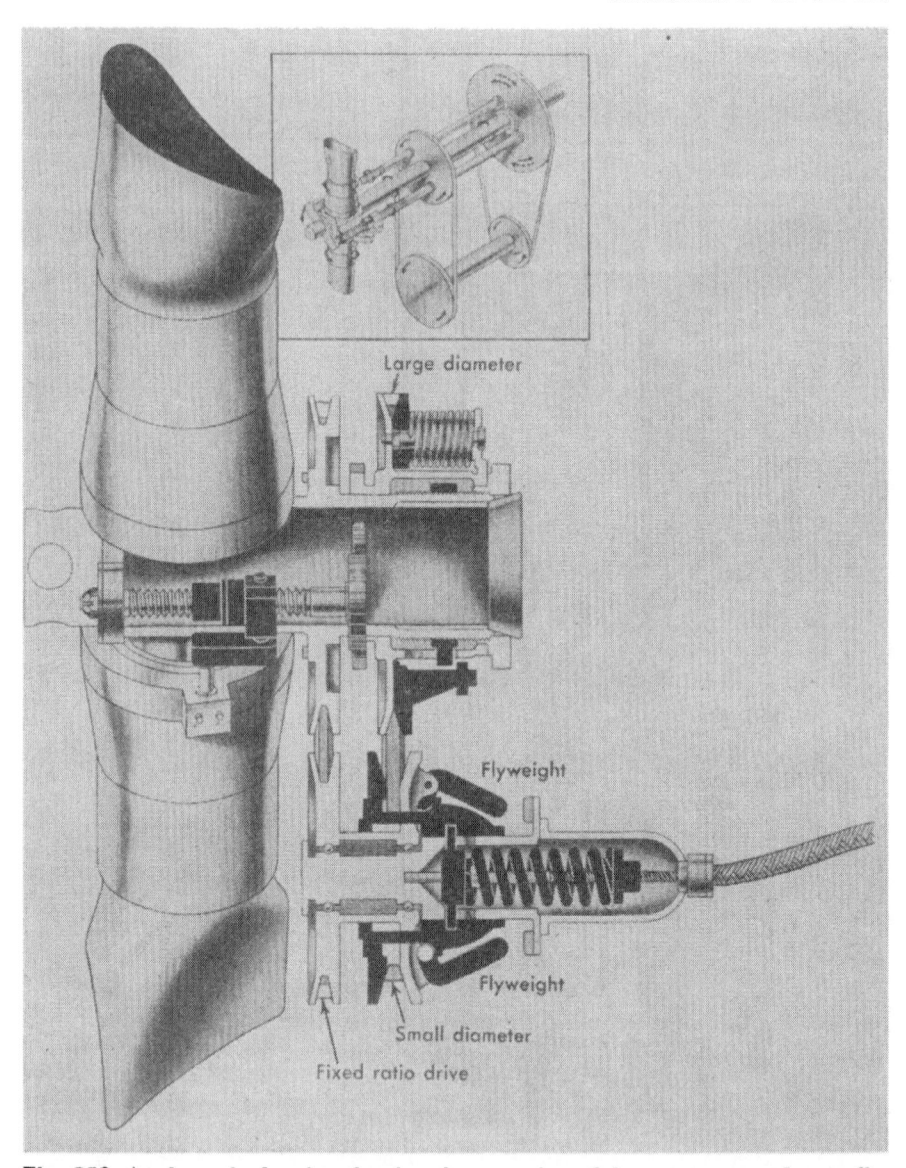

Fig. 252. A schematic drawing showing the operation of the constant-speed propeller for light airplanes. Note flyweights are in the under-speed position. (Courtesy Thomson Industries, Inc.)

shaft, is mounted on the engine nose. The governor shaft is driven by a fixed-ratio drive and is used to rotate the governor at a speed which corresponds to the engine speed. A sleeve which can rotate in either direction relative to the propeller hub is placed around the hub back of the fixed pulley. This sleeve is connected to the propeller blade by gearing in such a way that, when the sleeve is turned faster than the propeller,

CONTROLLABLE-PITCH PROPELLERS

it increases the pitch of the blades. When this sleeve rotates at a speed less than the rotation of the speed of the propeller hub, the pitch of the blades is reduced. When this sleeve rotates at the same speed as the propeller hub, there is no change in the pitch of the propeller blades. The variable-ratio drive connects this sleeve to the rotating propeller shaft.

Fig. 253. Front view showing the propeller in a full-feathered position. (Courtesy Thomson Industries, Inc.)

The ratio of the variable pulley is controlled by a speed-sensitive, flyweight governor. This flyweight governor is located on the rotating governor shaft. When the flyweights are displaced outward due to centrifugal force, the pulley halves on the governor shaft are brought closer together. This brings about an increase in the speed at which the sleeve rotates.

An increase in the speed of the governor sleeve increases the blade pitch. When the flyweights are in mid-position or neutral, no change in blade pitch takes place. If the speed of rotation decreases and the flyweights are displaced inwardly, the governor sleeve rotates more slowly than the propeller hub and the blade pitch is decreased.

The flyweights act against a spring, the tension of which may be

AIRCRAFT ENGINES

adjusted in flight. By adjusting the spring tension, the pilot can determine the speed to which he wishes the propeller to rotate. This brings about a constant-speed condition. The belt length remains constant, so any increase in the diameter of the variable pulley on the governor shaft must be accompanied by a decrease in the diameter of the variable

Fig. 254. Side view of the propeller in a full-feathered position. (Courtesy Thomson Industries, Inc.)

pulley on the propeller hub. A series of springs on the variable hub pulley maintains the proper belt tension.

As the variable pulley on the governor shaft changes its diameter, the spring tension on the propeller-hub, variable pulley allows the proper change in diameter of the propeller hub pulley.

Pitch limit stops are provided to prevent the blades from rotating beyond the desired limit of travel. The blades on this propeller may be rotated into the full-feathered position.

XVIII ENGINE INSTALLATION AND COOLING SYSTEMS

The engine mount is the arrangement by which the engine is attached to the aircraft. The engine mount is subjected to severe strains and stresses while the engine is in operation. Not only must the engine mount support the weight of the engine, but it must also resist twisting forces

Fig. 255. A five-cylinder, radial, air-cooled aircraft engine. Note the large number of cooling fins. (Courtesy Kinner Motors Inc.)

and transmit the thrust of the engine to the aircraft. The rear of the engine mount is usually fastened to main structural members of the aircraft. When the engines are mounted in the wings, or other than in the front of the fuselage proper, special reinforcing must be used.

In-line engines and opposed engines are usually mounted on a horizontal engine mount. In most aircraft, the mount is built up of welded steel tubing, one part of which is designed to be attached to the aircraft and the other part to the mounting pads on the engine. Extra reinforcment must be built into the engine mount to take care of torsional forces.

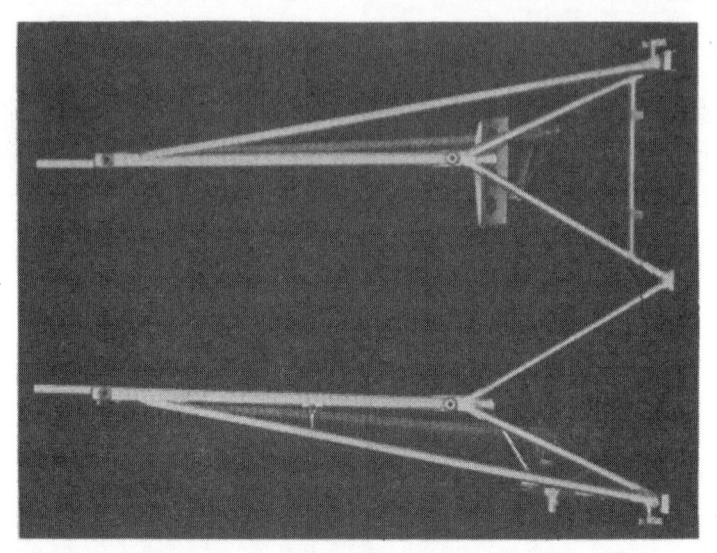

Fig. 256. An engine mount made from welded steel tubing. (Courtesy Fairchild Engine and Airplane Corporation)

When the engine mount is of the horizontal type, the engine usually rests on the top of the mount and is bolted firmly to it. The mount of a radial engine usually consists of a circle of tubing which is attached to the crankcase or other special attachments of the engine. The other end of the engine mount may be rectangular and is attached by means of special fittings to the aircraft. Much experimenting was done to design engine mounts which would not transmit the full force of the engine vibrations to the fuselage. Various methods of inserting rubber pads and vibration-absorbing devices between the engine and the aircraft have been tried with varying degrees of success. At the present time, some of the large engines are mounted by what is known as dynamic suspension. In this arrangement, the engine is suspended within an engine mount by shackles which absorb the vibration of the engine. This leaves the engine

Fig. 257. A rear view of a radial engine mount showing fuselage attachment. (Courtesy Douglas Aircraft Company, Inc.)

Fig. 258. A closely cowled, liquid-cooled engine. The radiators are located in the wing and cooled by air entering through openings in the leading edge of the wing near the fuselage. (Courtesy Bell Aircraft Corporation)

AIRCRAFT ENGINES

more or less free to vibrate in any direction without displacing the center of gravity of the engine from its desired location. With rubber pads and spring devices, the engine has a tendency to sag or tip downward.

An important part of the engine installation is the engine cowling. On the earliest aircraft, the engine was mounted out in the open. With this method of mounting, the engine offers excessive resistance to the

Fig. 259. A front view of a radial engine mount showing the ring to which the engine is attached. (Courtesy Douglas Aircraft Company, Inc.)

air flow and has no protection from the elements. As the speed of aircraft increased, this resistance to the air flow became more of a problem and the engine was enclosed in a cowling to decrease it. Since air-cooled engines were cooled largely by the blast of air from the propeller, much difficulty was encountered in designing a cowling which would not cause overheating. Various attempts were made to reduce the resistance of the air-cooled engine to the air flow. The ring cowling, which is simply a cambered ring of sheet metal, was mounted around the circumference of the cylinders. This ring decreased the drag of the engine to

ENGINE INSTALLATION AND COOLING SYSTEMS

such an extent that the speed of air craft was increased several miles per hour. The ring not only decreased the resistance of the engine to the airflow but also led to more efficient cooling.

As the horsepower of the engines increased, baffles were placed between the cylinders to direct the air flow between the cooling fins. In order to increase the efficiency of the baffle system, cowling was installed in back of the ring to direct the air flow where it was needed. With the

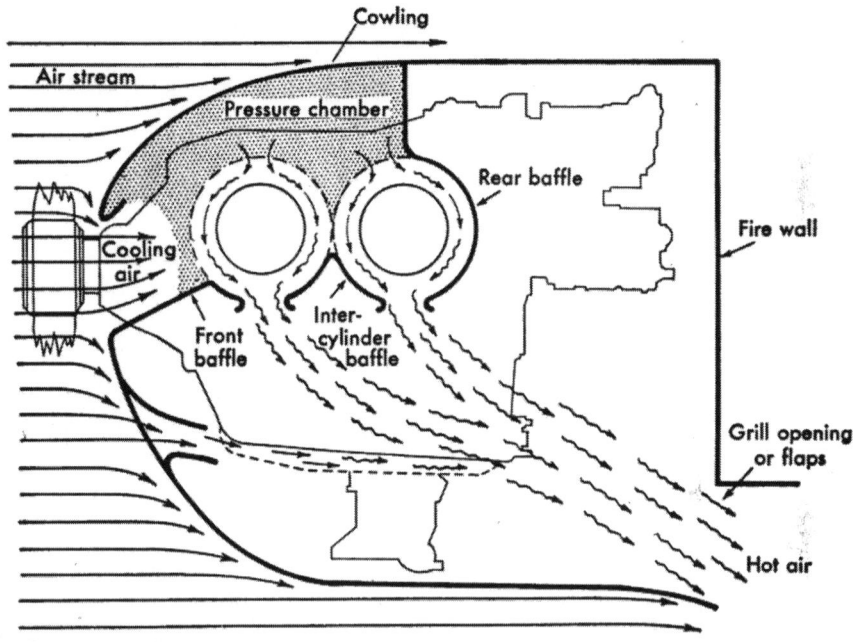

Fig. 260. A diagram showing the operation of pressure cowling in cooling an aircraft engine.

development of twin-row radial engines, the hot air from the front cylinder had to be deflected from the rear cylinders, and these had to be supplied with a stream of cool air. This led to the development of pressure cowling which is a tightly closed cowl having passages through which the air enters. The air is directed to the desired places about the engine and led out of the cowling in a smooth-flowing stream. When aircraft engines are not mounted in the fuselage, they are enclosed in a streamlined shelter called a nacelle. The removable part of this structure is the cowling.

With the old-style, wide-blade propeller, the part near the hub furnished sufficient pressure to force the air into the openings in the front of the cowling.

Fig. 261. A twin-engine aircraft with air-cooled engines showing cowling flaps open. (Courtesy Douglas Aircraft Company, Inc.)

Fig. 262. A twin-engine aircraft with air-cooled engines showing cowling flaps tightly closed. (Courtesy The Glenn L. Martin Company)

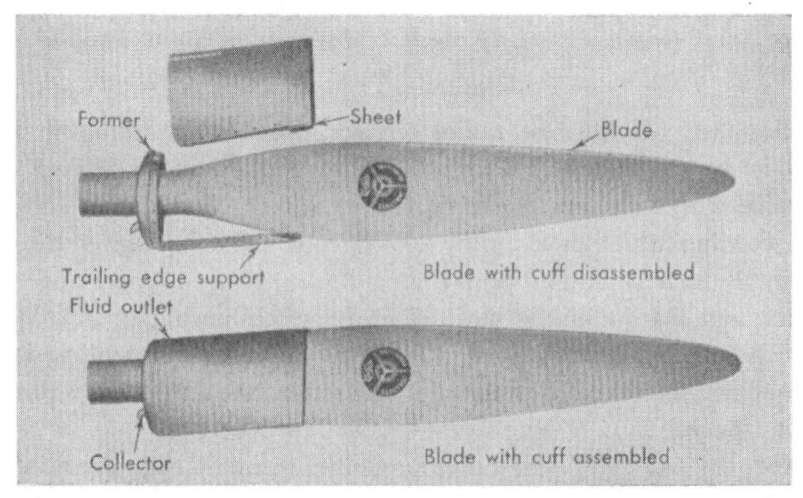

Fig. 263. Cuffs may be installed near the root of the propeller blade to assist in cooling the engine. (Courtesy Curtiss-Wright Corporation)

ENGINE INSTALLATION AND COOLING SYSTEMS

Metal blades have a round root which extends some distance from the hub, and decreases the slip stream near the propeller shaft. This type of blade may be equipped with cuffs to increase the air flow around the engine. In pressure cowling, however, cuffs are not usually needed as the air is supplied to the cowling from air scoops or openings located in the full blast of the air stream. Most pressure cowling is equipped with flaps which may be operated from the cockpit to regulate the flow of air about the engine. When engines are equipped with pressure cowling, cooling on the ground is not very effective.

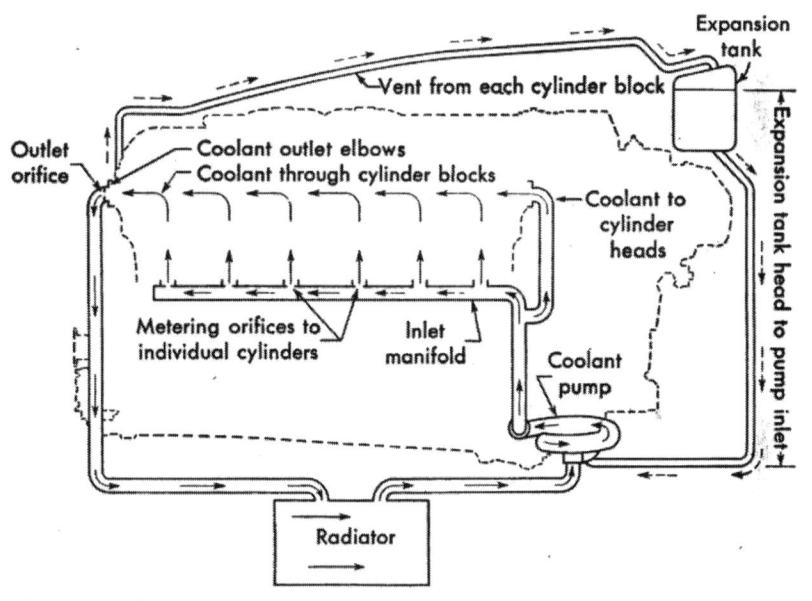

Fig. 264. A diagram showing a typical liquid-cooling system. (Courtesy Allison Division, General Motors Corporation)

Liquid-cooled engines may be closely cowled. On modern, liquid-cooled engines, the radiators may be located along the lower side of the wing, along the sides of the fuselage, or within the wing or fuselage. The liquid in the radiator is cooled by air led through the radiator by means of air scoops and air ducts. The liquid is pumped through spaces surrounding the cylinders and cylinder heads.

Oil is an important factor in the cooling of an engine. The inside of the piston is separated from the full heat of the combustion chamber by a comparatively thin layer of metal. The bottom of the pistonhead and the inside of the skirt are exposed to oil spray at all times. Some pistons are built with cooling fins on the inside of the skirt and underside

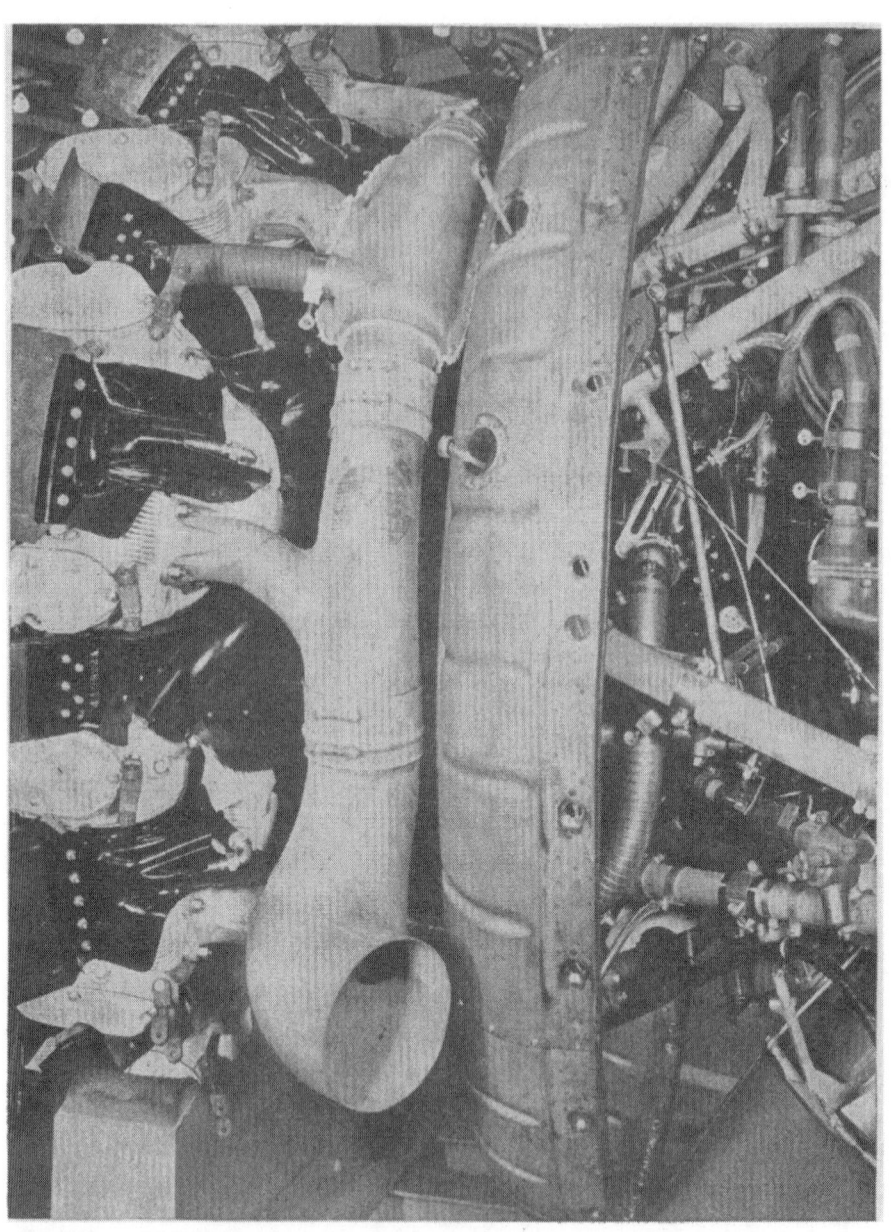

Fig. 265. A side view of engine and engine mount showing collector ring and inner cowling ring. (Courtesy Douglas Aircraft Company, Inc.)

ENGINE INSTALLATION AND COOLING SYSTEMS

of the head. The oil carries the heat away from the hot internal parts of the engine, and the heat is removed from the oil by means of oil radiators. These radiators may be equipped with a thermostatic control and/or with shutters which regulate the amount of oil and air flowing through the radiator. The amount of oil and air flowing through the oil radiator depends upon the temperature of the oil being returned from the engine to

Fig. 266. A radial engine with air deflectors or baffles between the cylinders. (Courtesy Jacobs Aircraft Engine Company)

the oil tank. A rapid increase in oil temperature is one of the first indications of excessive engine temperatures. During take-off and climb, under full throttle, most engines tend to overheat. Overheating under these conditions may be controlled somewhat by increasing the richness of the air-fuel mixture. The extra fuel is not completely burned and acts as a cooling agent in maintaining desirable engine temperatures. This method is known as fuel cooling. Proper cooling is important in prolonging the life of the engine, and the manufacturers' directions should be carefully followed at all times.

AIRCRAFT ENGINES

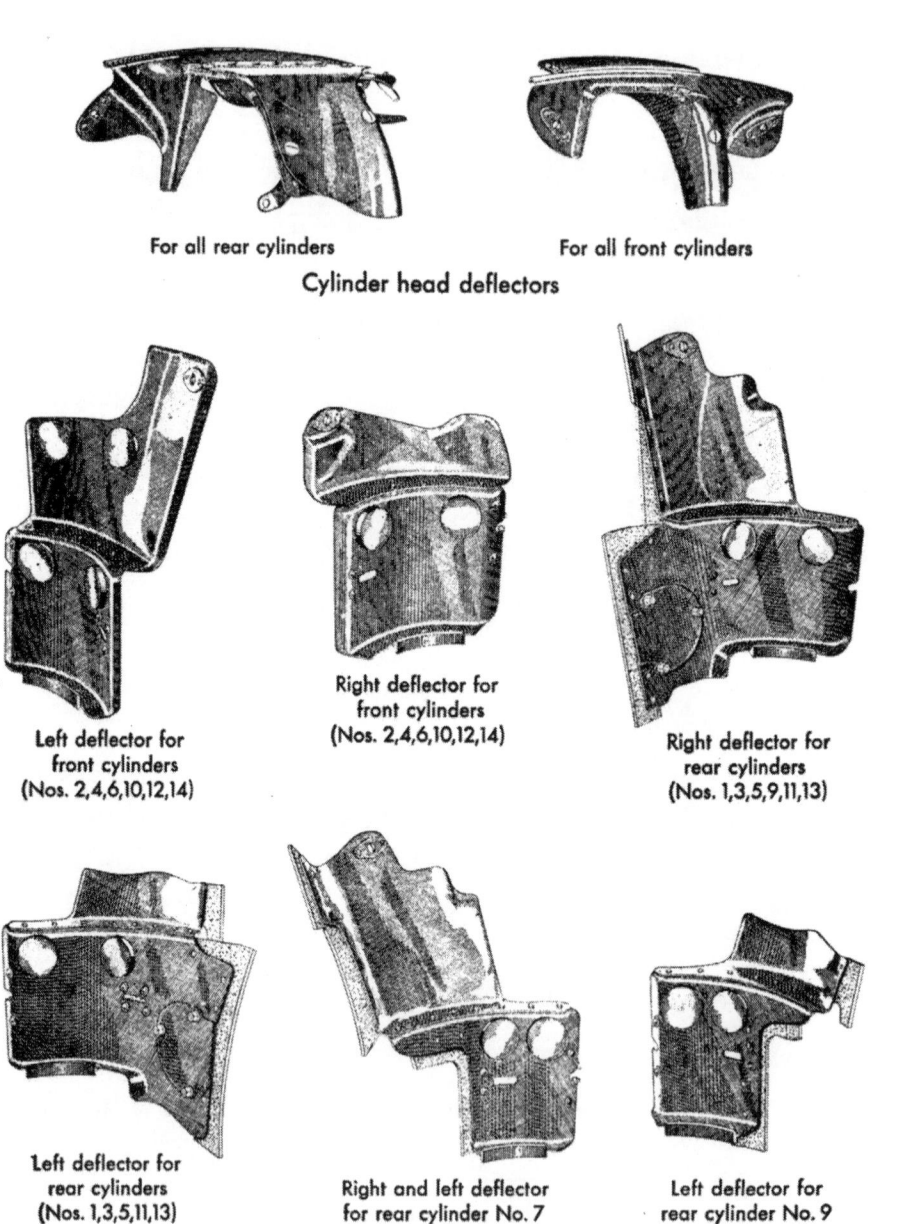

For all rear cylinders   For all front cylinders

Cylinder head deflectors

Left deflector for
front cylinders
(Nos. 2,4,6,10,12,14)

Right deflector for
front cylinders
(Nos. 2,4,6,10,12,14)

Right deflector for
rear cylinders
(Nos. 1,3,5,9,11,13)

Left deflector for
rear cylinders
(Nos. 1,3,5,11,13)

Right and left deflector
for rear cylinder No. 7

Left deflector for
rear cylinder No. 9

Fig. 267. Cylinder air deflectors or baffles for a twin-row, 14-cylinder, radial aircraft engine. (Courtesy Wright Aeronautical Corporation)

264

ENGINE INSTALLATION AND COOLING SYSTEMS

The assembly of different types of engines may vary somewhat but, in general, is the same. The assembly of an engine begins with the power transmission unit which consists of the crankshaft and crankcase. In the in-line engines, the crankshaft may be placed in one part of the crankcase, followed by the connecting rod and the other parts of the crankcase. In radial engines, the crankcase and connecting rods are assembled, followed by the crankcase parts. Connecting rods may already be fitted with the proper pistons and rings, which is usually the practice. The cylinders, with the valve mechanism properly assembled, are slipped over the pistons and fastened to the crankcase. Accessory drive gears and engine accessories are then attached, and the ignition system parts are installed along with the induction system and fuel supply accessories. After the engine is mounted in the aircraft, fuel lines, oil lines, controls, instruments, and electrical wiring should be properly connected. If the engine is liquid-cooled, the coolant lines must also be properly installed.

XIX ENGINE INSTRUMENTS

The earliest aircraft engines were operated entirely without instruments. Many aircraft after World War I were equipped with only a tachometer, oil temperature gauge, and water temperature gauge. These

Fig. 268. A typical instrument panel for a twin-motored aircraft showing the automatic pilot installed. (Courtesy Sperry Gyroscope Company, Inc.)

engines were of comparatively low horsepower. Their low compression ratios did not develop the high temperatures of the later engines. Most engines were operated at comparatively low altitudes, and supercharging was in its infancy. The throttle and ignition switch were practically the only engine controls with which the pilot was concerned. A few

ENGINE INSTRUMENTS

engines had a spark advance and a carburetor choke. Most modern engines have, in addition to the throttle and spark control, mixture controls, propeller controls, and an idle cutoff.

Engine instruments include the tachometer, oil pressure gauge, fuel pressure gauge, manifold pressure gauge, suction gauges, and temperature indicators for oil, air, carburetor, and cylinder heads.

Pressure Gauges. Pressure gauges may operate on the Bourdon tube principle, the aneroid cell principle, or by means of a diaphragm.

The Bourdon tube consists of a sealed metal tube which has been pressed into an elliptical shape. This flattened tube is bent into the form of an arc, as shown in Fig 277. One end of the tube is fastened securely to the end of the instrument case, while the other end is left free to move.

An opening in the fastened end of the tube allows liquid or gas to enter the tube under pressure. Pressure applied within the tube tends to straighten the tube. The free end of the tube is fastened by a suitable linkage to an indicating pointer which is rotated over the face of the dial by the movement of the free end of the Bourdon tube.

The aneroid cell is a flattened, metal, disclike cell. The sides of this cell are usually corrugated as shown in Fig. 272. Air is exhausted from the cell.

When this cell is mounted under spring tension, it becomes extremely sensitive to pressure changes. This cell is connected by a proper linkage to an indicating pointer designed to rotate over the face of a gauge. The diaphragm type of pressure gauge has a sealed-in diaphragm which is deflected by pressure against it.

OIL PRESSURE GAUGE. One of the most commonly used instruments that operates on the Bourdon tube principle is the oil pressure gauge. All aircraft engines obtain lubrication through the use of oil pumps. These pump the oil from either the wet sump within the engine or from a remote oil supply. The oil is delivered to the various engine bearings under pressure.

Fig. 269. An oil pressure gauge indicator. (Courtesy The Electric Auto-Lite Company)

Each engine lubricating system is designed for a pre-determined oil pressure. An oil pressure relief valve and surge chamber are commonly

267

AIRCRAFT ENGINES

installed to prevent excess pressure or sudden changes in pressure in the oil lines. The oil pressure gauge is usually a simple gauge of the Bourdon tube type and is usually calibrated for pressures from zero to approximately 200 lb. of pressure. Heavy oils, when cool, may show excess pressure. Therefore, gauge readings should not be depended upon until the oil has reached the correct operating temperature.

Fig. 270. An engine gauge unit indicating oil pressure, fuel pressure and temperature. (Courtesy The Electric Auto-Lite Company)

The surge chamber, which is a closed chamber in which air is trapped, prevents violent oscillation of the needle due to the action of the pump. The surge chamber is not as important to engines which operate under comparatively low oil pressures as it is to engines which operate under high oil pressures. A comparatively low oil pressure might be approximately 60 lb. per sq. in., while high oil pressures are in excess of 100 lb. per sq. in.

FUEL PRESSURE GAUGE. Most fuel pressure gauges operate on the Bourdon tube principle and are calibrated from zero up to approximately 25 lb. per sq. in. When fuel pressure gauges are used with pressure discharge carburetors, they are usually calibrated up to 25 lb. per sq. in., while gauges used on all other types of carburetors have ranges from zero to 10 lb. per sq. in.

ENGINE GAUGE UNITS. An engine gauge unit is simply a unit in which three instruments are combined. This instrument shows fuel pressure, oil pressure,

ENGINE INSTRUMENTS

and oil temperatures. The oil pressure and fuel pressure are the ordinary Bourdon tube type with the fuel pressure graduated from zero to 10 lb. per sq. in., and the oil pressure graduated from zero to 200 lb. per sq. in. The oil temperature is graduated from zero to 100° C. Each of the units operate entirely separately from each other. The thermometer unit is of a vapor pressure type. The fuel pressure and oil pressure Bourdon tube gauges are operated by the actual pressure of the fuel and oil on the Bourdon tube. The vapor pressure type of Bourdon tube instrument consists of a Bourdon tube which is connected by a capillary tube, that may be as much as 30 ft. in length, with a cell containing a highly volatile liquid. The cell and the Bourdon tube are hermetically sealed to opposite ends of the capillary tube. The cell is placed at the point where the temperature is to be measured. Any change in the temperature of the cell changes the vapor pressure within the capillary tube and the Bourdon tube, which causes a change in the curve of the arc. This change is indicated by the needle on the face of the instrument which is calibrated in terms of degrees. The degrees shown on the instrument face may be either in Fahrenheit or centigrade. This type of temperature indicator is called a vapor pressure thermometer.

MANIFOLD PRESSURE GAUGE. When an aircraft engine is not running, the pressure in the intake manifold and all parts of the induction system is equal to the atmospheric pressure. When a non-supercharged engine is in operation, however, the manifold pressure is less than the surrounding atmospheric pressure because of the suction caused by the air-fuel mixture being drawn into the cylinder. The particular function of the supercharger is to increase the pressure within the manifold until it equals or exceeds normal sea level pressure. The manifold pressure gauge is usually graduated from 10 to 50, the figures indicating inches of mercury pressure.

Fig. 271. A manifold pressure gauge indicator. (Courtesy Eclipse-Pioneer Division, Bendix Aviation Corporation)

The design of the combustion chamber in the engine determines the compression ratio which is most effective for the operation of the engine. The atmosphere grows less dense with altitude, and a non-supercharged enigne gets less air at greater altitudes. As the density of the air decreases, although the cylinder is filled with air, it is at lower pressure

AIRCRAFT ENGINES

and does not contain as much air by weight. By means of supercharging, the pressure is built up by forcing more air into the induction system. By means of proper supercharging, the manifold pressure may be maintained at 30 in. of mercury even at high altitudes. Since the horsepower of the engine depends upon the amount of oxygen burned with the gasoline in the combustion chamber, it will develop full horsepower only when the manifold pressure is maintained at 30 in., regardless of the

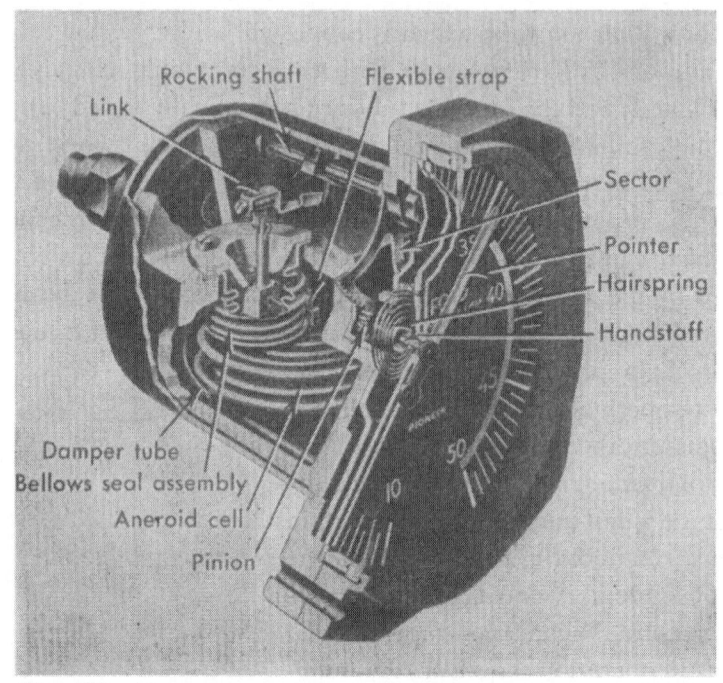

Fig. 272. A cutaway view of a manifold pressure gauge showing its construction. (Courtesy Eclipse-Pioneer Division, Bendix Aviation Corporation)

altitude. Without a manifold pressure gauge, the pilot cannot know the exact pressure within the manifold. A supercharged engine which would develop its maximum horsepower at 10,000 or 12,000 ft. at full throttle would, if the supercharger were left in operation, be seriously damaged by the increased compression near sea level. With the supercharger in operation, the pilot, by watching the indication of the manifold pressure gauge, reduces the manifold pressure by simply closing the throttle until the needle indicates 30 or whatever pressure is desired.

Engines may be operated at higher than 30 in. of mercury for short periods of time, such as take-off and climb, and the pilot must know how

ENGINE INSTRUMENTS

far the throttle may be opened with safety. For example, on the take-off, the engine manufacturer may recommend as much as 35 or 40 in. manifold pressure. It is only by means of the manifold pressure gauge that the pilot can tell just how far to open the throttle to obtain this pressure. Usually, the normal operating range is indicated on the instrument by a green band running from 30 to 36. A high-boost operating range is indicated by a yellow band extending from 36 to 42 on the instrument. Above 42 is a red band which indicates dangerously high pressures. The range covered by these bands varies with different engines in accordance with the manufacturer's specification.

Fig. 273. A manifold pressure indicator for a twin-motor aircraft. This indicator is an Autosyn remote indicating type. (Courtesy Eclipse-Pioneer Division, Bendix Aviation Corporation)

Fig. 274. A sensitive-type manifold pressure gauge. (Courtesy Kollsman Instrument Division of Square D Company)

In taking off, the pilot may open the throttle until the pointer on the gauge reaches the end of the yellow band. Shortly after the take-off, the pilot would close the throttle until the needle indicated the maximum manifold pressure for climbing. As he approached the cruising altitude, he would gradually throttle back until the needle indicated correct cruising pressure.

The manifold pressure gauge operates by means of an aneroid cell or double-metal diaphragm. This instrument is equipped with a two-cell diaphragm. The linkage system starts with a flexible shaft which connects the diaphragm to the bellows-seal assembly. A link ties the bellows to a rocking shaft which is connected with a pinion and a handstaff and pointer. A hairspring is installed to keep all parts of the linkage snugly in position. A sealed chamber in which the diaphragms are placed is con-

271

AIRCRAFT ENGINES

nected with the manifold by an airtight tube. If the pressure in the manifold exceeds 30 in. of mercury, the diaphragms tend to collapse, and this motion is transmitted through the linkage to the indicator needle, which will indicate a pressure above 30 in. of mercury. If the pressure within the manifold becomes less than 30 in. of mercury, the diaphragms expand and the needle indicates pressures below 30.

Fig. 275. An oil pressure warning unit. (Courtesy Eclipse-Pioneer Division, Bendix Aviation Corporation)

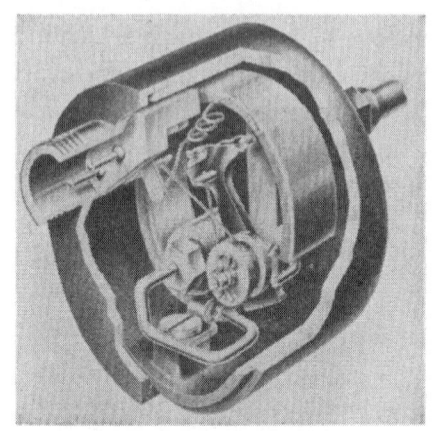

Fig. 276. A cutaway view of an oil pressure warning unit showing its construction. (Courtesy Eclipse-Pioneer Division, Bendix Aviation Corporation)

Pressure Warning Units. It is necessary that the pilot be warned immediately of any decrease in pressure in the oil or fuel system. However it is not possible for the pilot to watch all of the gauges continuously, and the oil or fuel gauge indication may drop suddenly and a considerable length of time could elapse before the pilot notices it unless some definite warning is given. Pressure warning units are designed either to operate warning lights or to sound signals to warn the pilot or flight engineer of this condition. Pressure warning units relieve the person operating an aircraft of the strain of continually watching instruments during flight. These instruments are designed to attract the attention of the operator as soon as there is any pressure drop below a minimum safe operating value. Warning units are installed in the modern airplane to indicate a number of different conditions.

The pressure warning units, of course, indicate low or high pressures. The oil pressure warning unit consists of a pressure-sensitive Bourdon tube and a contact assembly which completes an electric circuit when the oil pressure falls below a predetermined pressure. A fuel pressure

ENGINE INSTRUMENTS

warning unit consists of a fuel chamber, a pressure-sensitive diaphragm, a bellows, and two contacts. As the pressure drops below a predetermined point, two contacts are closed, completing an electric circuit which lights a warning light on the instrument panel. The bellows prevents the gasoline in the fuel chamber from reaching the electrical contacts should the diaphragm by any chance spring a leak. The oil pressure warning unit is a Bourdon tube which is normally extended by the oil pressure holding the electrical contacts apart. As the pressure falls, the Bourdon tube contracts, closes the contacts, and turns on a light on the instrument panel or sounds a warning. A switch opens an electric circuit to prevent the light from burning when the engine is not in operation.

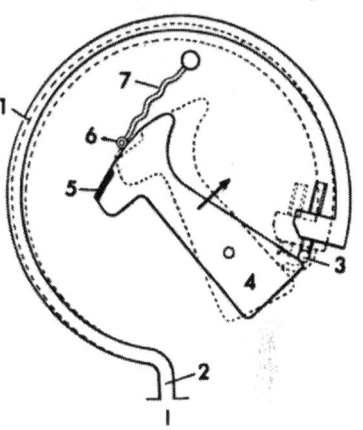

Fig. 277. A diagram showing the operation of a Bourdon tube connector. (1) Bourdon tube; (2) Bourdon tube connector; (3) adjusting pin attached to end of Bourdon tube; (4) commutator assembly; (5) contact surface; (6) contact part; (7) wiping spring which closes circuit. (Courtesy Eclipse-Pioneer Division, Bendix Aviation Corporation)

The fuel pressure unit containing a bellows and a diaphragm is operated by the fuel pressure which enters the instrument from the fuel line. The fuel enters a chamber, one wall of which is a diaphragm. The diaphragm expands under the high pressure of the liquid and, by pressing the electrical contacts apart, opens the electric circuit. If the fuel pressure falls, the diaphragm contracts, and this movement of the diaphragm is followed by the bellows which carries the lever to which is attached the multiple contact. As the pressure reaches the predetermined safe minimum, the contact closes and causes a lamp on the instrument panel to light. It is possible for this type of gauge to give warning of excessive pressure by use of a double contact. One contact closes with low pressure, and the other closes with high pressure, the contact point being carried by the expanding and contracting bellows.

Suction Gauges. A number of aircraft instruments, including the bank-and-turn indicator, contain a gyroscope which is driven by a stream of air produced by suction. This suction may be developed by either a vacuum pump, a Venturi tube, or the suction in the induction system of the engine. It is necessary to have a gauge which indicates the amount of suction being developed to be sure that the instruments are operating

AIRCRAFT ENGINES

properly. The suction gauge consists of a metal aneroid cell or diaphragm connected with the proper linkage to move the indicator needle over

Fig. 278. A fuel pressure warning unit. (Courtesy Eclipse-Pioneer Division, Bendix Aviation Corporation)

Fig. 279. A cutaway view of a fuel pressure warning unit, showing its construction. (Courtesy Eclipse-Pioneer Division, Bendix Aviation Corporation)

the face of the instrument. Most suction instruments are graduated in inches of mercury from zero to 10. The pressure-sensitive diaphragm cell measures variations in suction caused by the expansion of the cell under reduced conditions of pressure. It is similar to an altimeter working under

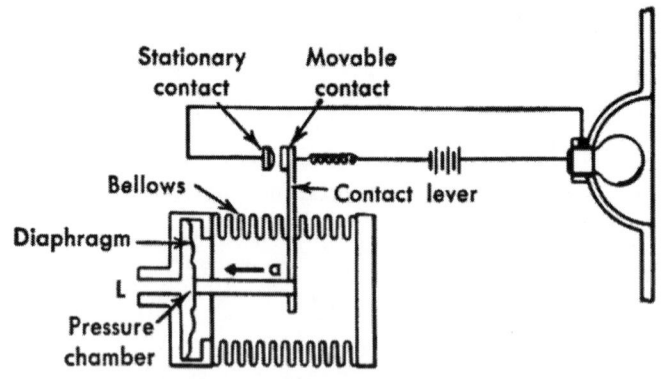

Fig. 280. A drawing showing the operation of a fuel pressure warning unit. a — Direction of movement to close contact. L — Fuel inlet to pressure chamber. (Courtesy Eclipse-Pioneer Division, Bendix Aviation Corporation)

artificial changes in pressure. The altimeter cell expands as it passes from low to high altitudes. The cell in the suction gauge expands due to decreased pressure caused by the vacuum pump or other sources of

ENGINE INSTRUMENTS

suction. The suction gauging cell is, of course, mounted in an airtight case connected by a tube with the suction-producing mechanism.

Fig. 281. An oil pressure indicator. (Courtesy AC Spark Plug Division, General Motors Corporation)

Fig. 282. An oil gauge engine unit which operates on a variable electrical resistance principle. (Courtesy AC Spark Plug Division, General Motors Corporation)

Tachometers. Tachometers are instruments which indicate engine r.p.m. They are operated by centrifugal force, electric or magnetic force, or by a timing arrangement such as that used in a watch.

It is necessary to have a tachometer for each engine. The tachometer informs the pilot whether or not the engine is functioning properly. Before taking off, the engine is run up a short time at full power to determine whether or not it is developing its required r.p.m. The tachometer is also necessary to determine whether or not the engine is being operated within its proper speed limits. The tachometer illustrated in Fig. 286 is calibrated from 500 to 2500 r.p.m. Other instruments may be calibrated within different ranges. When used in conjunction with a controllable-pitch or constant-speed propeller on a supercharged engine, a tachometer indication is combined with the manifold pressure gauge reading and the propeller pitch to arrive at the power developed by the engine.

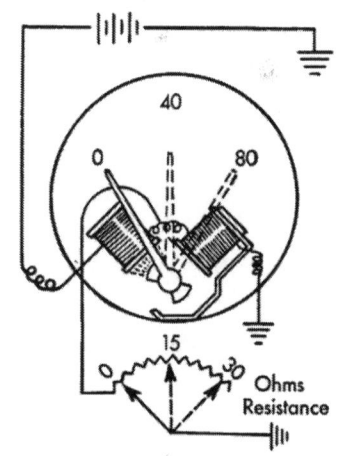

Fig. 283. A phantom view of an oil gauge indicator used in connection with a variable-resistance type engine unit. (Courtesy AC Spark Plug Division, General Motors Corporation)

275

AIRCRAFT ENGINES

CENTRIFUGAL TACHOMETERS. The centrifugal tachometer consists of a flyweight assembly connected to a linkage system which moves the pointer over the face of the instrument to indicate the engine r.p.m. The

Fig. 284. A suction gauge which indicates suction in inches of mercury. (Courtesy Eclipse-Pioneer Division, Bendix Aviation Corporation)

Fig. 285. A cutaway view of a suction gauge showing its construction. (Courtesy Eclipse-Pioneer Division, Bendix Aviation Corporation)

flyweight mechanism consists of three weights which are connected by links to a top and a bottom collar. This tachometer operates by means of these whirling weights. The rotating part is called a governor. The three weights are pivoted at each end, allowing them to move outward when acted upon by centrifugal force. As the shaft spins, these weights move outward. The more rapidly the shaft spins, the farther the weights move. As the weights move outward, the bottom collar slides upward and squeezes the governor spring together. The distance which the bottom collar moves as it compresses the spring measures the engine speed. This motion is transmitted to a pointer indicating the r.p.m. A hairspring maintains tension on the linkage forcing the indicator to follow all movements of the linkage. As the engine speed decreases, the centrifugal force becomes less, and the

Fig. 286. A centrifugal tachometer. (Courtesy Eclipse-Pioneer Division, Bendix Aviation Corporation)

ENGINE INSTRUMENTS

governor spring pulls the weights back toward the shaft. This action allows the governor spring to lengthen and, as the bottom collar moves downward under the pressure of the spring, a lower r.p.m. is indicated.

ELECTRIC TACHOMETERS. The electric tachometer is used in the same manner as the centrifugal tachometer. While the centrifugal tachometer is usually used on single motor installations where the flexible shaft is

Fig. 287. A cutaway view of a centrifugal tachometer showing its construction. (Courtesy Eclipse-Pioneer Division, Bendix Aviation Corporation)

comparatively short, the electric tachometer is used on multiengine installations, and the shaft is replaced by electric wires. It is necessary to have one tachometer for each engine and, in large multiengine aircraft, the length and weight of the flexible cable necessary for a centrifugal tachometer becomes excessive. Most four-motor aircraft have one tachometer for each engine in the pilot's cockpit and another set in the flight engineer's station. These aircraft, therefore, have eight tachometers. Such installations are much simpler with electric wires than with long flexible shafts.

The electric tachometer consists of two widely separated units. One unit is an alternator and is mounted on the engine itself and is driven by the engine. The other unit is the tachometer indicator and is mounted on the instrument panel. The panel unit contains a synchronous motor which receives its power and speed-controlling frequency from the alternator by means of wire connections. The synchronous motor driven by the alternator in turn controls the tachometer indicating mechanism.

Fig. 288. Two electric tachometer indicators. (Courtesy Eclipse-Pioneer Division, Bendix Aviation Corporation)

The alternator consists of a stator and a rotor. This unit is connected to the tachometer drive shaft of the airplane engine. As the engine rotates, the alternator drive shaft rotates with it, thus spinning the rotor. The rotor revolving inside the stator coil generates an alternating current, the frequency of which varies directly with the r.p.m. of the engine. Electric wires conduct this current to the synchronous motor mounted in the instrument panel. This motor, which operates the r.p.m. indicator, matches exactly the speed of the alternator on the engine. The self-synchronous motor also contains a stator coil and a rotor. A small permanent magnet and its return path rotate at the same speed as the rotor. The function of the whole system is to spin the magnet assembly at a speed equal to that of the engine tachometer drive.

The rest of the mechanism contained in the tachometer converts the motion of the magnet to a pointer indication of the engine r.p.m. The indicator mechanism consists of a drag cup which is mounted on the shaft connected to the pointer. A hairspring is mounted between the pointer and the drag cup. The drag-cup wall is between the magnet and

ENGINE INSTRUMENTS

its return path. There is no mechanical connection between the drag cup and the other two units. Lines of force set up by the field of the magnet loop from the magnet to the return path and back to the magnet. These lines of force leave the magnet at its north pole and reenter the

Fig. 289. A cutaway view of an electric tachometer showing its construction. (Courtesy Eclipse-Pioneer Division, Bendix Aviation Corporation)

Fig. 290. An electric indicator tachometer of the sensitive type. (Courtesy Kollsman Instrument Division of Square D Company)

magnet at its south pole, thus completing a magnetic circuit. As the magnet spins, these lines of force rotating with it sweep through the metal wall of the drag cup. A magnetic field moving through metal sets up eddy currents in the metal. These induced eddy currents in the drag cup

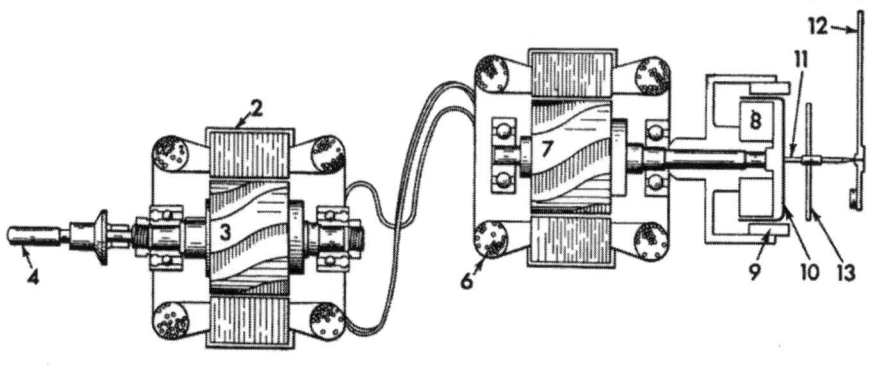

Alternator 1   Synchronous motor 5

Fig. 291. A diagram showing the operation of an electric tachometer. (1) The alternator unit; (2) stator; (3) rotor; (4) alternator drive shaft; (5) a synchronous motor; (6) stator coil; (7) synchronous motor rotor; (8) permanent magnet; (9) magnet's return path; (10) drag cup; (11) staff; (12) pointer; (13) hairspring. (Courtesy Eclipse-Pioneer Division, Bendix Aviation Corporation)

AIRCRAFT ENGINES

produce a rotative force causing the drag cup to follow the magnet around. The purpose of the hairspring is to resist the rotation of the drag cup. These two forces are balanced in such a way as to provide a pointer deflection which indicates the r.p.m. of the engine. The faster the engine rotates, the faster the magnet spins. As the magnet spins faster and faster, the eddy currents set up in the drag cup become greater, and the drag cup twists with a greater force against the hairspring. This causes the hairspring to be twisted into a tighter spiral, moving the pointer on the dial to indicate the higher engine r.p.m.

CHRONOMETRIC TACHOMETERS. A chronometric tachometer is an instrument used to indicate engine r.p.m. The instrument may be used in the same manner as other tachometers, except that it is usually limited to single-engine airplanes as it is shaft-driven. Standard types of chronometric tachometers usually have a range of from zero to 3500 r.p.m. This type of tachometer is designed to rotate at one half the crankshaft speed, and is provided with a reversing mechanism which prevents damage due to engine kickback and allows the drive unit to be driven in either direction.

There are four principal parts of the chronometic tachometer: (1) the driving mechanism, (2) the counting mechanism, (3) the watch mechanism, and (4) the synchronizing cams which time the counting arrangement. The watch mechanism, which times the action of the synchronizing cams, and the counting mechanism, which counts the revolutions during one-second periods, are connected with a pointer gear which shows the results on the dial.

The chronometric tachometer actually totals the revolutions which occur during each alternate second. These revolutions are automatically measured by a special watch escapement. A chain of gears driven by the shaft from the engine operates the escapement cam and counting system. The escapement cam is operated by a friction drive which causes a gear to become meshed with the counting gear intermittently for one-second intervals. During the one second that these gears are in mesh, the large gear rotates a distance proportionate to the speed of the drive system. This movement is transmitted to the pointer gear which rotates the indicator hand through the same distance as the counting gear. At the end of the second, the counting gear is disconnected and returned by a spring to its starting position. The indicator hand is held stationary during this interval.

If, during the next second, the engine has run faster than before, the

ENGINE INSTRUMENTS

counting gear pushes the indicator pointer farther around the dial. If the engine has run slower than it did during the preceding second, the pointer is released and a spring drops it back to the position of the counting gear at the end of the second. The pointer moves by jerks. The indicated reading at any instant is the speed of the engine during the previous second. This cycle of operation continues as long as the drive shaft continues to rotate. After the engine is stopped, a ticking may be heard which is caused by the escapement continuing to run because of the tension of the main spring until this spring loses its stored-up energy. This ticking may last as long as 30 sec.

Temperature Indicators or Thermometers. Temperature is indicated by means of a thermometer. Any device for indicating temperature may be classed as a thermometer.

The ordinary thermometer depends upon the expansion of a liquid contained in a bulb which is connected to a capillary tube. The capillary tube is graduated to indicate temperature degrees. Changes of temperature of the liquid in the bulb cause the liquid to expand or contract, extending or retracting a column of liquid in the capillary tube.

Vapor pressure thermometers depend upon changes in vapor pressure within a sealed system due to changes in temperature.

Electric thermometers usually depend upon the variation in electrical conductivity of conductors due to changes in temperature. The electrical conductivity varies with temperature, being inversely proportional to the temperature. At high temperatures, most conductors carry less current than they carry at low temperatures.

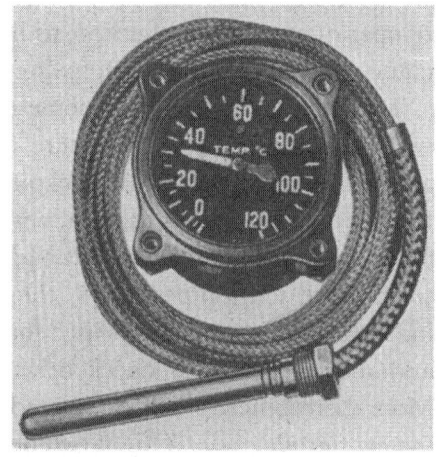

Fig. 292. A vapor pressure thermometer. (Courtesy The Electric Auto-Lite Company)

Another type of electrical thermometer is the thermo-couple. When two wires of different composition are joined at one end and this junction is heated, an electric current is generated in the wires which may be measured at the opposite ends of the wires by means of a galvanometer. This electric current varies directly with the temperature of the heated junction at the ends of the wires.

Thermometers are used to determine the temperature of engine lubricating oil and may measure the oil temperature as it enters the engine or as it drains from the engine either into a crankcase sump or into a sump from which it is removed for return to the storage tank.

Thermometers are used to measure the temperature of the air both inside and outside of the cockpits and cabins and in the carburetor. The measurement in the carburetor is usually made in the carburetor throat just before the air is mixed with the fuel. The temperature of the liquid in liquid-cooled engines is taken to determine the operating temperature under various flight conditions. Before taking off, the temperature of the cooling liquid, if the engine is liquid cooled, and the temperature of the oil are always checked. The cooling liquid should reach a temperature of approximately 70° C. or 160° F. The oil temperature should be approximately 30° C. or 90° F. before the take-off is begun.

Thermometers enable the pilot to adjust the oil radiator shutters or other controls to maintain the proper oil temperature. Carburetor thermometers are used to determine when the temperature is such that ice might form in the carburetor. When carburetor heat is used, the temperature must be checked, as hot air admitted to the carburetor may cause detonation in the engine's cylinders.

The term "free air thermometer" is used to indicate thermometers which measure the temperature of the atmosphere either outside the aircraft or in the cabin or cockpits. The outside temperature is important because it largely determines the conditions under which ice may form on the various parts of the airplane. When thermometers are used to indicate ice temperatures, they are sometimes called "ice-warning indicators." Thermometers used for this purpose are accurately calibrated and may be used to check other thermometers used in the aircraft. Most thermometers are calibrated either to the Fahrenheit scale or to the centigrade scale. Thirty-two degrees on the Fahrenheit scale is equal to zero degrees on the centigrade scale and indicates the freezing point of water; 212° on the Fahrenheit scale corresponds to 100° on the centigrade scale and is the temperature at which water begins to boil under normal conditions of pressure. Oil thermometers are usually calibrated from 0° to 100° C., which corresponds to 32° to 212° F.

The thermometers used to indicate the temperature of the liquid coolant are usually calibrated from 0° to 200° C., which corresponds to 32° to 390° F. Free air thermometers are usually calibrated to indicate approximately minus 40° C. to plus 50° C., which corresponds to approxi-

ENGINE INSTRUMENTS

mately minus 40° F. to plus 155° F. Since it is necessary to have the temperature indicated on the instrument board for the convenience of the pilot, it is often necessary to transmit to the instrument board, the temperature at a point considerably remote.

VAPOR PRESSURE THERMOMETERS. The vapor pressure thermometer is a remote indicating type. This thermometer consists of the indicating instrument mounted on the instrument board, a long capillary tube, and a metal bulb located at the point where the temperature is to be measured. The bulb is usually a hollow brass cylinder having an outside

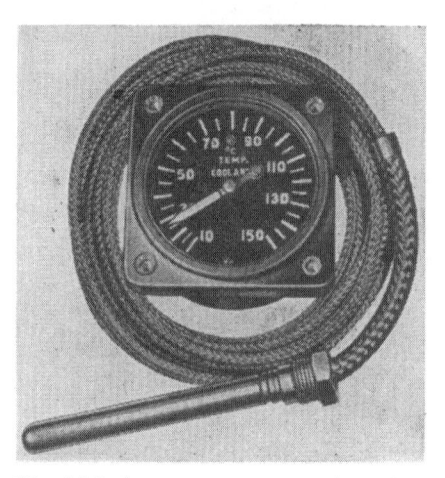

Fig. 293. A vapor pressure coolant thermometer. (Courtesy The Electric Auto-Lite Company)

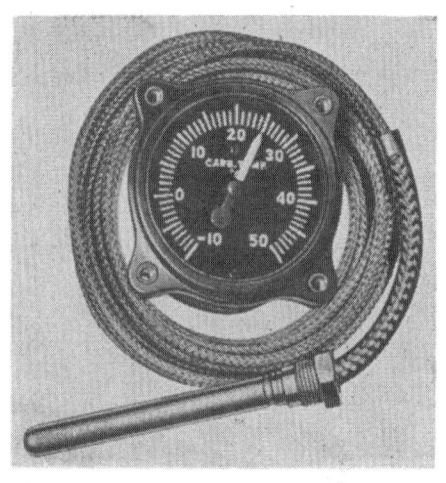

Fig. 294. A vapor-pressure type carburetor thermometer. (Courtesy The Electric Auto-Lite Company)

diameter of approximately ½ in., and it may vary from approximately 2 in. to approximately 4 in. in length. The indicator itself consists of a Bourdon tube connected by a linkage system to the indicator needle. The capillary tube is fastened to the stationary end of the Bourdon tube with an airtight joint. The other end of the capillary tube forms an airtight joint with the bulb which contains a highly volatile liquid, such as methyl chloride.

As the temperatures of the liquid in the bulb change, the vapor pressure in the entire system changes and affects the Bourdon tube. Vapor pressure does not vary directly with temperature, and a series of stops are arranged to regulate the movement of the Bourdon tube. The capillary tube is usually of copper protected by a braided metal covering. When carefully calibrated, this type of thermometer gives quite accurate temperature indications. The vapor pressure thermometer may be used

283

AIRCRAFT ENGINES

to measure the temperature of free air, cabin air temperature, cylinder head temperatures, cylinder base temperatures, oil temperatures, coolant temperatures, carburetor air temperatures, or carburetor mixture temperatures. The capillary tube may be any length up to more than 20 ft.

ELECTRICALLY OPERATED THERMOMETERS. These thermometers consist of the temperature sensitive element or bulb and wires connecting the bulb with the instrument which is installed on the instrument panel. The indicating ranges on these thermometers are similar to those on the

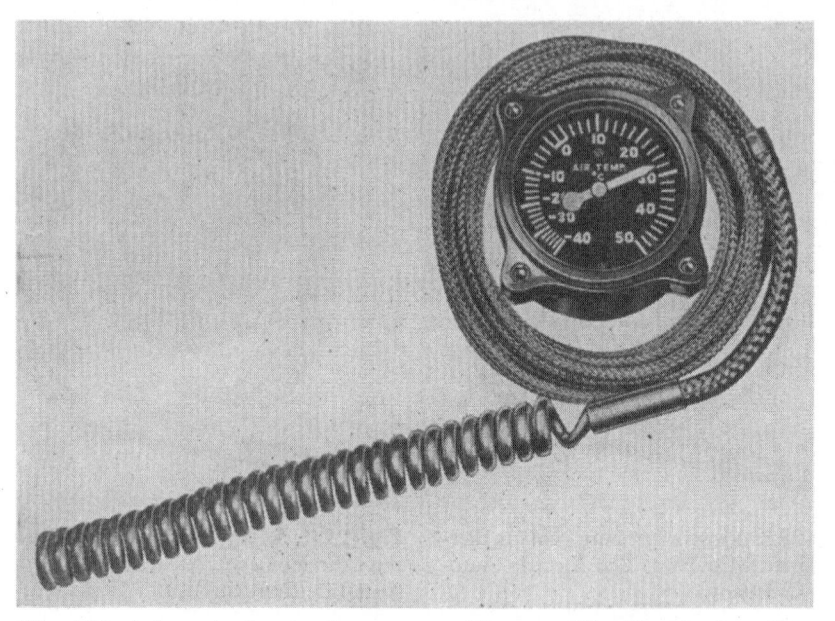

Fig. 295. A free air electric thermometer. (Courtesy The Electric Auto-Lite Company)

pressure type. These thermometers may be used to measure any of the temperatures desired in the same manner as the pressure type.

Electrical thermometers are of two general types. One type makes use of the Wheatstone bridge arrangement and the other type works similarly to a regular voltmeter having a permanent magnet and a moving coil.

The Wheatstone bridge type contains three resistance coils having a resistance of 100 ohms each. These coils are of manganin wire. The three coils form three arms of the Wheatstone bridge. The sensitive element or bulb is a coil of pure nickel wire wound on an anode-treated aluminum tube. This coil has exactly 100 ohms resistance at a temperature of 0° C. or 32° F., depending on the scale used. This coil is inserted into a tube of

ENGINE INSTRUMENTS

Monel metal and is soldered into place. The two ends of the nickel coil are soldered to a Bakelite plug having molded into it two silver-plated, brass, female inserts. These inserts are arranged to receive two split pins made up of silver-plated brass. These split pins are connected with the wires leading to the indicator part of the instrument on the instrument panel. This sensitive bulb forms the fourth arm of the Wheatstone bridge. Metals, as a rule, are better conductors of electricity when cold than when hot. The indicating instrument, which is similar to a voltmeter, is connected across the Wheatstone bridge. When the temperature is at $0°$ C., the sensitive element and one of the 100-ohm coils are balanced against the other two 100-ohm coils, and no current flows through the circuit in which the instrument is inserted. With any change in the temperature of the sensitive element, this balance is upset. If the temperature rises, the resistance increases in the sensitive element, and part of the current flowing through the bridge arrangement passes through the indicating instrument. This causes a deflection of the indicating needle toward the right, and the deflection indicates accurately the temperature of the bulb. If the temperature falls below $0°$ C., the resistance in the bulb is decreased, and current flows through the instrument circuit in the opposite direction deflecting the needle to the left. The instrument itself is electrically shielded to protect it against magnetic effects of other instruments in the electrical system of the aircraft.

One type of electrical thermometer contains a movable coil and a permanent magnet and is called the ratio-type thermometer. This thermometer is an electrical arrangement made up of a circuit having two parallel branches. One branch has a fixed resistance in series with a coil of known resistance. The other branch has the sensitive element or bulb in series with a second coil of known resistance. As the current varies in the circuit, due to changes in the resistance of the sensitive element because of changes in temperature, the current flows through the instrument in one direction for temperatures above zero, and in the other direction for temperatures below zero, deflecting the indicating needle either to the right or to the left depending upon the direction of current flow.

CYLINDER TEMPERATURE INDICATORS (THERMO-COUPLE). Another type of temperature indicator or thermometer is based on the thermo-couple. This instrument consists of a sensitive indicator of the voltmeter type, the wires connecting with the thermo-couple, and the thermo-couple itself. When two wires of different composition are joined together at

AIRCRAFT ENGINES

one end and this junction is heated, an electric current is set up in the wires which may be detected by connecting the other ends of the wires across a sensitive voltmeter or galvanometer. The number of the volts generated depends on the difference in temperature between the hot ends of the wires and the cold ends of the wires. The ends of the wires heated are sometimes called the "hot junction," and the opposite ends are called the "cold junction." One wire or lead of the thermo-couple is usually iron, and the other an alloy of copper and nickel called constantin. These two metals are selected because they develop a definite electromotive force per degree change in temperature.

The two wires in close contact are brazed to the surface of a solid copper gasket which is designed to replace a standard spark-plug gasket. This gasket is inserted in place of a spark-plug gasket and indicates accurately the temperature of the cylinder head at that point. A thermo-couple may have any desired form of ending and may be inserted under a crankcase stud nut or clamped to any part of the aircraft engine to indicate the temperature. The cold junction must have a compensating device to cancel out the effect of temperature changes of the instrument itself. A bimetal spiral spring is attached to one of the controlled springs of the instrument for this purpose. The indicator needle is moved not only by the voltage of the thermo-couple, but also by the temperature surrounding the instrument itself. When a thermo-couple is disconnected, the instrument indicates the temperature of the surrounding air. An arrangement is also included to compensate for the temperature of the moving coil.

Air-Fuel-Mixture Indicator (Exhaust Gas Analyzer). The air-fuel-mixture indicator or exhaust-gas analyzer is an instrument to indicate the ratio between air and fuel in the mixture being used by the engine. The air-fuel mixture is determined by the composition of the exhaust gases. When air and fuel are burned together, a number of by-products are given off in the exhaust gases. The exhaust gas from the engine contains water vapor, carbon dioxide, carbon monoxide, oxygen, hydrogen, nitrogen, and a small percentage of other gases. The ratio between the air and fuel burned in the engine cylinder affects the amount of these gases in the exhaust gas.

Without a fuel-mixture indicator, the pilot usually determines, as nearly as possible, the correct mixture by moving the mixture control toward the rich position until the r.p.m. of the engine begin to fall off. The control is then moved toward the lean position until the r.p.m. of

ENGINE INSTRUMENTS

the engine again begin to decrease. He then picks out a position between these two points where the engine develops the maximum r.p.m. With the fuel-mixture indicator, he can determine, by reading the indication on the gauge of the instrument, when the mixture is in the proper ratio.

The instrument consists of a Wheatstone bridge arrangement, an indicator of the sensitive voltmeter type, a current supply, and a ballast tube. The ballast tube is merely a resistance coil sealed in a hydrogen-filled glass tube to protect it from temperature, pressure, and altitude effects. The ballast tube maintains the current supply at 4.1 v. Two arms of the Wheatstone bridge are arranged so that they are exposed to the exhaust

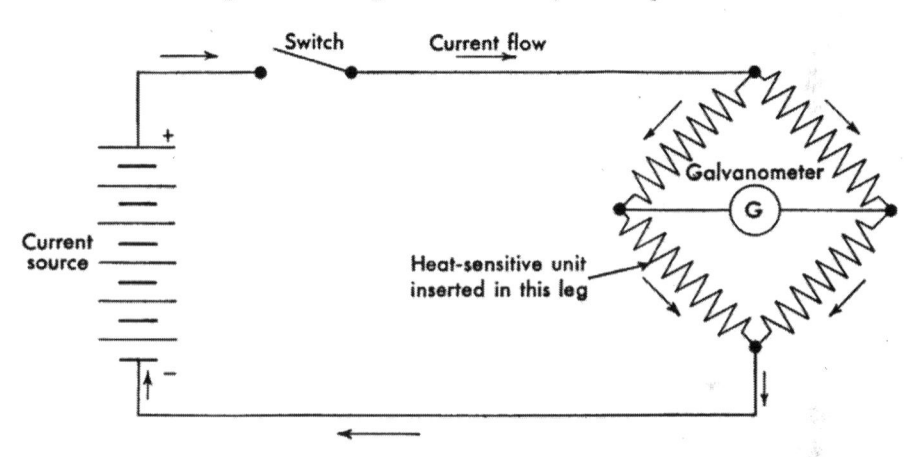

Fig. 296. A diagram showing the Wheatstone bridge circuit.

gases from the engine. The other two arms are sealed in a cell filled with air which is kept saturated with moisture by means of a removable wick.

The principle of operation is based upon the difference in the heat conduction of hydrogen and carbon dioxide. Hydrogen conducts heat about six times as readily as does air, while carbon dioxide has a heat conductivity of approximately one half that of air. Hydrogen, therefore, conducts heat about twelve times as readily as does carbon dioxide.

Enough current is allowed to flow through the Wheatstone bridge to maintain the resistors at a temperature of about 260° F. A continuous sample of gas from the exhaust manifold is carried through the cell containing two arms of the Wheatstone bridge. The rate at which the heat is carried away from these two arms indicates the composition of the exhaust gases.

The indicator needle is so set that, with the proper mixture, it indicates zero or remains at the center of the dial. A rich mixture, which is one

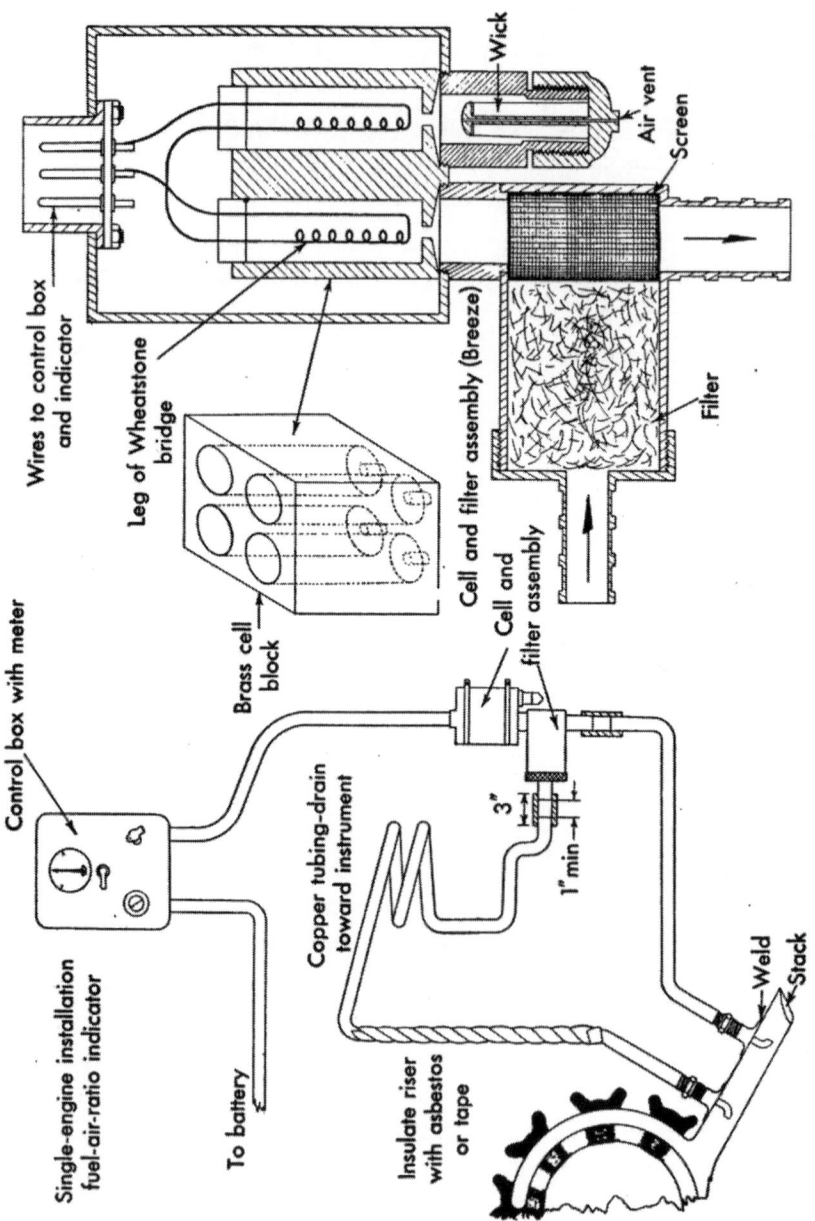

Fig. 297. A diagram showing the construction and installation of an air-fuel-mixture indicator, or exhaust gas analyzer.

ENGINE INSTRUMENTS

having more fuel than is necessary for the correct air-fuel ratio, produces a larger proportion of hydrogen, and the carbon dioxide content is reduced. Hydrogen carries away the heat from the coils more rapidly than carbon dioxides. This rapid carrying-away of heat lowers the temperature of the coils, and the needle on the indicator swings toward a rich mixture. The richer the mixture, the farther the needle will swing as the richer mixtures increase the content of hydrogen in the exhaust gas. As the mixture becomes lean, the amount of carbon dioxide increases and the amount of hydrogen decreases. This combination of gases removes less heat from the coils, and the needle on the indicator swings to the lean position. As the mixture is made more lean, the amount of hydrogen decreases and less heat is carried away from the coil and the indicator swings farther to the lean side.

There is one condition which the pilot must keep in mind which is that, when detonation in the cylinder occurs, the amount of hydrogen in the exhaust gas is greatly increased. Detonation occurs with a lean mixture and the needle, under these conditions, will indicate an overrich mixture. As the mixture is leaned still further, more detonation occurs and the needle indicates a still richer mixture. If, at any time, as the mixture is leaned, the needle suddenly indicates rich, detonation is indicated and the mixture lever should be moved toward the rich position.

Self-Synchronous Instruments. Self-synchronous instruments are usually of the remote indicating type. The transmitting part of the instrument transmits the desired information to a remotely located instrument panel. These instruments may indicate such conditions as fuel pressure, oil pressure, manifold pressure, engine r.p.m., landing-gear position, tail-wheel position, flap position, temperatures, liquid flow, or fuel levels. By means of selector switches, one instrument on the panel may be used in connection with several transmitters. For example, the temperature of four engines may be indicated on a single instrument by using a selector switch in one of four positions. Some indicators may be equipped with more than one indicating needle. One needle indicates a condition in the right engine, while the other needle indicates the corresponding condition in the left engine.

There are several satisfactory, self-synchronizing, remote indicating systems in common use. The trade names of three systems are: the Selsyn, the Telegon, and the Autosyn. The operation of these systems is similar but they are designed to operate on different electric currents. The Selsyn systems are designed to operate from a standard 12- or 24-volt storage battery; the Telegon instruments are designed to operate on a 110-volt alternating current; while the Autosyn instruments are designed to operate on a 26-volt, 400-cycle current. Some of these instruments may, by means of an adapter, use currents from batteries or from a source having a different phase from that for which the instrument is designed.

The Autosyn System. The Autosyn system is a system of remote indication. This system makes it possible to transmit to the pointer of an instrument on the instrument panel the actions of a distantly placed measuring instrument or mechanism. While this system may be used on small aircraft, it is generally used on large multiengine aircraft in which a considerable distance separates the point where the measurement is to be made and the instrument panel.

The system is composed of two units — a transmitter and a receiver or indicator. Electric wiring joins these two basic points of the Autosyn remote indicating system. Both transmitter and indicator units contain an Autosyn. The transmitter contains a measuring device and an Autosyn. The indicator consists of an Autosyn plus a pointer. The pointer is attached to the Autosyn rotor shaft which rotates it about the instrument dial by means of the shaft.

The Autosyn represents an adaptation of the synchronous motor principle. The self-synchronous motor principle consists of two separated motors which operate in exact timing with each other. The rotor of one motor turns exactly the same distance as the rotor of the other motor. In the Autosyn, the rotors neither spin nor produce power. The electrical design of the Autosyn is such that the rotors of two connected Autosyns match each other as to position when energized by means of an electric current. The rotor of one Autosyn moves only the distance necessary to match any movement of the rotor of the first Autosyn.

The transmitter and indicator Autosyns are much alike both in construction and in electrical characteristics. Each has a rotor and a stator. When a rotor is energized by an alternating electric current, a transformer action causes three distinct voltages to be induced in the secondary stator windings. The values of the voltage vary with the position of the rotor in relation to the stator. Any change in position of the rotor causes a new and completely different combination of the three voltages which are induced.

When two Autosyns are connected, the rotors of both Autosyns occupy exactly the same position in relation to their stators. Both sets of induced voltages are then equal and opposite, and no current flows in the inter-

ENGINE INSTRUMENTS

connected stator leads. Under this condition, both rotors remain stationary. When the two rotors do not have exactly the same position in relation to their stators, the voltages in the two stators are not alike. When the voltages are not alike, rotation of the second rotor occurs which continues until both rotors are in identically the same position in their electric fields. When they reach the same relative positions, the induced voltages are in balance and the rotation stops.

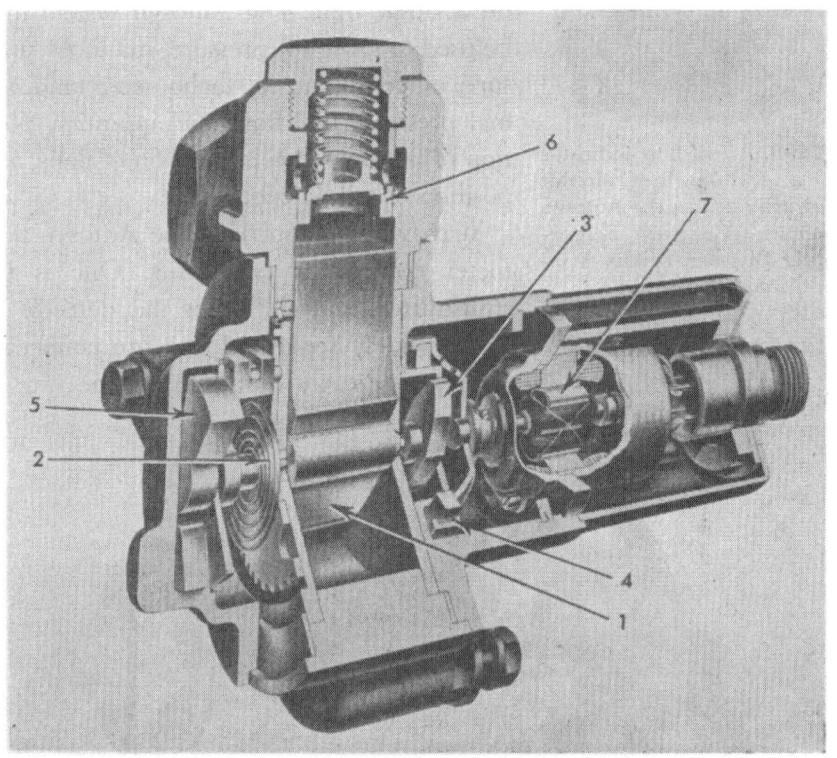

Fig. 298. A cutaway view of an Autosyn fuel flow meter showing its construction. (1) Vane; (2) spring; (3) bar magnet; (4) ring magnet; (5) damper vane; (6) relief valve; (7) transmitter rotor. (Courtesy Eclipse-Pioneer Division, Bendix Aviation Corporation)

When a measuring element is used to turn a rotor of a transmitter Autosyn to a certain position, the rotor of the indicator Autosyn takes the same identical position in its field. The indicator Autosyn remains fixed until a change takes place in the transmitter. The Autosyn system moves the indicating pointer of an aircraft instrument by direct action. The measuring part of the transmitter unit is connected to any device which will cause the rotor to move indicating the position of the mechanism. The measuring device may be a Bourdon tube, a pressure dia-

AIRCRAFT ENGINES

phragm, a flyweight governor, or any type of prime mover. Instead of moving the pointer by means of a linkage system, the gauge element is moved by the indicating rotor. The movement of the transmitter rotor causes a movement of the Autosyn indicator rotor which, in turn, moves the pointer.

Two or more indications may be shown on a single dial. The Autosyn system may be used to show oil pressure, manifold pressure, oil temperature, tachometer readings, fuel pressure, fuel flow, fuel quantity, position of landing wheels, position of flaps, and position of tail wheels.

Fig. 299. A fuel flow indicator for a twin-engine aircraft which operates on the Autosyn principle. (Courtesy Eclipse-Pioneer Division, Bendix Aviation Corporation)

AUTOSYN FLOW METER. The Autosyn flow meter consists of two units. One is the transmitter, and the other the detector or indicator. The two units may be widely separated and are connected by electric wires. This instrument indicates at all times the rate at

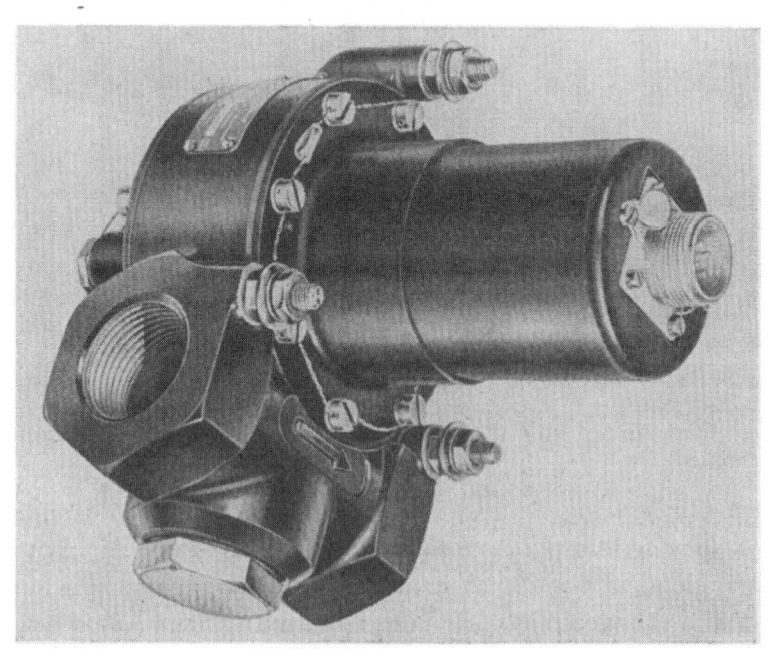

Fig. 300. An Autosyn fuel flow meter. (Courtesy Eclipse-Pioneer Division, Bendix Aviation Corporation)

which gasoline flows from the tanks to the engine. Some instruments indicate how many gallons per hour are flowing, but most instruments

ENGINE INSTRUMENTS

indicate the number of pounds per hour of gasoline that the engine is using.

Flow meters are primarily used to get the most economical performance of the engines. An overrich mixture is indicated by a higher rate of flow than should be used to develop the indicated horsepower. An overrich mixture is corrected by adjusting the mixture controls. The transmitter

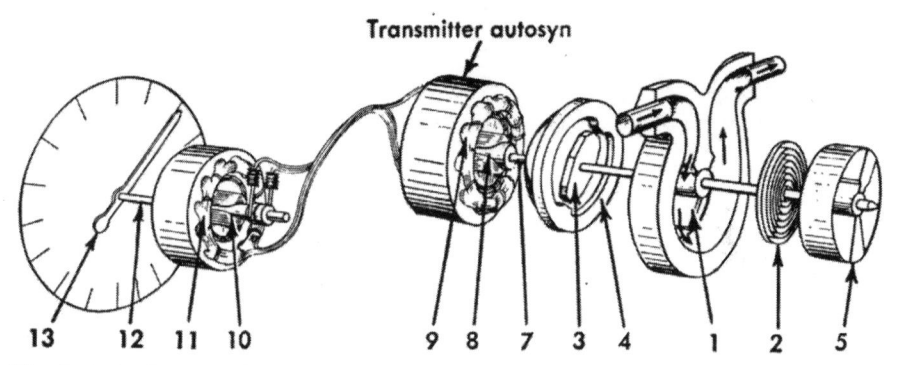

Fig. 301. A schematic drawing of the Autosyn flow meter transmitter. (Courtesy Eclipse-Pioneer Division, Bendix Aviation Corporation)

OPERATION. Gasoline entering the inlet port strikes the vane (1). Impact of the fuel plus pressure drop across the vane, causes the vane to move against the restraining force of a calibrated spring (2). The two forces are so balanced that any static position assumed by the vane represents a measure of the rate at which fuel is passing through the metering chamber.

The bar magnet (3), attached to the vane shaft, moves with the vane. The ring magnet (4) repeats motions made by the bar magnet, since the two are magnetically coupled. The ring magnet is attached to the rotor shaft (7) of the transmitter Autosyn; therefore, motions of the ring magnet cause the rotor (8) to change position relative to the stator (9).

In accordance with the Autosyn principle of operation, the rotor (1) of the panel-mounted indicator Autosyn assumes an identical position relative to its stator (11). The pointer (13) is attached to the rotor shaft (12) of the indicator Autosyn. Motion of this rotor causes the pointer to deflect on a calibrated dial.

The damper vane (5) cushions wide, misleading fluctuations sometimes caused by large air bubbles passing through the metering chamber. Should the flow exceed the capacity of the instrument, the relief valve automatically opens and allows fuel to by-pass the metering chamber, and closes again when the flow drops to within the instrument's range. (10) Indicating unit rotor.

unit is mounted between the fuel tank and the carburetor on the engine side of the fuel pump. The gasoline flowing to the engine passes through the metering chamber in the transmitter. In the metering chamber of the transmitter is a movable vane. This vane extends into the stream of gasoline flowing through the transmitter. The more rapid the flow of gas, the greater the distance the vane is moved out of the neutral position. The vane moves the rotor in the transmitter. This movement of the rotor causes the rotor in the indicating unit to place itself in the same

AIRCRAFT ENGINES

relative position to its stator as the rotor in the transmitter is to its stator.

The indicating needle is attached to the shaft of the rotor in the indicating unit. As in all Autosyn units, the rotor in the indicator matches exactly the position of the rotor in the transmitter. As the pivoted vane in the metering chamber moves back and forth moving the rotor in the transmitter, the rotor in the indicating unit matches this movement. The farther the pivoted vane moves, the farther the indicating needle moves over the face of the indicator dial.

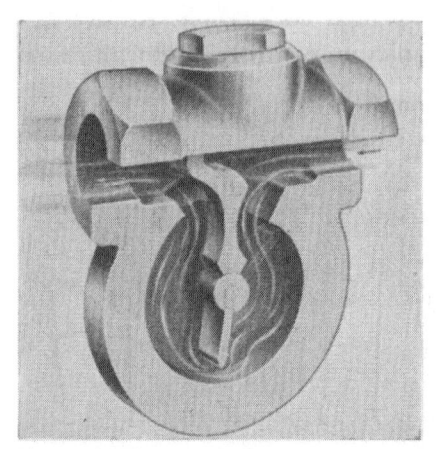

Fig. 302. The vane of an Autosyn fuel flow meter. (Courtesy Eclipse-Pioneer Division, Bendix Aviation Corporation)

Engine Synchronizers. The engine synchronizer is an instrument by which the pilot may, by operating the throttles or propeller controls, cause the propellers to rotate at exactly the same speed. When two or more propellers are operating at different speeds, vibration and throbbing noises may be set up. This instrument consists of a very sensitive indicator of the voltmeter type which is adjusted to measure a difference in voltages between two similar electric tachometers. The indicator instrument is usually calibrated to

Fig. 303. An engine synchroscope for a twin-engine aircraft. (Courtesy Kollsman Instrument Division of Square D Company)

Fig. 304. An engine synchroscope for a four-engine aircraft. (Courtesy Kollsman Instrument Division of Square D Company)

ENGINE INSTRUMENTS

indicate a variation in engine r.p.m. of approximately 40 for each engine. The instrument is connected into the circuit with the tachometer indicators and generators. There is usually a three-position control switch having an OFF position, a tachometer position, and a synchronizer position. The tachometer indicators cannot be used when the synchronizer is in operation. This instrument will show a difference in engine revolutions of approximately 2 per min. The engine throttles should be so adjusted that the pointer remains at zero.

ENGINE SYNCHROSCOPE. Another type of engine synchronizing instrument is called the engine synchroscope. This instrument consists of an indicator of the voltmeter type. A condenser unit is installed to smooth out the pointer movement. Coils of different resistance may have to be installed depending upon the type of engine magneto used. The synchroscope, with the proper resistance coils, is connected between the primary circuits of two magnetos. One magneto for each engine is used. When it is connected to the two magneto switches, two alternating voltages, the frequencies of which depend on the engine r.p.m., are passed to the instrument. When two currents having different frequencies are applied, a series of beats or recurring excessive waves appear at regular intervals. If the difference in r.p.m. of the two magnetos is great, these beats cause a wide swing of the synchroscope pointer. If the magnetos are rotating at exactly the same speed, the steady alternating current causes no beats and the pointer remains stationary. The pilot adjusts the throttles to get the slowest swing possible on the needle, as it is not always possible to get the magnetos to rotate at exactly the same speed. The greater the difference in the speed of the magnetos, the more rapid the oscillations of the synchroscope indicator needle.

Fuel Level Indicator. Fuel level gauges vary from the simple sight-reading gauge, consisting of a float which operates an indicator in a glass tube projecting from the bottom of the fuel tank, to very complex remote

Fig. 305. A combination tachometer and synchroscope for a twin-engine aircraft. (Courtesy Kollsman Instrument Division of Square D Company)

AIRCRAFT ENGINES

Fig. 306. Nose of a large aircraft engine showing the torque meter, or horsepower indicator. (Courtesy Wright Aeronautical Corporation)

indicating devices. The simplest form is a graduated sight glass mounted directly on the fuel tank. In this glass tube is an indicator connected with the float in the tank which indicates to the pilot by its position in the glass tube the fuel level in the tank.

One type of indicator consists of hydraulically operated bellows. This system consists of a closed hydraulic system having four bellows connected by tubing. One pair of bellows is connected with a float lever in

ENGINE INSTRUMENTS

the fuel tank. The other pair of bellows is connected with a movable arm which indicates on a scale the fuel level or the contents of the fuel tank. As the float rises in the tank, it releases the pressure on one bellows and compresses the other. This causes a movement of the two bellows connected with the indicator, moving the indicator over the dial. When the tank is located conveniently, a rotating needle may be arranged under a glass dial which is rotated directly through a gear train attached to a float in the tank.

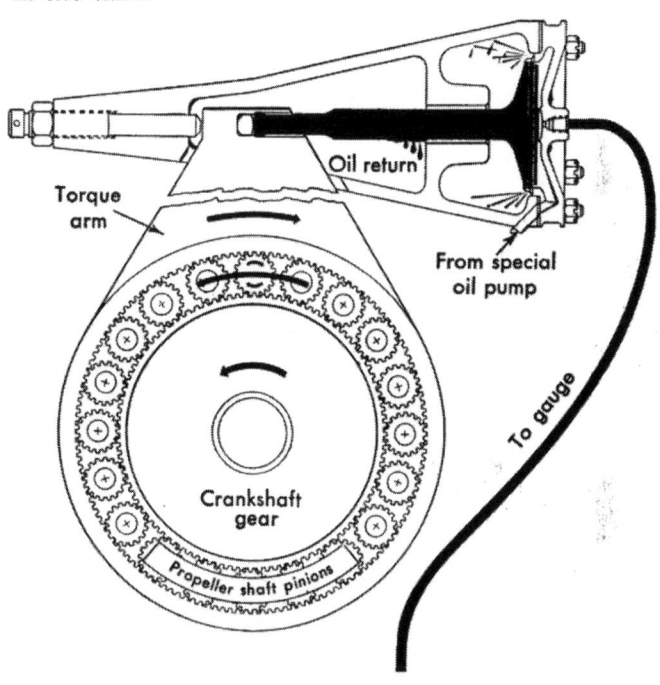

Fig. 307. A diagram showing the operation of the torque meter, or horsepower indicator. (Courtesy Wright Aeronautical Corporation)

Remote indicating instruments usually consist of a float arrangement in the tank which moves a contact over a resistance coil, causing a variation in the flow electric current which is measured by a voltmeter type of indicator on the instrument panel. This type is commonly used in automobiles. Another fuel indicator operates on the hydraulic pressure developed in the tank by the fuel. The greater the depth of the fuel, the greater the pressure on the instrument, which causes a deflection of a bellows or a diaphragm which is transmitted to the instrument board by means of one of the remote indicating devices. The remote indicating type of instrument, such as the Selsyn, may be used with this

AIRCRAFT ENGINES

arrangement. The float causes the rotor in the transmitter to turn, causing a corresponding turn of the rotor in the indicator.

Torque Meter or Horsepower Indicator. A torque meter measures the torque reaction on the stationary gear by the propeller reduction gear. While this instrument measures the torque reaction, it is used to indicate engine brake horsepower. The effective power of the engine is used to turn the propeller. A measure of the force exerted by the engine in turning the propeller is a close indication of the engine's power output. This instrument is mounted on the front section of the crankcase.

The instrument consists of a torque arm which has a ball race on its inside diameter. The torque arm is assembled to a support by means of 86 balls inserted between the races permitting relative motion between the torque arm and the support, and acting as an antifriction device. The torque arm at its outer end acts on a balance valve. The valve is actuated in one direction by the torque reaction of the stationary gear and is balanced in the other direction by the oil pressure on the valve head. Oil is supplied to the torque meter by a small booster pump which also serves to drive the governor gear shaft. When the engine is operating, the torque reaction changes the twisting force of the stationary gear to a slight outward movement of the valve. This results in an increase in the oil pressure on the valve head. The oil pressure increases until it is just sufficient to overcome the torque reaction and thus causes a slight inward movement of the valve. This inward movement of the valve opens metering slots in the sleeve in which the valve moves. The oil escaping through these openings is led to the crankcase. The oil pressure on the head of the valve then decreases until the torque reaction is not sufficient to overcome the oil pressure. When another outward movement of the valve is produced, a balance of oil flow is quickly reached.

Fig. 308. A horsepower indicator for an aircraft engine. (Kollsman Instrument Division of Square D Company)

An oil pressure gauge line is connected into the valve housing and is led to a special oil pressure gauge calibrated to read "brake mean effective pressure." Since engine brake horsepower is proportional to engine

ENGINE INSTRUMENTS

revolutions per minute and to brake mean effective pressure, brake horsepower may be calculated by multiplying the engine revolutions per minute by the brake mean effective pressure registered on the gauge and divided by a constant value for the engine upon which the instrument is installed.

XX STARTING AND STARTING SYSTEMS

Most early aircraft engines, although of comparatively low horsepower and compression, were rather difficult to start. The magneto with which these engines were equipped delivered a weak spark due to the low speed with which the propeller was "pulled through." Priming systems and accurately adjusted carburetors had not come into use. Some early aircraft were equipped with a booster magneto located in the cockpit. As the propeller was pulled through, the pilot turned this magneto by means of a hand crank.

Fig. 309. A light aircraft engine equipped with two magnetos, a generator and a starter. (Courtesy Continental Motors Corporation)

STARTING AND STARTING SYSTEMS

Later, magnetos were equipped with an impulse coupling. This consists of a spring arrangement which prevents the magneto's turning as the propeller first starts to rotate. At the time the spark is to be delivered to the cylinder which is under compression, the spring is suddenly released, allowing the armature to snap forward at high speed. This action delivers a hot spark to ignite the mixture in the cylinder.

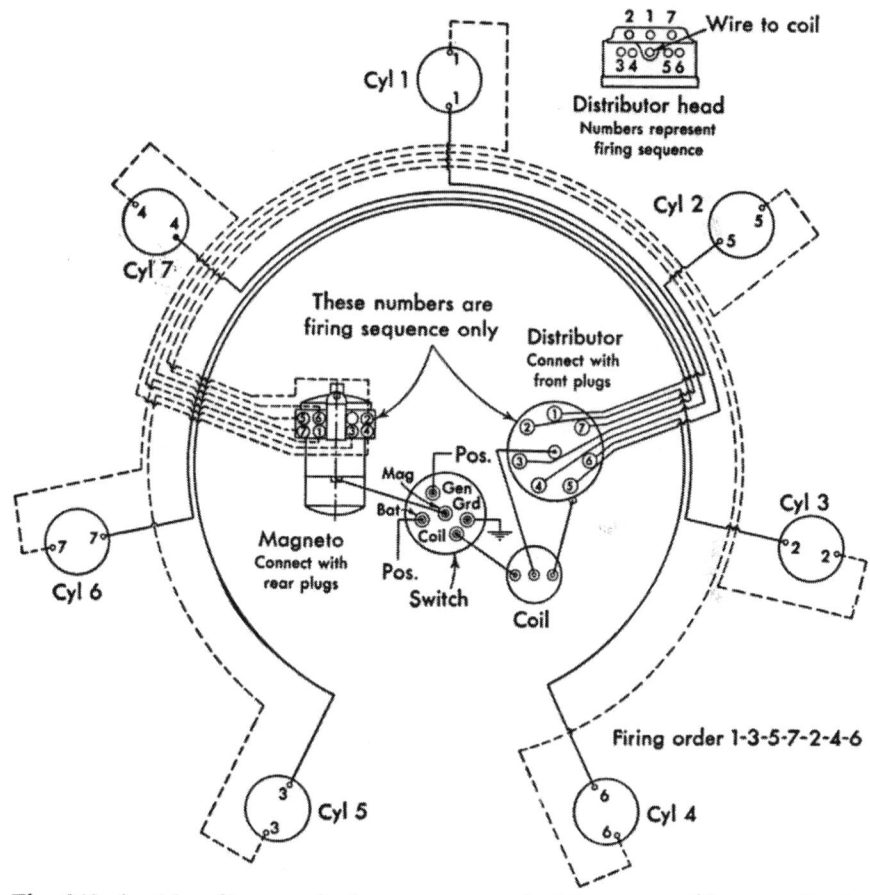

Fig. 310. A wiring diagram of a battery-magneto ignition system. (Courtesy Jacobs Aircraft Engine Company)

On many engines at the present time, one set of plugs is connected with a battery system which delivers a hot spark at low speeds. The ignition switch connects the battery system with one set of spark plugs. As soon as the engine starts, the switch is turned to the regular magneto ignition system.

On modern engines there are a number of different ways by which the engine may be started other than by pulling through. These methods

AIRCRAFT ENGINES

include direct hand cranking, hand inertia starters, electric inertia starters, direct electric starters, air injection starters, and the cartridge starter.

The direct, hand-crank arrangement consists of a shaft equipped with a worm gear which engages a gear on the crankshaft when the starter

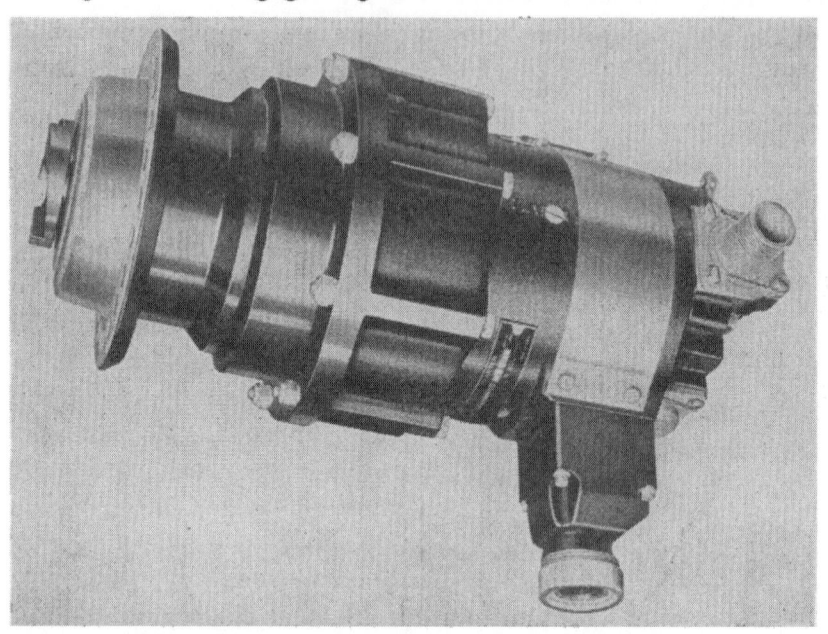

Fig. 311. An electric inertia or hand-cranking starter. (Courtesy Eclipse-Pioneer Division, Bendix Aviation Corporation)

shaft is turned. This shaft is turned by means of a detachable hand crank and is equipped with a ratchet device which disengages the starting mechanism when the engine begins firing. A self-locking device is installed to prevent injury to the person handling the crank if the engine backfires.

Hand inertia starters make use of the energy stored in a small, heavy flywheel which rotates at high speeds. This flywheel speeds up to 10,000 to 12,000 r.p.m. through a gear train operated by a hand crank. When the proper speed of the flywheel is obtained, the crank is removed, and the pilot engages a clutch which transfers the energy of the flywheel to the crankshaft. On some aircraft the engaging clutch is located close to the hand-crank opening, and the person doing the cranking removes the crank and engages the clutch. Usually, the clutch arrangement is such that some slipping is allowed to overcome the inertia of the engine. This clutch also prevents damage in case of kickback.

STARTING AND STARTING SYSTEMS

Electric inertia starters are operated on the same principle as the hand inertia starter, the only difference being that the flywheel is rotated by means of an electric motor instead of a hand crank.

Direct electric starters are similar to those used in automobiles and consist of an electric motor driven from a battery system acting either

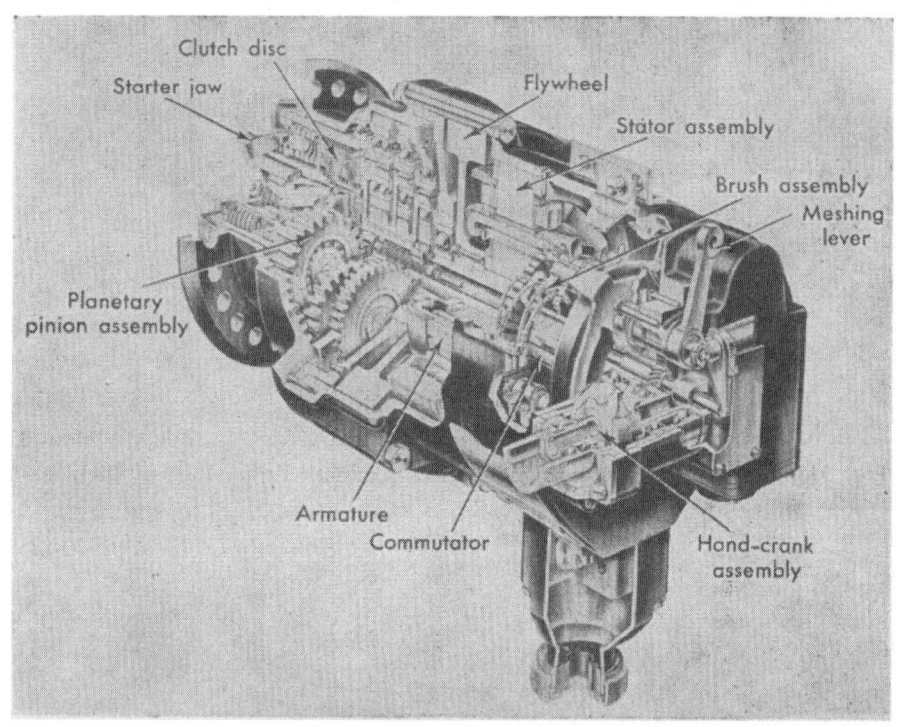

Fig. 312. A cutaway view of an electric inertia or hand-cranking starter. (Courtesy Eclipse-Pioneer Division, Bendix Aviation Corporation)

directly upon the crankshaft or turning the crankshaft by means of a gear train. This system is also equipped with a safety clutch and an automatic disengaging device.

The air injection type of starter is not in common use. It consists of an air tank and a high efficiency air compressor operated by the engine. The air pressure which is built up to approximately 450 lb. per sq. in. by the compressor is regulated by automatic valves which prevent excessive pressure. A control located in the cockpit allows the air to be released to the cylinders in their regular firing order. This is done by alternately pulling out the starter and then pushing it back. The air lost from the supply tank is replaced by the compressor while in flight. This starting system was used on light airplanes and weighs approximately 30 lb.

Fig. 313. An electric direct-cranking starter. (Courtesy Eclipse-Pioneer Division, Bendix Aviation Corporation)

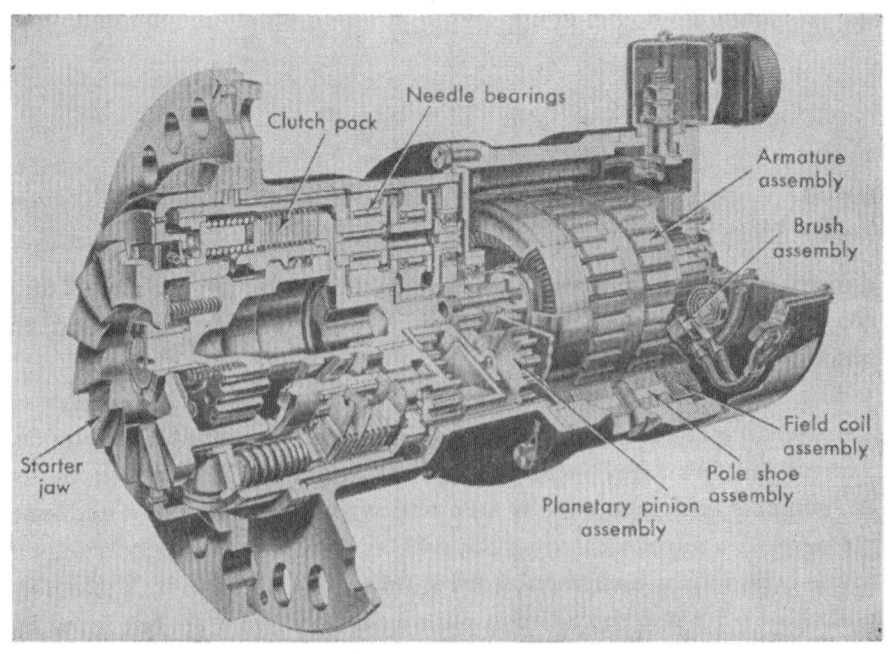

Fig. 314. A cutaway view of an electric direct-cranking starter showing its construction. (Courtesy Eclipse-Pioneer Division, Bendix Aviation Corporation)

STARTING AND STARTING SYSTEMS

A cartridge starter is sometimes used to start an engine. This starter consists of a chamber in which a cartridge is placed. The cartridge is ignited by electric contact. The explosion of the cartridge in the chamber of the starter forces a cylinder to move along spiral grooves, rotating the crankshaft. These starters may be used when extremely cold weather makes starting difficult. A special cartridge is necessary, and ordinary shotgun shells should not be used. These starters are equipped with a safety diaphragm which blows out in the event that the starter builds up excessive pressures.

Most starting systems make use of either a booster coil or a vibrating coil to furnish a hot spark to start the engine. The booster arrangement cuts out automatically when the engine starts.

Fig. 315. The jaws of an aircraft engine starter. (Courtesy Jack & Heintz, Inc.)

With a battery system, a vibrating coil may be used which furnishes a series of hot sparks to the magneto distributor while starting the engine.

Fig. 316. An electric inertia and direct-cranking starter. (Courtesy Eclipse-Pioneer Division, Bendix Aviation Corporation)

AIRCRAFT ENGINES

Before starting an aircraft engine by hand cranking, the pilot should be sure that the wheels are chocked or that the parking brakes are firmly set. The oil and fuel supply should also be checked. With the throttle closed and the switch off, the proper number of priming strokes, if the engine is equipped with a primer, is given on the priming pump. The engine is then rotated by hand through four or five revolutions in

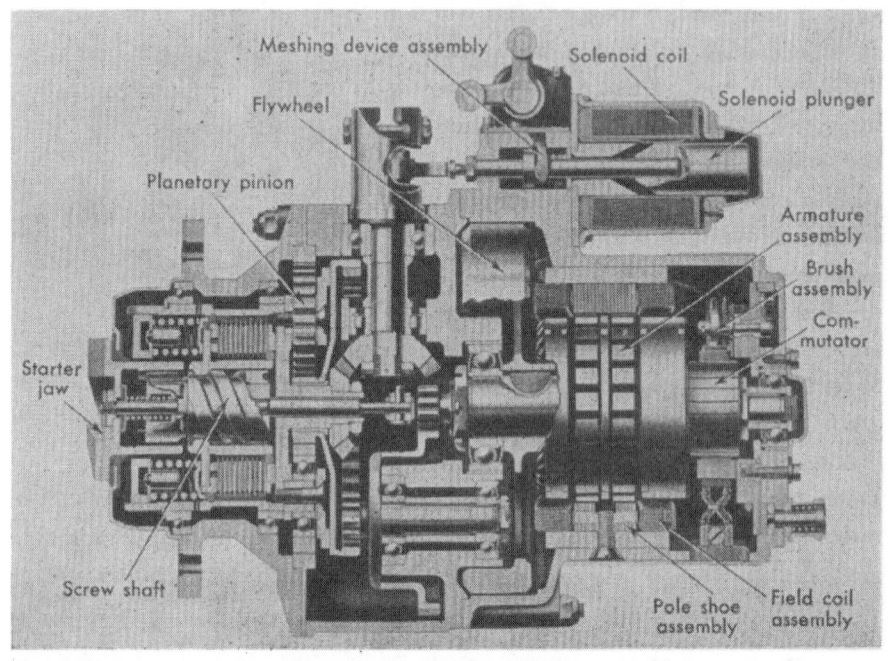

Fig. 317. A cutaway view of an electric inertia and direct-cranking starter. (Courtesy Eclipse-Pioneer Division, Bendix Aviation Corporation)

the direction of normal rotation. The propeller is put in the proper position to pull through, and the ignition switch is turned on. The throttle is then cracked, which means that it is opened slightly, and the propeller is pulled through as rapidly as possible. If the engine fails to start, the switch is turned off and the engine rotated as before, but without additional priming. The switch is then turned on and the engine again pulled through. If the cylinders of the engine become flooded with fuel, the engine is rotated backward through five or six complete revolutions with the switch off and the throttle fully opened. Then, with the throttle closed, the starting procedure is repeated.

On large engines equipped with a starter, the engine is primed and, with the switch off, the engine is started to rotate by means of the starter. As soon as the engine begins to rotate, the ignition switch is turned on.

STARTING AND STARTING SYSTEMS

This procedure assists in preventing kickback which might cause damage or place excessive strains on the structure.

Before starting a newly installed engine, a thorough inspection of all parts and their installations must be made. The ignition switch on an aircraft works somewhat differently than does the ignition switch on an automobile. On the automobile, the ignition switch when off simply

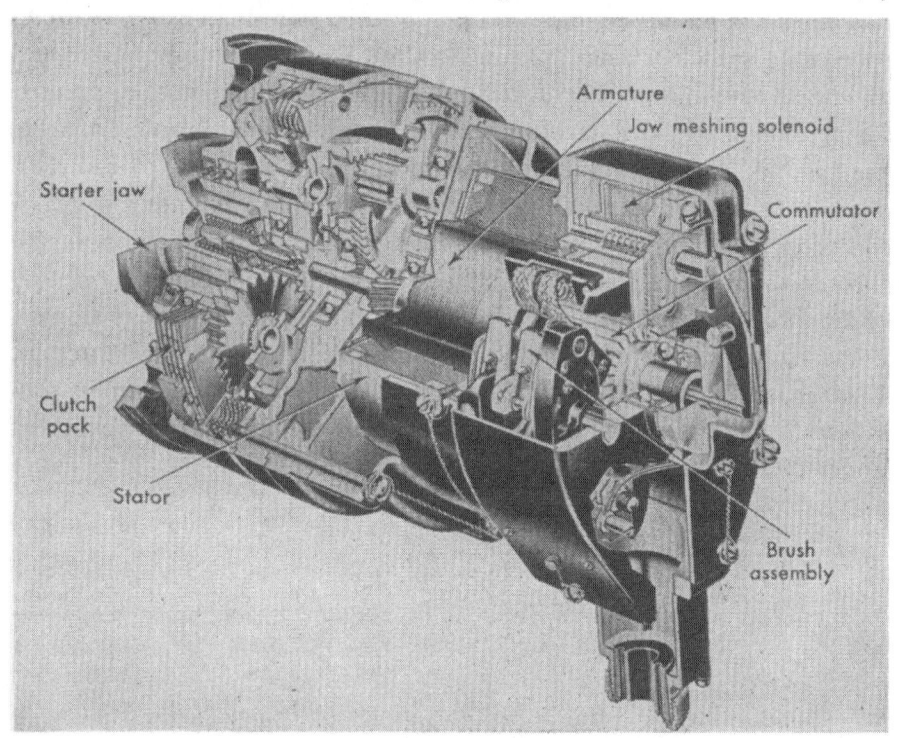

Fig. 318. A cutaway view of a direct-cranking electric starter. (Courtesy Jack & Heintz, Inc.)

opens the primary circuit. On aircraft the ignition switch or magneto switch grounds the magneto by means of a ground wire. It is important that the ground wire be securely in place, for if it becomes loose the ignition system will operate and the engine will continue to run although the ignition switch is in the OFF position. If the ground wire is loose or disconnected, the engine may start if, for any reason, it is rotated. In starting the engine after the proper priming charges have been drawn into the cylinder, unless the aircraft is equipped with a battery ignition on one set of plugs, the ignition switch is placed in the BOTH ON position. As soon as the engine starts, the oil pressure gauge should be inspected and, unless the oil pressure begins to register within a few seconds, the engine should be immediately stopped. The oil system should be checked before any further attempt is made to start the engine. After the engine is started, the ignition switch should be turned to the right and left positions, thereby testing each magneto and its corresponding set of plugs separately. Operating on each magneto separately should not cause a loss of more than 50 to 150 r.p.m.

The engine should be warmed up at a speed slightly in excess of idling speed for a sufficient length of time to allow the oil to become warm and all engine parts thoroughly lubricated. The engine should not be operated for periods longer than approximately 30 sec. at full throttle while on the ground. This "revving up" should be done, however, to ascertain whether or not the engine is reaching its maximum r.p.m. and to check the proper operation of all engine instruments. All movements of the throttle should be smoothly and gradually performed. Sudden opening of the throttle should be avoided, as this may cause the engine to stop or, if the engine takes hold, will cause excessive strain on the structure. Engines should not run for extended periods on the ground unless provisions for extra cooling are made, as most engines will seriously overheat under these conditions.

XXI ENGINE THEORY AND OPERATION

The mechanic, pilot, and operator of aircraft engines should know more about the engine than just how it is constructed. While it is necessary to know the various parts of the engine, how they are constructed, and where they fit into the structure, it is also necessary to understand how the engine performs and how it develops power. The entire function of the aircraft engine is to develop power. To understand this production of power, a clear understanding of what happens when an engine is started and what makes it continue to operate and produce power should be had by all who have a part in its care, maintenance, repair, or operation.

No internal-combustion engine will start without the help of some external force which turns the crankshaft.

As the engine is first turned over, the cylinder in which the exhaust stroke has just been completed is ready for the intake stroke. The intake valve in this cylinder is opened by means of the cam and the valve-operating mechanism. The rocker arm, which is pivoted in the middle, operates the valve. One end of this arm is pressed upward by the cam action, while the other end depresses the valve against the valve springs, opening the valve port. As the intake stroke is started, the exhaust valve closes and the air-fuel mixture flows into the cylinder through the intake valve port. The air-fuel mixture is supplied to the engine through the intake manifolds which, in the carburetor fuel system, is connected with the carburetor. As the piston starts downward on the intake stroke, the partial vacuum left in the cylinder causes the air to move through the intake manifold and the carburetor. The carburetor may be open to the atmosphere or connected with a duct leading to an air scoop or a supercharger. The air stream is speeded up in the carburetor by passing through a Venturi. This rapidly moving stream of air draws a spray of fuel from the spray nozzle in the carburetor into the air stream, thus forming the air-fuel mixture. During the compression stroke, both valves are closed and, near the end of this stroke, the mixture is ignited. The burning mixture transfers its energy to the piston which it forces downward in the cylinder. The piston, being connected to the crankshaft by means of the connecting rod, transmits this force to the crankshaft, causing it to rotate. The power developed by this first power impulse is usually great enough to complete a compression stroke in another cylinder. In a single-cylinder engine, this first power stroke starts a heavy flywheel turning, the momentum of which carries the piston through the succeeding strokes until another power stroke adds to its momentum. Enough momentum must be built up not only to turn the crankshaft but also to operate all moving parts of the engine between power strokes. The crankshaft is not only connected with the propeller but, through gearing, also drives the camshaft which operates the valves, the oil pumps, the magnetos, the superchargers, and all other moving parts. Most of these parts are driven by means of gear trains which are arranged to give the proper rate of rotation to each part.

The oil pressure pump draws oil from the oil supply tank or from the engine sump and supplies it to the engine bearings under pressure, lubricating all moving parts through the proper oil passages. The inside of the cylinders is lubricated by sprays of oil and oil vapors within the crankcase. If the engine is the dry-sump type, the oil is collected in a sump and is returned to the oil supply tank by scavenger pumps. In the wet-sump type, the oil simply drains back into the sump which is usually a part of the crankcase. A certain amount of oil always works past the pistons and is burned in the combustion chamber. This burning accounts for most of the normal oil consumption of an engine. All of the power of an engine is developed in the combustion chamber and cylinders. All power from the engine is transmitted to the crankshaft which is held in its proper position by the crankcase.

Only about 25 per cent of the energy contained in the fuel is put into useful work. Work is measured in terms of brake horsepower. About 50 per cent of the energy is lost in the form of heat in the exhaust gases. The cooling system removes about 20 per cent of the energy in the form of heat from the engine, and the other 5 per cent of the total energy in the fuel is used up by friction in the engine itself. Much of the frictional loss is due to the fact that there is only one power stroke for each four strokes of the piston. During the three idle strokes, power is not only used to complete these three strokes but also to continue the movement of all the moving parts of the engine. Assuming that the air-fuel mixture

ENGINE THEORY AND OPERATION

is in the proper proportion, the power developed by any engine operating on any particular grade of fuel depends almost entirely on piston displacement. Piston displacement depends upon the bore and stroke of the cylinder.

The bore of the cylinder is the diameter of the cylinder barrel. The distance that the piston travels from top dead center to bottom dead center is called the stroke. Piston displacement is the volume of a cylinder having a diameter equal to the bore of the engine cylinder and a height equal to the stroke of the piston. Total piston displacement of the engine is determined by the formula, $D^2 0.7854$ times the stroke times the number of cylinders. The part of the formula, $D^2 0.7854$, is used to find the area of a circle having a diameter equal to the bore of the cylinder. This figure in square inches, multiplied by the length of the stroke in inches is equal to the piston displacement of a single cylinder in cubic inches. If the letter B equals the bore and S equals the stroke and N the number of cylinders, the formula for total engine displacement becomes $B^2 0.7854 S N$. The volume of the combustion chamber and piston displacement is usually given in cubic inches or cubic centimeters. When the volume is given in cubic centimeters, all measurements, such as bore and stroke, should be in centimeters. The total volume of the charge in any one cylinder at the end of the intake stroke is the piston displacement plus the volume of the combustion chamber.

Compression ratio is the ratio between the total volume of the air-fuel-mixture charge with the piston at bottom dead center before the compression stroke and the volume of the charge when the piston is at top dead center at the end of the compression stroke. For example, the piston displacement of a given engine having a bore of 4.125 in. and a stroke of 5.5 in. is found to be approximately 73.75 cu. in. If the combustion chamber in this cylinder has a volume of approximately 12.35 cu. in., the compression ratio is approximately 6.9:1. While this compression ratio is rather high for aircraft engines, it is within safe limits as some engines operate with a compression ratio of approximately 7:1. The compression ratio of the modern aircraft engine varies from approximately 6:1 to 7:1. The compression ratio largely determines the type of fuel used. The pressure developed in the cylinder follows the standard gas laws. Boyles law states that, "The volume of a given weight of gas when confined varies inversely as the pressure exerted upon it, the temperature remaining constant." The law of Charles and Gay-Lussac states that, "The volume of a given weight of gas when confined

AIRCRAFT ENGINES

varies directly with the absolute temperature, the pressure remaining constant." As stated in these two laws, the gaseous air-fuel mixture becomes heated upon being compressed. This increases the pressure on the piston head to a somewhat greater extent than would be done by the compression alone. As the air-fuel mixture is burned, the high temperatures developed greatly increase the pressure.

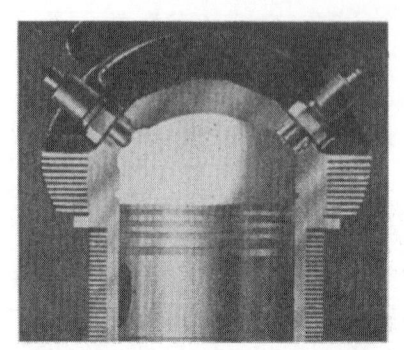

Fig. 319. A cutaway view showing flame in the burning air-fuel-mixture charge advancing with a clearly defined front. At about this point when the mixture is ignited at one point only, the traveling flame front slows down. (Courtesy Wright Aeronautical Corporation)

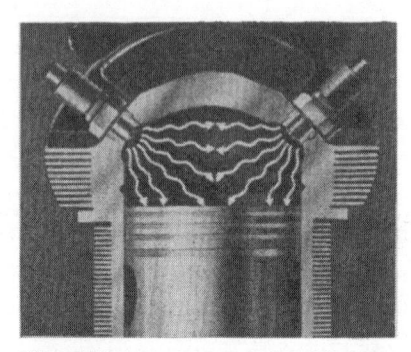

Fig. 320. A cutaway view showing where the two flame fronts meet when the mixture is ignited at two places at the same time. (Courtesy Wright Aeronautical Corporation)

The high performance of the modern aircraft engine depends upon successful combustion. As the spark jumps between the electrodes of the spark plug, a flame starts at this point and spreads rapidly through the air-fuel mixture. This flame advances with a definite "front." There seems to be a well-pronounced boundary line separating the burned from the unburned portion of the charge. The flame slows down somewhat as it approaches the walls of the combustion chamber. This flame travels through the charge normally at the rate of about 100 ft. per sec. With dual ignition, the charge is ignited at two points on opposite sides of the combustion chamber at the same time. Since the flame travel is at the normal rate of speed, it takes only about one half the time for the charge to burn as it would when ignited at only one point. The flame fronts meet at approximately the middle of the combustion chamber. When detonation occurs, the flame advances at the normal rate up to a point where it normally tends to slow down, and then the remainder of the charge explodes. When a charge explodes, the unburned portion of the charge burns instantly. An exploding charge develops excessive

ENGINE THEORY AND OPERATION

temperatures and pressures. A detonating charge exerts a hammer-like blow on the cylinder structure. This blow strains the metal and produces a metallic ping or knock. A detonating charge develops high cylinder-head temperatures very rapidly. Detonation may be caused by a fuel having a too low antiknock rating or the failure of one spark plug. Detonation may be caused by compression pressures in excess of those for which the engine was designed. These pressures may develop because the engine is overheated or the intake air-fuel supply is too hot.

Preignition takes place when the air-fuel mixture is ignited before the proper time. This may be caused by wrong timing, such as excessive spark advance, or by incandescent particles of carbon, overheated spark plugs, or excessive compression. Preignition may cause the fuel to detonate. It is the fuel itself which detonates, not the engine. Detonation may cause pressures in the cylinder up to three times that of the normal combustion. Pressures may become so great as to cause engine failure.

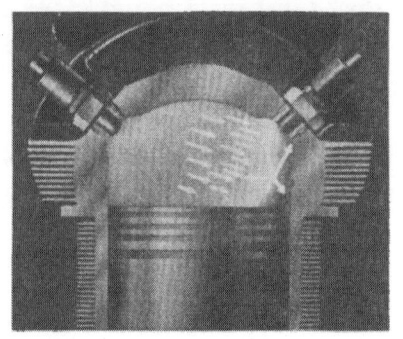

Fig. 321. A cutaway view showing where detonation takes place. The flame front advances normally as indicated by the group of arrows. Instead of slowing down at this point, the remainder of the charge explodes. This action is called detonation. (Courtesy Wright Aeronautical Corporation)

The time when the valve opens and closes, which is known as valve timing, lends much to the efficient operation of the engine. The valves do not close at the exact end of any of the four strokes. For example, the intake valve may open several degrees of crankshaft rotation before the piston reaches top dead center on the exhaust stroke. The exhaust valve does not close at the exact moment that the piston reaches top dead center at the end of the exhaust stroke. The intake valve may open as much as 25° or more ahead of top dead center, and the exhaust valve remain open as much as 30° or more after top dead center. This means that, at the end of the exhaust stroke, there is a considerable length of time when both the exhaust and intake valves are open. This condition is known as valve lap. For example, one modern engine opens the intake valve 15° before top dead center and closes the exhaust valve 30° after top dead center, giving a valve lap of 45°. Another modern engine opens the intake valve 20° before top dead center and closes the exhaust 20° after top dead center, giving a valve lap of 40° of crankshaft rotation.

During the power stroke, the exhaust valve opens from 60° to 70° before bottom dead center. The main force of the burning charge has already been expended when the piston reaches this point in its downward stroke. The exhaust valve then remains open during the exhaust stroke and for a considerable number of degrees after top dead center. On one engine, the exhaust valve remains open 280° of crankshaft rotation. On this same engine, the intake valve opens 15° before top dead center and remains open 65° after bottom dead center. This valve remains open 260° of crankshaft rotation. In this engine, power is developed on the piston through only 110° during the 720° of rotation necessary to complete the four cycles. Power is developed less than one sixth of the time in each cylinder. With this valve timing arrangement, six and a fraction cylinders would be required to produce a continuous flow of power. If power is exerted through 120° of crankshaft rotation, as is true in some engines, a six-cylinder engine theoretically produces a continuous flow of power. This is not true, however, as the power developed in the cylinder builds up and dies down and is not continuous during the power stroke. When more than six cylinders are used, the power impulses overlap each other. This condition is power overlap. The larger the number of cylinders operating on a single crankshaft, the smoother the flow of power.

Valve lead is the term used to describe the number of degrees a valve opens before top or bottom dead center. Valve lag is the term used to describe the number of degrees a valve remains open after top or bottom dead center.

The time at which the spark occurs affects the time when the maximum pressure is developed in the combustion chamber. In most engines, the spark occurs before the piston reaches top dead center on the compression stroke. The number of degrees before top dead center at which the spark occurs is ignition advance, and may be as great as 25° or more. When dual ignition is used, the spark advance is not usually as great as when single ignition is used. The spark timing is adjusted to ignite the charge at the proper time so that the maximum pressure will be developed at the beginning of the power stroke.

The opening and closing of the valves are regulated by the timing of the camshaft. The camshaft is usually gear driven from the crankshaft. When the valves are opened by cam plates, the plates are usually fastened to a flange on the crankshaft or are a part of the crankshaft. The size of the valve opening is determined by valve lift which varies with the

ENGINE THEORY AND OPERATION

size of the valve. On one large engine, the lift is slightly over ½ in. On most engines, the valve lift is approximately one fourth the diameter of the valve port. Valve clearance is the amount of clearance between the valve stem and the valve operating mechanism. This clearance varies when the engine is hot or cold. The valve stem expands, and the clearance is less when the engine is hot. If there is not enough valve clearance, the valve may be held open slightly when hot. This condition would allow the escape of gases which would burn the valve face and seat, and might cause backfiring. Valve lash is due to clearance in the valve operating mechanism.

The cams which operate the valve tappet must be carefully designed to prevent their bouncing the cam follower. The lift of the follower must start gradually but build rapidly to a maximum and then allow a rapid, smooth closing of the valve. Snappy valve action is necessary to good engine performance. The maximum opening of the valves should last as long as possible, but the valve should open and close promptly.

The proper timing of the ignition and valves is one of the most important factors in efficient engine performance.

The induction system and exhaust system have considerable influence as to whether or not a proper-sized charge is drawn into the cylinder on the intake stroke and whether or not the exhaust gases are efficiently disposed of. When the engine is operating at 2000 r.p.m., each valve opens and closes 1000 times a minute. Any back pressure in the exhaust manifold would tend to prevent the free flow of exhaust gas from the cylinder through the exhaust valve opening. Any drop of pressure in the intake manifolds or induction system would prevent the maximum amount of air-fuel mixture from entering the cylinder. Supercharging, air scoops, and streamlining assist in maintaining proper manifold pressures. Without supercharging, the intake manifold pressure cannot exceed outside atmospheric pressure, except in flight when this pressure may be built up to some extent by the ramming effect of air through the scoop. With superchargers, the manifold pressure may be considerably in excess of the atmospheric pressure. The higher the manifold pressure, the greater the charge of air-fuel mixture delivered to the cylinder on the intake stroke. Engine power depends largely upon the amount of air-fuel mixture in the cylinder. All parts of the modern aircraft engine are carefully coordinated, and each part is designed to give the maximum efficiency with the lowest possible weight necessary for the required strength.

The operation of the Diesel engine is, in general, similar to the operation of the standard internal-combustion engine. The principle difference is in the way in which the fuel is induced into the cylinder and ignited. The method of introducing fuel makes it possible to operate a Diesel engine with but one valve per cylinder. This acts both as the intake and exhaust valve. However, better performance is obtained by having both an intake valve and an exhaust valve. In the Diesel engine, air alone is drawn into the cylinder through the valve opening on the intake stroke. This charge of air is compressed to a much higher degree than in engines using an air-fuel mixture. Compression ratios in the Diesel engines may be as high as 14:1 to 16:1. This high compression raises the air charge to a temperature high enough to ignite the fuel without an electric ignition system. With a compression ratio of 16:1, the temperature of the compressed air is approximately 1000° F. At the proper time, which is slightly before the piston reaches top dead center, the exact fuel charge needed is sprayed into the compressed air through a fuel nozzle which is inserted through the cylinder head. The fuel is supplied to the nozzle by a high-pressure plunger pump. There is usually one pump plunger for each cylinder. Just before the end of the power stroke, the exhaust valve opens, and the waste gases are pushed out of the cylinder during the exhaust stroke. This engine operates on a four-cycle principle. When a single valve is used, the valve remains open during both the exhaust and intake stroke. When two valves are used, the valve timing is similar to that of a standard internal-combustion engine.

The power output of any engine is measured in terms of horsepower. Horsepower does not represent a fixed quantity. Horsepower is the rate at which power is being developed. For example, if 100 gal. of water will flow through an opening per minute, at any given instant the quantity of water leaving the opening is zero. In order that a quantity of water leave the opening, a time element must enter. It requires a full minute before the entire 100 gal. of water have passed through the opening. The rate of flow is 100 gal. per min. The horsepower of an engine is the rate at which power is being delivered. A horsepower is defined as that power necessary to raise 1 lb. 33,000 ft. in 1 min., or that power which will raise 33,000 lb. 1 ft. in 1 min. Horsepower has to do only with the rate at which the power is delivered. Whenever 33,000 ft.-lb. of work are accomplished in 1 min., 1 hp. has been used.

The indicated horsepower of an engine is determined by formula. One method of figuring indicated horsepower is by means of the formula,

ENGINE THEORY AND OPERATION

$PLANK/33,000$, where P is equal to the pounds pressure per square inch developed in the cylinder, L is the length of the stroke in feet, A is the area of the piston in square inches, N is the number of power strokes per minute, and K is the number of cylinders. The power strokes per minute for any one cylinder are equal to one half the revolutions per minute of the engine crankshaft of a four-cycle engine.

Brake horsepower is found by connecting the engine with an apparatus which actually measures the power output. Such an instrument is called a dynamometer. Another way of determining brake horsepower is by arranging a brake by which the engine may be slowed down and figuring the power necessary to force the engine to develop its full power.

Friction horsepower is the amount of power lost to friction in the engine. Friction horsepower is equal to the indicated horsepower minus the brake horsepower. Horsepower is always stated at a definite number of revolutions per minute. The brake horsepower is the true measurement of the power of the engine and is given at the least number of revolutions per minute at which the engine will continue to operate.

The mechanical efficiency of an engine is found by dividing the brake horsepower by the indicated horsepower, and is stated as a percentage. This figure for the average aircraft engine is approximately 90 per cent. Thermal efficiency is the actual amount of heat energy furnished in the form of fuel divided into the useful work performed at the propeller shaft. The thermal efficiency of an aircraft engine is approximately 25 per cent. The heat furnished by the fuel is lost in a number of ways: approximately 50 per cent in the hot exhaust gases; 20 per cent in the cooling system; and 5 per cent in friction within the engine itself. One pound of aircraft fuel furnishes approximately 20,000 B.t.u. The B.t.u. is a unit of heat measure. The B.t.u. is that amount of heat necessary to raise the temperature of 1 lb. of water 1° F. Volumetric efficiency is based on piston displacement. Horsepower may be figured by using the piston displacement of an engine. Some of the modern aircraft engines develop approximately 1 hp. for each 2 cu. in. of piston displacement. The volumetric efficiency depends upon how much of a full charge at atmospheric pressure is drawn into the cylinder on the intake stroke. By supercharging, the volumetric efficiency may possibly exceed 100 per cent.

The engine controls are the various mechanisms used to control the operation of the engine or its accessories. The engine controls are located in the cockpit within reach of the pilot. The control handles and levers are usually marked to distinguish them or indicate their use. The control

AIRCRAFT ENGINES

handles or levers are connected with the various parts by means of push or pull rods, ball joints, linkages or steel wires, or rods. The controls should be so adjusted that they remain at any setting and yet move freely. There should be no play or lost motion between the lever and the part operated. If any play or slack is felt when the controls are operated, the control mechanism should be examined to determine the cause, and the slack should be eliminated. Ball joints, as well as other joints, should be frequently inspected to see that they are properly adjusted and safetyed.

The engine controls consist of a throttle, mixture-control lever, spark control, and heat control. Propeller controls are often included with the engine controls.

The operation of the throttle is used, with engines equipped with superchargers, to regulate manifold pressure. On small engines the speed of the engine is kept within the desired limits by direct action of the throttle. On airplane engines, the throttle is pushed in to open and pulled out to close it.

In cockpits where the pilot and copilot are side by side, the throttle lever is located between the pilots, within easy reach of either. In tandem airplanes, the throttle is usually located at the left side of the cockpit. In some airplanes, there is a throttle lever on each side in order that the pilot may operate the stick with his left hand, leaving the right hand free. In multiengine planes there is, of course, a throttle for each engine, and they are usually so located that the pilot can operate all of the throttles with one hand at the same time.

In multiengine aircraft, the throttles are usually located at the left of the control group levers, with the propeller controls in the middle and the mixture controls at the right. These controls are usually located close together in a single group. The operation of the throttle should always be gradual, as sudden opening or closing imposes excessive strains on the engine structure. This is particularly true with heavy propellers. Every manufacturer furnishes with each aircraft manufactured a set of operating instructions which should be carefully followed. These instructions describe the proper method of starting, primer operation, proper warm-up, and checks to be followed when operating the aircraft.

There are certain checks which the pilot should always make after starting the engine. After the warm-up period, he should check all instruments to be sure that they are operating properly. Before starting the engine, he should check the oil, fuel, and coolant, if the engine

ENGINE THEORY AND OPERATION

is liquid-cooled. He should also check the movement of all controls, both engine and aircraft. Fuel gauges, oil pressure gauges, manifold pressure gauges, thermometers, tachometers and all other instruments or gauges should be checked for proper operation.

There are a number of operating limits with which the pilot should be familiar. Nearly every instrument indicates a maximum and minimum limit.

Such items as cylinder head temperature, manifold pressure, carburetor air temperature, oil temperature, oil pressure, fuel pressure, air-fuel-mixture ratio, crankshaft speed, and engine r.p.m. should be carefully watched.

On light aircraft, the instruments are neither complicated nor many. A tachometer and oil temperature and oil pressure gauge may complete the engine instrument-board layout. On multiengine planes, the instruments and other gadgets may number well in excess of 100. On most instruments, maximum and minimum indications are shown by means of differently colored lines. A safe operating range is usually indicated by means of a green or yellow line, while danger areas are indicated by red lines.

Air speed, while not directly connected with engine operation, has its effect on proper cooling, in that an engine may overheat at excessively low air speeds when operating under full throttle. The most efficient cooling air speed should be marked on the face of the air-speed indicator. Excessive air speed often leads to excessive engine speed, particularly when a fixed-pitch propeller is used.

When the engine is equipped with a supercharger, care should be taken that excessive manifold pressures are not developed. Under conditions of excessive manifold pressure, the limits for the octane rating of the fuel and the compression ratios of the engine may be exceeded. Under these conditions, detonation is almost sure to occur, which may cause failures of piston heads, cylinder heads, or other parts of the engine. Excessive manifold pressure is usually indicated by a rapid rise in engine temperatures. Excessive temperatures due to this cause usually result in damage to the piston rings and, possibly, in scoring of the cylinder walls. The manifold pressure is regulated by the throttle setting, but may vary with the engine speed because the supercharger speed varies directly with the engine speeds. A throttle setting which maintains a safe manifold pressure while in level flight may become excessive when the aircraft is put into a dive, unless the throttle setting is changed.

Aircraft engines operate most efficiently within a comparatively narrow temperature range. Any indication of a rapid change in oil temperatures or cylinder-head temperatures, where the engine is equipped with cylinder-head-temperature indicators, should be a cause for concern on the part of the pilot. Usually, cylinder-head temperatures should be maintained below approximately 450° F. The temperature of the air entering the carburetor may have considerable influence on the engine operating temperatures. When operating under excessively high air temperatures, it may be necessary to increase the richness of the air-fuel mixture or take other precautions to maintain the engine temperature within the desired limits. Decreasing the throttle setting or opening the cowling flaps may also assist in maintaining a desired engine temperature.

The oil temperature gauge, which usually indicates the temperature of the oil at the point where it is removed from the engine, is generally the first to indicate a rise in engine temperature. A proper minimum temperature of the oil is necessary to proper lubrication. Even though plenty of oil is available and the oil pressure gauge indicates a maximum pressure, if the oil is cold, the viscosity may be increased to such an extent that the engine is not properly lubricated. With excessively high temperatures, the viscosity may be lowered to such an extent that not enough oil is retained in the bearings to prevent metal-to-metal contact. Oil operating temperatures vary from approximately 150° F. to slightly above 200° F.

The manufacturers' instructions concerning oil temperatures should be carefully followed. Many engines are equipped with an automatic oil temperature control which maintains a fairly constant temperature or viscosity. One type maintains a constant temperature while another type maintains a constant viscosity. Excessively thin oil may allow a "blow-by," which will increase the oil consumption and cause serious overheating of the engine. It is important during the warm-up period of the engine, that a low engine speed be maintained until the oil temperature has reached a safe minimum operating temperature. The fuel burned in the cylinder during a given length of time largely determines the amount of heat developed in the engine. The farther the throttle is opened, the hotter the engine tends to become. The throttle setting should be the minimum setting which will give the desired performance under any given condition.

Lean air-fuel mixtures tend to cause overheating. At low altitudes, particularly during take-off and landing, the mixture controls should be

ENGINE THEORY AND OPERATION

set at full rich. With an increase in altitude, a certain amount of leaning of the mixture may become necessary.

On light aircraft, leaning of the mixture is not usually necessary at altitudes below approximately 5000 ft. At higher altitudes, if the engine seems to be operating on a too-rich mixture, the mixture controls should be moved towards the lean position until the maximum r.p.m. are obtained. It is often well, if a constant altitude is to be maintained, to advance the mixture control towards the lean position until the maximum r.p.m. are reached and then move it a short distance towards the rich position. If the engine temperature rises, the mixture controls should be moved towards the rich position. As the altitude increases and the density of the atmosphere decreases, the propeller speed tends to increase with any given throttle setting. This increase is gradual and, unless the pilot checks the tachometer at frequent intervals, an engine may be exceeding the safe operating r.p.m. without the pilot's becoming aware of this condition. This is particularly true with fixed-pitch propellers.

With most liquid-cooled engines, there are manually operated flaps on the radiator by which engine temperatures may be controlled. Many air-cooled engines are enclosed in cowling equipped with manually controlled flaps which may be used to regulate the flow of air around the cylinders. Usually, carburetor heat is used only to prevent ice formation in the carburetor. Indications of carburetor icing are loss of power and irregular operation of the engine. The supply of coolant in liquid-cooled engines, grade of oil, oil supply, and atmospheric conditions are items which influence engine operating temperatures. Low-grade fuels lead to detonation or slow burning of the mixture which results in loss of power and increased operating temperatures. It is the pilot's responsibility to determine that proper grades of fuel and lubricants are always used. If the fuel which has been placed in the tank tends to detonate, the pilot should use the lowest possible throttle setting to maintain safe operating conditions. A poor grade of oil causes an increase in friction which, in turn, causes an increase in engine operating temperature. Any indication of failure in the lubricating system should be a reason for landing as soon as possible to determine the cause. An engine may be completely ruined in a few minutes of operation unless it has sufficient lubrication.

Any variation from the prescribed limits of fuel pressures may result in improper functioning of the carburetor. Excessive fuel pressures may cause an overrich mixture, while low fuel pressures may cause a lean mixture with a resultant loss in power. A low fuel pressure may also cause the engine to stop.

XXII JET PROPULSION

Jet propulsion, since it converts "a physical force such as heat into mechanical power and motion," falls within the scope of engines as defined by Webster.

The rocket, while not made use of in propelling aircraft, except to lend additional power for short periods of time, is so similar to the action of the jet-propulsion engine that a thorough understanding of its principles assists greatly in understanding the operation of the jet engine. The rocket, as well as the jet-propelled aircraft, depends directly upon the fundamentals set forth in Newton's third law of motion. These three laws, which were originally written in Latin, stated:

1. A body remains in a state of rest, or of uniform motion in a straight line, unless compelled by an external force to change that state.

2. If a given body is acted upon by a force which gives motion to the body, the movement is proportional to the force and in the same direction in which the force acts.

3. For every action there is an equal and opposite reaction.

If a person standing in a light boat, which is free to move in the water, attempts to jump ashore, the force exerted pushes the boat backward, and the person may fall into the water between the boat and the shore. The force which shoved the boat backward was the reaction to the force necessary to propel the person from the boat to the shore. As a person walks along the street, each step forward is made possible by the force in the form of friction on the ground which allows his feet to push him forward, step by step. It would not be possible to walk even on a level surface were it not for this reactive force. On a frictionless surface, or if the bottom of the feet were resting on ball bearings, the feet would slide backward with each step and, because of the inertia of the body, there would be no forward movement. The experience of trying to walk forward on roller skates with the skates pointed directly forward is an experiment which illustrates this action.

JET PROPULSION

A toy balloon, when filled to capacity with air and the opening closed, has no tendency to move of its own accord. The pressure of the gas in the balloon is equal in all directions. If, however, after being fully extended the balloon is let loose without closing the opening, it immediately begins to travel rapidly in a direction away from the opening. This movement

Fig. 322. The photograph shows the evolution of the fighter-type airplane. The older type is at the top. The bottom airplane is a propellerless jet-driven airplane. (Courtesy Bell Aircraft Corporation)

is not caused, contrary to common conception, by the air pushing against the outside air, but by decreasing the force on a part of the inside of the balloon. The force is exerted within the balloon, literally by the air kicking itself out through the opening. The force with which the air moves through the opening has a reactive force equal to it acting in the opposite direction within the balloon.

The rocket operates on this principle. The burning powder within the rocket develops a comparatively high pressure. This pressure is exerted equally in all directions within the rocket, except toward the open end. In the direction of the opening, no force is exerted on the rocket proper.

AIRCRAFT ENGINES

The equal opposite force within the rocket is exerted on the rocket itself and forces it in a direction opposite to the opening in its base.

The action of an aircraft propeller in pulling the aircraft through the air is also dependent upon Newton's third law. The propeller displaces large volumes of air backward. In this case the reactive force which pulls the aircraft forward is called "thrust." In the case of a "pusher" propeller, the aircraft is pushed forward. A fish swimming through the water

Fig. 323. A three-quarter rear view of a jet-propelled airplane. Because there is no propeller, this airplane rests closer to the ground than the conventional type. (Courtesy Bell Aircraft Corporation)

displaces water backward and, in so doing, forces itself forward. Many sea animals propel themselves by taking in water through the front of their body and expelling it backward with considerable force in the form of a small jet. This action forces the animal through the water.

Jet propulsion, as applied to aircraft, is developed by several different methods, all of which depend upon the same action, namely, that of discharging to the rear large volumes of gases at high velocities. Technical details recently released show that the modern jet engine operates on principles similar to the types already known, the operation of which is fairly well understood.

Jet-propulsion units fall, roughly, into two classes. The first is the true rocket type which is self-contained and carries its complete fuel supply. This fuel supply may be in the form of a dry mixture, such as gun cotton or other explosives, or in the form of a liquid fuel which is burned with oxygen carried within the unit itself. When liquid fuel is used, the oxygen is usually in the form of liquid oxygen. The second group consists of those jet-propulsion units classified as "air-stream engines." This

JET PROPULSION

type obtains its oxygen from the air. The oxygen from the air is burned with fuel supplied to the engine, and the hot gases thus formed are allowed to escape through a properly designed opening at high velocities. In this group are the intermittent firing engine, the engine developed by the Germans in their V-1 robot bomb, the gas-turbine jet engine, the gas-turbine engine, and the supercharged internal-combustion engine having an exhaust jet.

The main difference between the two types is the source of oxygen. The rocket type is independent of the oxygen in the atmosphere from which the air-stream engines obtain their supply. All of these engines, with the exception of the internal-combustion engine, are operated on the reaction principle. With the internal-combustion engine, the exhaust gases are directed to the rear, giving some propulsive force.

The use of the solid-fuel rocket as a military weapon goes back many hundreds of years. Rocket action was discovered by the Chinese and, in the 12th century, was used as a military weapon. A rocket is formed of a heavy cardboard case pointed at one end to decrease air resistance and is attached to a stick or piece of bamboo to give direction to the flight. The burning explosive in the form of powder inside the rocket develops the reaction necessary to push the rocket forward in one direction, and the burned gases escape in the opposite direction through the open end of the rocket case.

The first rocket was fired at the enemy more or less as a projectile. Later, an explosive charge was included in the front part of the rocket to make the projectile more effective against the enemy. At first, this charge was simply enclosed in a cardboard container, but later it was enclosed in a metal container to increase the explosive force.

Modern weapons, such as antiaircraft rockets, aircraft rockets, and bazooka projectiles, are simply a modification of the skyrocket. Instead of the trailing stick to give direction to the projectile, metal fins are used.

A true rocket plane is one which makes use of a self-contained rocket. A rocket of this type carries its own fuel, usually in solid form.

Oxygen, usually in the form of liquid oxygen, is burned with almost any inflammable liquid fuel, and the hot gases rushing out through the rear of the rocket give it forward force. As the fuel in the self-contained rocket burns, the rocket becomes lighter and increases in speed. Such fuels as kerosene, alcohol, liquid hydrogen, and gasoline have been used. A disadvantage in using liquid oxygen is that it boils at minus 297° F. and cannot be closely confined.

AIRCRAFT ENGINES

The thermo-jet, air-stream engine makes use of air obtained from the atmosphere to combine with the fuel carried either by the engine itself or in separate tanks. The load any rocket motor must lift is equal to its fuel plus the weight of the rocket itself and any attachments which may be added. If the oxygen itself must be carried, it is a dead load until consumed. If air is used, some method to force large quantities into the rocket

Fig. 324. Front view of a jet engine mounted in the test stand. This is the first picture of the G. E. jet engine. (Courtesy General Electric Company)

engine is necessary. This principle was used in the German V-1 bomb which made use of the intermittent-firing duct engine. The air is forced into the combustion chamber by ram effect. In the front end of the motor is a grill covered by a number of shutters which open inwardly against spring pressure.

In order to launch this type of motor, sufficient speed must be attained on the launching device to open the shutters by means of air pressure. As the shutters are forced open, air rushes into the combustion chamber into which fuel is continuously supplied. The mixture is ignited by an electric spark to start combustion. As the pressure builds up, the shutters

JET PROPULSION

are forced shut and the burned gases discharged through a suitable tube and nozzle at the rear. This action creates a suction at the end of the combustion period. This suction causes the shutters to open, allowing more air to enter, and the cycle is repeated.

In the V-1 bomb, explosions occur approximately forty times a second. The thrust on this type of engine is not steady, but consists of a series of impulses. These rocket bombs attain speeds of 250 to approximately 400 m.p.h.

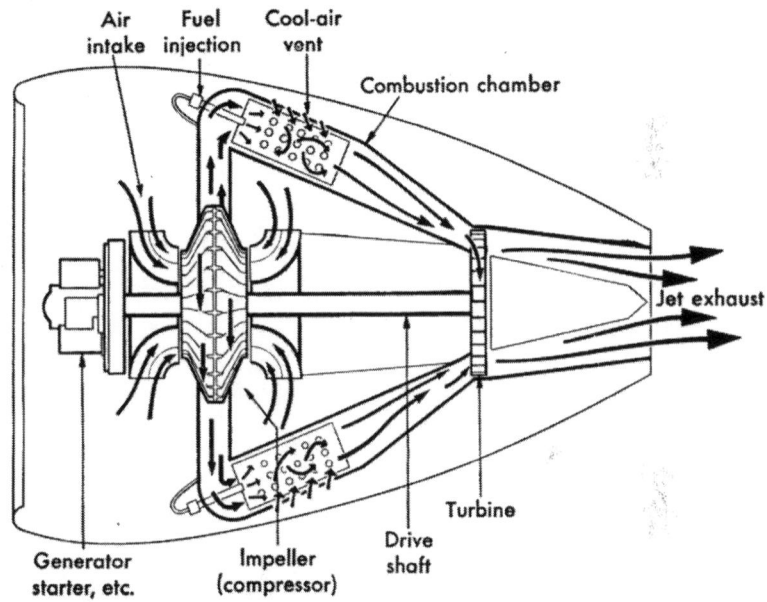

Fig. 325. A schematic drawing showing the basic principles of the turbo-jet engine which drives propellerless airplanes through the air faster than man has ever flown before. Air, sometimes as cold as 75° F. below zero, is compressed by the impeller and fed into combustion chambers. Fuel, such as kerosene, forms an air-fuel mixture which is burned developing temperatures approximately 1,500° F. A stream of hot air and gasses passing through the turbine spins it approximately 10,000 r.p.m. The turbine is connected by a shaft with the impeller and furnishes the power for the compression of the air. The air and gasses sweep out through the jet exhaust producing the reactive force which drives the airplane forward. (Courtesy General Electric Company)

About 1913 there was developed in France a continuous-firing duct engine. This engine was developed in England and was given the name, "Athodyd." This engine looks something like a long barrel with both ends removed. The fuel is discharged through small openings in a ring somewhat ahead of the middle of the barrel-shaped body. As the air enters the front end of the engine, it expands and is speeded on its way by the burning of the fuel. This increased velocity develops enough jet reaction to keep the whole device up to the speed necessary to force the desired quantity of air into the fuel chamber. Enough power is developed to drive forward an aircraft to which it may be attached. While this type of air duct is simple in its construction, enough experimenting has not yet been done with it to determine all of its possibilities.

Another kind of jet-propulsion engine is the gas-turbine jet. This type of engine may drive a propeller. The engine itself is an elongated tube open in the front and with the rear end forming a discharge jet. In the front part of the tube is a compressor fitted with blades which suck in and compress the air that enters through the open front part of the engine. This compressed air is fed into a combustion chamber to which fuel is continuously supplied. The highly heated gases which result from the combustion are expanded through the turbine before they are discharged from the jet. This gas turbine rotates the compressor unit in the front part of the engine and may drive a propeller through reduction gears. This type of engine may be used alone or as a combined jet and propeller engine.

The true rocket type of propulsion is not satisfactory for most aircraft, but may be used to assist in take-off or as the propelling unit for flying bombs. Speeds reached by this type of projectile may be above 600 m.p.h. The Athodyd may reach speeds of 750 to 1000 m.p.h. and may be used for flying bombs.

The gas-turbine jet engine without a propeller is used for aircraft propulsion and is expected to develop extremely high speeds. The gas-turbine jet engine operating in combination with a geared propeller is being used for aircraft propulsion and is expected to develop speeds of from 300 to more than 600 m.p.h.

Supercharged internal-combustion engines, using a geared propeller and an exhaust jet to take advantage of the discharged gas, develop speeds of from 150 to 450 m.p.h.

The first successful application in this country of a gas turbine to an aircraft power plant occurred in September, 1918, when Dr. Sanford A. Moss, with a group of Army Air Corps representatives from McCook Field, took the first turbo-supercharger, installed on a Liberty motor, to the top of Pikes Peak for tests under altitude conditions. On this occasion it was demonstrated that the output of the Liberty engine, which was calibrated at 350 hp. at sea level, was not only maintained but actually increased to 380 hp. at an altitude of 14,000 ft.

JET PROPULSION

At 35,000 ft. altitude, the power developed by the turbine is actually equal to the output of the engine without the turbo-supercharger. At higher altitudes it is even greater.

Its use for ground-boosting engines to increase the power for the take-off has long since become an everyday occurrence, as has its use

Fig. 326. A three-quarter front view of the turbo-jet engine. In starting, the impeller shaft is rotated by means of a small electric motor, and the air-fuel mixture is ignited by an electric spark. When once ignited, the burning of the fuel in the combustion chamber is continuous. (Courtesy General Electric Company)

under combat conditions to obtain emergency power far beyond the normal output of the engine. Perhaps the most important secondary function which has been developed is its contribution to extending the operating range of aircraft.

It has long been realized that the fuel economy of an aircraft engine is greatly improved when operating at from 30 per cent to 50 per cent of its normal rated power, at low engine speed but with the highest intake manifold pressure it is possible to hold. The turbo-supercharger, on account of its favorable characteristics and ease of control, is more adaptable to this mode of operation than the superchargers driven by gears from the crankshaft. In fact, it involves nothing more complicated than closing the waste gate in the exhaust line and directing a greater proportion of the exhaust gases through the wheel.

For the lower altitudes, the engine without a turbo-supercharger but using exhaust jets seems to enjoy an advantage. At the higher altitudes, the situation is reversed, and the turbo-supercharger comes into its own.

Beyond this usage lie three major developments, each of which it is believed will have a decided effect on future aircraft design. These are:

1. The compounding of an internal-combustion engine with a gas turbine.
2. The use of a gas turbine for straight jet propulsion.
3. The use of a gas turbine for combined propeller and jet propulsion.

The nation's first jet-propelled plane is known to be extremely fast and has already been rated at well over 400 m.p.h. Weighing better than 5 tons, it has a slender fuselage, a long upswept tail, and a wing span of 49 ft.

In appearance, this plane largely resembles any mid-wing plane of its type. The chief distinction is the lack of a propeller. This allows it to be built lower to the ground, since there is no prop clearance to be considered, and thus makes it an easier plane for mechanics and ground crews to work on.

It has an extended nose and a tricycle landing gear. It employs a laminar-flow type of wing which, combined with the lack of a propeller, cuts down air resistance and ensures higher speeds. In conventional planes, compressibility effects from the propeller occur at speeds approaching 450 m.p.h., long before the breakdown of the air flow over the wings.

The jet plane project was one of the most remarkably guarded secrets of the war. It had existed for almost two and a half years before it was publicly announced on January 7, 1944. The nation has now seen pictures of the planes and has become familiar with jet propulsion and its principles and promise.

The world's fastest jet airplane of its time was unveiled on August 1, 1945. Photographs of the propellerless plane disclosed droppable fuel tanks mounted on inner shackles and faired into the extreme tips of the wings, giving the plane the out-of-this-world appearance of a Buck Rogers space ship.

JET PROPULSION

To reach its terrific speeds, faster than the top speed of any other plane in the skies and nearer to the speed of sound than any other airplane ever has been able to achieve, the plane uses a superjet-propulsion turbine, the most powerful aircraft engine in the world.

Fig. 327. View of the nose of a jet engine in the parted fuselage of a single-seated airplane. This airplane has a high-powered conventional engine driving a propeller mounted in the nose. This airplane has both conventional power and jet propulsion. (Courtesy General Electric Company)

Problems of range, which limited German jet planes to flights of a few minutes' duration, have been solved in this plane by economy of jet and aerodynamic lines that are the cleanest of any airplane in the skies. With its jettisonable tanks, this plane is capable of carrying out missions now handled by long-range planes. Its ceiling is well above 45,000 ft., a mile higher than the rated top altitudes of first-line, reciprocating engine planes.

This plane has a wing span of 38 ft., 10½ in. Its total empty weight is approximately 8000 lb. The gross take-off weight with maximum fuel capacity is about 14,000 lb. The aerodynamic sleekness of this new plane

AIRCRAFT ENGINES

is unscarred by the exterior attachments of conventional planes. The air scoops and canopy are the only protuberances on the all-metal, semi-monocoque fuselage.

The center line of the sharp, laminar-flow, low wing is 2 in. behind the mid-point of the fuselage. The flat wing tapers at both leading and trailing edges. The control surfaces of both wing and empennage are much smaller than those of conventional planes. With no propeller slipstream or torque to overcome, rudder tabs are eliminated.

Fig. 328. Right-side view from above shows a jet-propelled airplane. This airplane will exceed 600 m.p.h. with a ceiling above 40,000 feet. The top speed of this airplane has not been revealed. (Courtesy General Electric Company)

The cockpit, topped by a plastic bubble canopy and located forward of the wing in this plane's long slender nose, gives the pilot excellent visibility for every maneuver. The cockpit is pressurized from the jet to give the pilot comfort in the substratosphere. The plane's tricycle landing gear is unusually short and light. With no necessity for propeller clearance, the low wings of the airplane skim the ground to assure a high safety factor in both take-off and landing.

The engine is hidden within the fuselage directly behind the cockpit. The jet engine actually has but one moving part, an impeller and turbine connected by a shaft. The turbine and impeller spin at more than

JET PROPULSION

10,000 r.p.m. Air compressed by the impeller frequently is $-50°$ F., while that pouring from the combustion chambers is blazing hot, $1500°$ F. or more.

The impeller whips air into the combustion chamber from the two intakes molded into the fuselage at the wing roots. Kerosene injected into the combustion chamber burns fiercely in the compressed air. The velocity of the air and gases is increased by the heat before they strike the turbine buckets. These gases turn the turbine at great speed and then pass out through the jet exhaust nozzle directly under the tail assembly.

Fig. 329. Looking upward at a jet-propelled airplane. The tanks at the ends of the wings contain fuel and may be dropped by the pilot. Note the clean lines of this airplane. (Courtesy General Electric Company)

Unlike conventional types of engines, the jet requires no warm-up for take-off. The airplane is under way 60 sec. after the engine starts. The efficiency of the engine increases greatly with speed and altitude. The engine output is rated in pounds of thrust. The superjet develops a maximum thrust several times more powerful than certain reciprocating engines. Thrust exists at all times when the velocity of the jet exceeds the speed of the airplane, and it increases as jet velocity and air-mass flow increase. The only engine control is the throttle. The tachometer that shows the revolutions per minute of the gas turbine is the principal instrument necessary to determine thrust output.

AIRCRAFT ENGINES

The airplane is extremely maneuverable. Even at great speeds the hydraulic aileron boost and balanced empennage controls give the plane a maneuverability limited only by the pilot's ability to withstand the forces of tight turns and pull-outs. This plane has the fastest roll of any plane in the world. Stall characteristics are good. Hard to get into a spin, this plane recovers in one-fourth to one-half turn. The fuselage flaps are arranged to form an unbroken line with the wing flaps across the plane's undersurface, when extended. Since there is no propeller to create air drag, the fuselage flaps help slow the plane down for landings at conventional speeds. These flaps beneath the fuselage may be operated together with or independently of the conventional split-type of wing flaps.

Without the conventional plane's propeller, carburetor, ignition system, generators, oil system, radiators, air scoops, and all the complex controls for such items, this plane and its jet engine are exceptionally easy to build, service, repair, and fly.

The plane is constructed in four major assemblies — nose section, wing section, center fuselage section, and aft fuselage and tail group. Both nose and aft fuselages are secured by quick-action tension fittings and may be removed for service and maintenance within a matter of minutes. With the aft fuselage, including the nozzle, removed by detaching three fittings and the tail-pipe clamp, the simplified jet engine is accessible for maintenance and may be run up without removing it from its position in the fuselage. A complete engine change may be finished in less than 20 min.

The nose section contains compartments for oxygen, radio equipment, and the adjustable landing light. Two fluorescent lights are mounted on the sides of the cockpit. In the center fuselage section are the cockpit, fuel tanks, and power plant. The space beneath the cockpit holds hydraulic, fuel, and radio equipment.

The aluminum alloy wing of full cantilever design is built in one unit containing retractable sway braces for carrying the droppable fuel tanks on the wing-tip shackles. The ailerons are conventional. The main landing gear, operated hydraulically, folds inboard into the wing. A glass-smooth "piano" finish adds greatly to its overall speed and performance. To attain this surface, the rivets are cut and surface ground. A zinc chromate primer is applied. All butt joints are cement filled, and the flexible joints are covered with organdy mesh tape. The finish is waxed and highly polished.

JET PROPULSION

There is no ignition system to cause radio interference. There is no propeller to hamper forward-firing armament or to provide a clearance or torque problem. The engine has a self-contained lubricating system that eliminates oil coolers, piping, and tanks.

Fig. 330. First propeller-drive gas turbine on the test stand. This new engine has been named the Propjet. (Courtesy General Electric Company)

The combustion in the engine chamber is complete. There is no trail of flame or smoke emerging from the nozzle, and there is no danger to the pilot of carbon monoxide poisoning. Although present engines are designed to operate on kerosene, which reduces fuel problems and fire hazards, modifications in the fuel system would enable them to operate on gasoline of any octane rating with approximately the same efficiency.

Compact and superpowerful gas turbines which drive propellers may have a wider application than jet propulsion for the big long-range air-transport planes of the future, according to engineers who have had a prominent part in the development of the jet engines. The most likely fields for the different engines and combinations of engines, according to some engineers, are listed below.

AIRCRAFT ENGINES

1. For the utmost in speed, disregarding other considerations, jet propulsion is by far the best performer.
2. For operation at extreme ranges, the internal-combustion engine combined with an exhaust gas turbine gives by far the best performance. This compound engine combined with water injection provides a power plant with a tremendous reserve of power at sea level and at all altitudes up to the critical, or the point where the turbine nozzle pressure starts to fall off. This combination gives it a very high rate of climb, exceeding that of any other power plant.

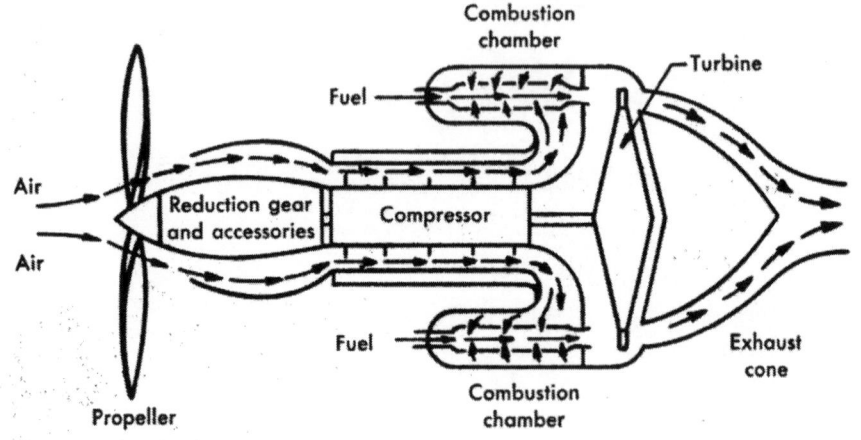

Fig. 331. A schematic drawing showing the basic principles of the propeller-drive gas-turbine engine. The principle is basically the same as the pure jet engine, the only addition being a reduction gear and accessories which drive a propeller. (Courtesy General Electric Company)

3. The gas turbine-propeller combination provides a power plant which, because of its low specific weight, gives a performance approaching that of the jet unit except at very high speeds, but with much better range.
4. For operation at extreme altitudes, the present internal-combustion engine, equipped with a modern turbo-supercharger and properly utilizing the exhaust gas through a jet, is nearly equal to the jet-propulsion unit in speed and far surpasses it in climb.

According to some engineers, all of the design and operating advantages obtained in the jet engines will carry over in the case of a gas turbine driving a propeller. Such features as simplicity and minimum of vibration, for example, will also characterize the gas-turbine power plants.

Performance of the gas turbine does not decrease at high altitudes as much as might be expected. This is because, while the power or thrust output does decrease with decreasing air density, the cold air at high altitudes has a favorable effect on the over-all gas turbine output and partially makes up for the lack of supercharging.

JET PROPULSION

Some engineers have stated that most propeller-type turbines, particularly those designed for use in high-speed planes where a jet can be used effectively, will get their power both from the propeller drive and the jet. They said that a favorable ratio for the two is about 75 per cent propeller and 25 per cent jet, although the amount delivered by the jet could be adapted to suit the requirements of the plane on which the power plant is installed.

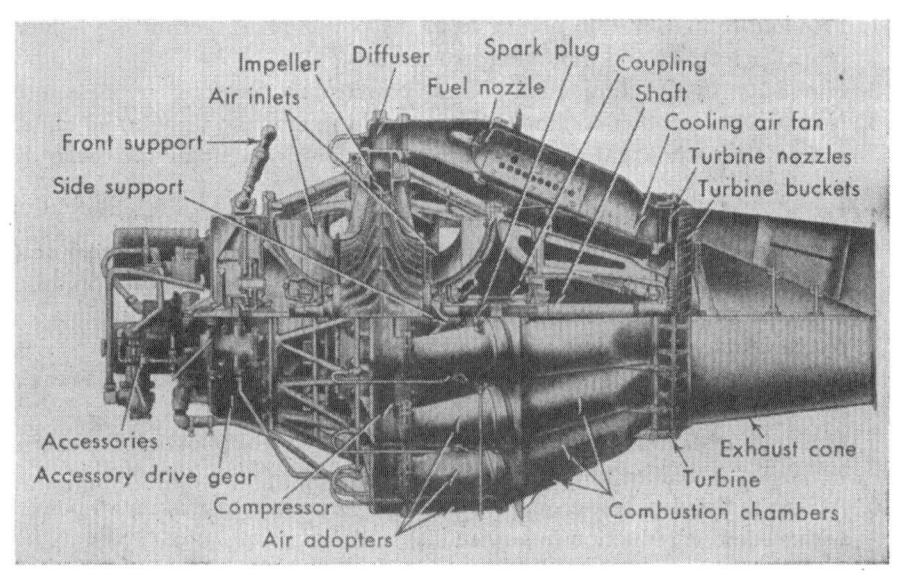

Fig. 332. A cutaway view of an aircraft gas-turbine engine. (Courtesy General Electric Company)

The jet generates virtually no vibration. This eliminates a factor which causes much pilot fatigue. Comfort for passengers in transport planes also will be increased by this when jet engines and gas turbines are harnessed to propel the larger aircraft.

This plane is far lighter than those with conventional reciprocating engines of anything like comparative power. Jet propellerless planes also function smoothly at altitudes higher than those at which planes propelled by conventional reciprocating motors can operate efficiently. No new flight problems are presented to a pilot by the jet engine. Any competent pilot can fly a jet-propelled plane.

The most attractive uses for the gas turbines are in the field of direct propulsion rather than units used as accessories to some other type of power plant. It is in this field that the greatest advances have been made during the war and have exerted the greatest influence on aircraft design.

AIRCRAFT ENGINES

In any gas turbine the net shaft power available is the difference between the power required to compress the air and that delivered by the turbine. In most gas turbines between 65 and 75 per cent of the total turbine output is required to drive the compressor, leaving only from 25 to 35 per cent of the output as netpower.

In its present simple form, the aircraft gas-turbine jet has numerous attractive features as follows:

1. Freedom from vibration arising from the power plant, permitting lighter propeller sections and mounting structure.
2. Simplicity, by the elimination of some power-plant controls.
3. No radiators or other cooling surfaces to add weight and drag on jet units, except that when gas turbines drive a propeller, a small oil cooler is required.
4. Small cooling-air requirements and consequent reduction in drag.
5. Potentiality of very high outputs, well above the range which has been developed by reciprocating engines.
6. Reduced nacelle diameter in comparison with radial air-cooled engines.
7. No carburetors, hence no icing, and no mixture controls.
8. An available supply of compressed air, not contaminated by oil or gasoline, for supercharging the cabin.
9. Quietness of operation, which means less fatigue for the pilot and less cabin insulation to deaden noise.
10. Ability to develop best economy at a high percentage of maximum power without undue reduction in service life.
11. Lower specific weight.
12. Less consumption of lubricating oil.

The actual operation of the jet is this simple:

1. The air to form the air-fuel mixture is picked up by a compressor at the nose of the engine.
2. From the compressor it passes to a chamber where fuel and air burn, increasing the velocity of the air and gases.
3. The hot air and gases then sweep through a turbine, spinning it at a tremendous speed. This turbine, in turn, furnishes the power for the compressor. The turbine and the compressor unit are connected by a shaft, and they rotate as a single unit.
4. After swirling through the turbine, the hot air and gases funnel at high pressure through a nozzle, or jet, in the rear of the engine. The velocity of this air and the gas gives the reactive thrust which drives the plane forward.
5. The gas and hot air pass out of the jet in a steady stream. The thrust power of this stream of hot air and gas is controlled by the fuel throttle.
6. When properly regulated, no flame, glow, or stream of smoke emerges from the jet nozzle.

JET PROPULSION

7. The noise made by the jet is a rumbling roar.
8. The jet is started by an electric starter by current from batteries in the plane or from an outside source. The starter is a small motor mounted in front of the compressor. The motor spins the compressor, which forces air into the combustion chamber. The mixture of air and fuel is ignited and burned. Expanding gases reach a velocity high enough to turn the turbine faster than the starting motor turns it, and the starter motor automatically cuts off. The turbine continues the job of running the compressor.

A new type of aircraft power plant has been developed which makes two uses of a single gas turbine. This power plant, by means of the gas turbine, drives a propeller and boosts with jet thrust at the same time. This marks a decided advance in the development of aircraft power units. This power plant is called the "Propjet." It was primarily designed to drive large, high-speed military transports and bombers. The Propjet has been subjected to rigid tests on test stands and has been installed in experimental airplanes of advanced design. This power plant was designed for installation in the wings of multiengine aircraft or in the nose of a single-engine airplane.

The air rams into the nose of the Propjet through air ducts and scoops opening forward. The air is compressed by axial flow units in the forward part of the engine and is then forced into combustion chambers. The fuel is injected into these combustion chambers and burns with intense heat. This burning raises the temperature and velocity of the gases. The great energy developed is transmitted to the blades of the turbine wheel. This turbine turns more than 10,000 r.p.m. The temperature developed is over 1500° F. The turbine absorbs the greater part of the energy in the gases.

The turbine not only turns the compressor, but through reduction gears drives a propeller. The reactive thrust created by the energy remaining in the gases after they have passed through the turbine wheel is used as jet propulsion to assist in driving the airplane forward. The Propjet combines both propeller and jet power.

The power generated by these new units is great already, and engineers see no difficulties in increasing the output of this engine to almost any force necessary to drive the huge airplanes of the future. This engine opens a new and unexplored field in aircraft propulsion. It is generally agreed by the engineers that the size and power of conventional aircraft engines is approaching the peak which can be reached without making the engine too complex.

Fig. 333. The first propeller-drive gas turbine for aircraft. This light engine is the number one gas turbine both to drive an airplane propeller and boost with jet propulsion. (Courtesy General Electric Company)

JET PROPULSION

Some of the advantages of this new engine are given below.

1. The Propjet type of engine will make it possible to power airplanes by smaller, lighter-weight engines, thus enabling them to carry heavier loads at greater ranges.
2. The Propjet is simple and compact, and the power is developed by a single high-speed rotor.
3. This power plant develops practically no vibration even when operating at maximum power. The ordinary reciprocating engine produces high vibration when operating at full power.

Fig. 334. A drawing to show how the new Propjet engine may be mounted in a slender torpedo-shaped nacelle in the wings of the airplane of the future. (Courtesy General Electric Company)

4. This engine operates most efficiently and economically under full power conditions during long periods of time. The reciprocating engines are most economical when operated at cruising speeds which are much below their maximum power.
5. The high-speed turbine drives the propeller through reduction gears. The development of a suitable reduction gear was one of the main problems in the development of this engine.
6. The Propjet can be developed to operate efficiently on almost any liquid fuel. Kerosene has been the fuel used thus far in the actual tests.
7. Airplanes powered by the Propjet power plant will have a long range of operation. Transcontinental and transoceanic flights will be possible with airplanes powered by this engine. When operating at low speeds and at low altitudes, the gas turbine uses more fuel than the conventional engine when cruising under similar conditions. However, the Propjet will cruise at full

AIRCRAFT ENGINES

power with less fuel than the reciprocating engine under similar conditions. The Propjet power plant is most effective at high altitudes where the air is colder.
8. The speed limit of airplanes powered by this engine apparently will be limited only by the compressibility barriers which are already reached by propeller-driven airplanes. This point is reached somewhere slightly above 500 m.p.h. Airplanes powered by the Propjet can operate over long ranges at speeds close to this compressibility wall, while planes driven by the conventional engine must cruise at much reduced speeds on long flights.

Airplanes powered by the Propjet will probably not reach the extreme speed attained by the new pure-jet airplanes. The fastest airplane in the air today is powered by the pure-jet engine. On propeller-driven aircraft, the propeller blades spinning at high speeds are the first part of the plane to be affected by compressibility. The pure-jet planes have no propeller, thus eliminating this problem, and their top speed is closer to the compressibility barrier. It is the opinion of engineers that while the new Propjet may not power our faster planes, it will give ranges greater than those attained by means of pure-jet propulsion.

There is much research to be done with this new engine before it can be generally used for aircraft propulsion. Practical developments of great importance to swift and economical transportation are looked forward to through improvements of the application of this new type of engine, particularly in the field of large, civilian aircraft.

INDEX

Accelerating device, automatic . . 152
   pump 152, 156
Acceleration 25
Air bleed 149
   deflectors, cylinder 264
   -fuel mixture . . . 11, 12, 143
   -injection starter 303
   -stream engines 324
Aircraft 25
   rockets 325
Airfoil 25
Airplane 25
Alcohol 138
   blends 138
Alloys 3, 18, 25
   properties 19
Alternating current (ac) . . . 25
Alternator 25
Altitude 25
Aluminum 17
   strong alloys of 20
Ammeter 127
Ampere 25, 103
Aneroid 25
Annealing 22, 25
Anode 25
Antiknock value 25, 141
Antimony 21
Arcing 25, 116
Armature 25, 128
Articulated rod (link rod) . . . 25
Athodyd 327
Atmospheric pressure 25
Atom 100
Austenite 20
Autosyn 25
   flow meter 292
   system 290
Axial motion 25

B.t.u. (British thermal unit) . 26, 317
Baffle 259
   cylinder 26

Bar magnet 99
Battery 26, 104
   charging system 125
   ignition system 113
   storage 26, 105
Bazooka projectiles 325
Bearing 174
   ball 175
   roller 174
Benzine 138
Benzol 139
Blade-angle 204
Boiler, steam 8
Booster segments 121
Brass 21
Breaker points 114
Breather 75
British thermal unit (B.t.u.) . 26, 317
Bronze 21

Cadmium 21
Cam 26, 86, 93
   follower 90
   plate 26, 95, 96
Camber 26
Camshaft 26, 91, 97
Carbon 18
Carburetor 26, 148
   accelerating pump 152
   downdraft 149
   float-type 150, 151
   injection-type 161
   pressure-type 155
   types of 148
   updraft 149
   variable Venturi . . . 159, 160
Carburizing 23
Cartridge starter 305
Case hardening 23
Castings 19
Cathode 26
Cell, dry 25, 105
   wet 26, 105

343

INDEX

Cementite	.	20
Centrifugal force	.	26
pump	.	26
Chamber, combustion	.	12, 26
compression	.	12, 26
Circuit, electric	.	104
parallel	.	104
primary	.	114
secondary, voltage	.	116
series	.	104
Coefficient of expansion	.	18
Combustion chamber	.	12, 26
Commutator	.	128
Compression chamber	.	12, 26
ratio	.	26, 142, 316
Condenser	.	101, 102, 114
Conductivity, electric	.	18
Conductors of electricity	.	100
Congeal	.	26
Connecting rod	.	13, 52, 60
Control panel	.	127
Controls, engine	.	317
Coolant	.	261
Cooling systems	.	40, 255
Copper	.	18
Core, soft iron	.	108
Corrosion	.	18
Coulomb	.	26, 103
Counterweight	.	13, 26
Cowling	.	258
pressure	.	259
ring	.	258
Crank	.	13
cheek	.	64, 66
throw	.	66
Crankcase	.	14, 26, 75
Crankpin (wrist pin)	.	26
Crankshaft	.	13, 26, 64
Cuno cleaner	.	192
filter	.	184
Current, induced	.	28, 108
limitator	.	127
Cutout, reverse current	.	127
Cyaniding	.	24
Cycle	.	26
Cylinder	.	26, 43
barrel	.	43
head	.	45, 46
types	.	49
skirt	.	46
Dampers, dynamic	.	27, 74
Dead center, bottom	.	12
top	.	12

Density	.	17, 26
Detonating characteristics	.	139
Detonation	.	27, 312
Diesel engine	.	27, 316
Difference in potential	.	102
Diffuser	.	27
Direct current (dc)	.	27
electric starters	.	303
hand-cranked starters	.	302
Distill	.	27
Distributor	.	114
head	.	117
Dry-sump	.	186
Dynamic damper	.	27, 74
Dynamotor	.	27
Economizer	.	153, 156
Efficiency, mechanical	.	317
thermal	.	317
volumetric	.	317
Electric circuit	.	104
conductivity	.	18
conductors	.	100
nonconductors	.	100
Electrical fundamentals	.	99
Electrode	.	27
Electron	.	27, 101
Element	.	17, 27
Engine	.	8
classification	.	33
controls	.	318
dry-sump	.	38
external-combustion	.	8, 27
four-cycle	.	11
fundamentals	.	8
gauge units	.	268
in line	.	32
installation	.	255
internal-combustion	.	8, 9, 27
jet	.	42, 331
mount	.	256
multiple-cylinder	.	32
radial	.	32
steam	.	8
supercharged	.	40
suspension, dynamic	.	256
synchronizers	.	294
synchroscope	.	294, 295
theory and operation	.	309
two-cycle	.	10
types	.	32
V-type	.	33
W-type	.	39

344

INDEX

wet-sump	38
X-type	39
Ether	138
Exhaust gas analyzer	286
port	27
valve	27, 86
Feathering	27, 199
Ferrite	20
Field switch	127
Fins, cooling	90
Firing order	133
Flame hardening	24
Flow meter, Autosyn	292
Flywheel	12, 27
Foot-pound (ft.-lb.)	27
Forgings	19, 27
Four-cycle engine principles	11
Fractional distillation	139
Fuel	27, 137
gauge	145, 146
hydrostatic	146
heat value	141
-injector system	162, 163
level indicator	295
mixture	11, 12, 143
pressure gauges	268
supply pump	144
systems	41, 137
types	137
Fuselage	27
Gas turbine, propeller-drive	339, 340
engine	325, 337
jet engine	325
Gasoline, weight	15
Gauge, fuel	145, 146
hydrostatic	146
pressure	268
manifold pressure	269
oil pressure	267
pressure	267
suction	273
units, engine	268
Generator	27, 107, 127
simple shunt-wound	107
Glide	27
Gold	17
Governor	27
Graduations	27
Ground wire	119
Gyroscope	27

Hardening	22
Heat treatment	21
value of fuels	141
Horsepower	28, 103, 316
brake	317
friction	317
indicator	298
Hydraulic	28
Hydrocarbon group of fuels	138
Hydrometer	28
Idle cutoff	151, 155
Idling	28
system	151
Ignition	28
dual	124
systems	112
parts of	114
timing	133
Impeller	28
Inconel	21
Indicator, air-fuel mixture	286
cylinder temperature	285
Induced current	28
Induction coil	28, 108, 115
system	158
Inertia	28, 322
starters, electric	303
hand	302
Inflammable	28
Instruments, engine	266
Intake manifold	12, 159
valve	28
Intermittent-firing duct engine	326
Iridium	17
Iron	19
gray cast	20
wrought	19
Jet engine	42, 331
-propelled airplane	330, 332
propulsion	28, 322
-propulsion engine	328
Jig	28
Junction, cold	286
hot	286
Kilowatt hour	104
Lapping	28
Lead	17
Leyden jar	101
Linear	28
Link rod	28

345

INDEX

Lithium	17
Loading, power	28
Lobe	28
Lodestone	28, 99
Lubricants	28, 174
classified	174
properties of	176
Lubricating systems	174
Magnesium	21
Magnet, bar	99
Magnetic field	28, 100, 115
lines of force	106
Magneto	28, 120
ignition system	117
Main bearing	28
Manifold	28
intake	12, 159
Martensite	20
Meshes	28
Metallurgy	2, 28
Metals	17
hardness	17
melting points	17
physical properties	17
Metering jet	28
Mixture control	154
Molecule	100
Molten	28
Momentum	28
Monel metal	21
Nacelle	28
Naphtha	139
Neutron	29
Newton's laws of motion	322
Nitriding	24
Nonconductors, electric	100
Normalizing	22
Nozzle, discharge	152
Octane	29, 141
Ohm	29, 102
Oil, asphalt-base	182
color	179
fire point	179
flash point	177
gravity	177
mineral	139
paraffin-base	182
pressure gauges	267
pump	183, 186
radiators	263
rings	192
sump	84
weight	15
Otto cycle	10
Oxide	17
aluminum	18
iron	18
Oxygen	325
per cent in atmosphere	143
p.s.i. (pounds per square inch)	30
Parallel circuit	104
Pearlite	20
Petroleum	137
fuels	138
Phosphorus	18
Piston	13, 29
assembly	52
head types	55
parts	57
pin	52
rings	52, 192
types	58
Pitch angle of propeller blade	193
Plate	19
Points, breaker	114
Precision instrument	29
Preignition	29, 138, 313
Pressure gauges	267
Primary circuit	114
Primer	147
Propeller	29, 193
adjustable-pitch	195, 197
area	29
constant-speed	210
controllable-pitch	197, 205
counterweight-type	202, 203
cuffs	260, 261
effective pitch	194
efficiency	29
electric adjustable	224
feathering	197
fixed-pitch	197
high pitch	198
hydromatic quick-feathering	212
light airplane controllable pitch	249
low pitch	198
pitch	29
radius	29
reduction gear	196
self-contained controllable	241
slip	29, 194
theoretical pitch	194
thrust	29
tipping	29

INDEX

two-pitch	197
two-position	206
wooden	199
Properties of metal, physical	17
Propjet	335, 339
Proton	30, 101
Pulsating direct current	30
Pumps, scavenger	186
Push rod	90
Radiator	261
Ram effect	30
Rarefied atmosphere	30
Relay current control	120
Resistor	30
Reverse current cutout	127
Rheostat	30
Rocker arm	30, 86, 90
Rocket	30, 323
plane	325
solid-fuel	325
Rod, connecting	13, 60
Rotor	115
shaft	116
S.A.E. (Society of Automotive Engineers) or Saybolt numbers	30, 176
Safety gap	117
Safetied	30
Scavenger oil (or scavenged oil)	30
pump	30
Secondary circuit voltage	116
winding	115
Self-induction	116
Self-synchronous instruments	289
Selsyn	297
Servo unit	30
Sheet	19
Shielding, radio	30, 135, 136
Silicon	20
Sodium	30
Solders	21
hard	21
soft	21
Solenoid	30
Solution, storage battery	106
Sorbite	20
Spark gap	116
plug	30, 111, 121
bushings	47
types	123
Specific gravity	17, 30
Spinner	30
Splines	30

Starter, air-injection	303
cartridge	305
direct, electric	303
hand-cranked	302
inertia, electric	303
hand	302
Starting systems	300
Steel	18
cast	20
chromium	19
chromium molybdenum	19
chromium molybdenum vanadium	19
chromium vanadium	19
copper nickel	19
manganese	19
manganese vanadium	19
molybdenum	19
molybdenum nickel	19
nickel	19
nickel chromium	19
stainless	19
tungsten	21
Stellite	30
Stroke, compression	12
exhaust	12
intake	12
power	12
Suction gauges	273
Sulphur	18, 20
Sump, dry	30, 38
wet	30, 38
Supercharged	40
Supercharger	30, 148, 169, 170
Surging	30
Switch, main	114
Synchronize	30
Synchronizers, engine	294
Synchronous motor	278, 279
Tachometer	31, 275
centrifugal	276
chronometric	280
electric	277
Tappet	31, 90
Temperature indicators	281
Tempering	22
Tension	31
Tetra-ethyl-lead	141
Thermal circuit breaker	31
Thermo-couple	285
Thermo-jet air-stream engine	326
Thermometers	281
electrically operated	284
vapor pressure	283

347

INDEX

Timing	133
Torque	31
meter	298
Torsional force	31
Transformer	31, 108
simple	109
Troostite, primary	20
secondary	20
Turbo-jet engine	327, 328
Turbo-supercharger	173, 328
Two-cycle engine principle	10
V-1 robot bomb	325
Vacuum	31
Valve	31, 86
butterfly	149
by-pass	186
cooling	87
exhaust	27, 86
face, angle of	88
guide	88, 174
heads	87
intake	86
lead	314
lifter, hydraulic	91, 92
parts	87
poppet	87
pressure relief	186
retainer	90
seat	88
angle of	88
sliding	87
springs	90
throttle	149
train	90, 94
types	87
Vapor pressure	31
Vaporization	138
Venturi tube	31, 148
Vibrator	110
Viscosity	177
Volatility	31
Volt	31, 102
Voltage booster	31
regulator	127
Voltmeter	127
Volumetric efficiency	31
Warming up	308
Warning units, pressure	272
Watt	31, 103
Weight per horsepower	15
Wet-sump	185
Wheatstone bridge	31, 284
Windings, secondary	115
Windmilling	31
Wobble pump	31, 145
Wright brothers	2
Wrist pin (*See* crankpin.)	31
Zinc	21

348

The Aviation Collection by Sportsman's Vintage Press

www.SportsmansVintagePress.com

Aircraft Construction Handbook	by Thomas A. Dickinson
Aircraft Sheet Metal Work	by C. A. LeMaster
The Aircraft Apprentice	by Leslie MacGregor
Aircraft Woodwork	by Col. R. H. Drake
Aircraft Welding	by Col. R. H. Drake
Aircraft Sheet Metal	by Col. R. H. Drake
Aircraft Engines	by Col. R. H. Drake
Aircraft Electrical and Hydraulic Systems, and Aircraft Instruments	by Col. R. H. Drake
Aircraft Engine Maintenance and Service	by Col. R. H. Drake
Aircraft Maintenance and Service	by Col. R. H. Drake